Topology and Geometry for Physicists

CHARLES NASH

Department of Mathematical Physics
National University of Ireland
Maynooth, Ireland

SIDDHARTHA SEN

School of Mathematics
Trinity College
Dublin, Ireland

DOVER PUBLICATIONS, INC.
Mineola, New York

Bibliographical Note

This Dover edition, first published in 2011, is an unabridged republication of the work originally published in 1983 by Academic Press, Inc., New York.

International Standard Book Number
ISBN-13: 978-0-486-47852-4
ISBN-10: 0-486-47852-1

Manufactured in the United States by LSC Communications
47852109 2023
www.doverpublications.com

Preface

One noticeable feature of theoretical physics of the last decade or so has been rapid growth of the use of topological and geometrical methods. This book is intended to teach physicists these methods. No previous knowledge of topology or geometry is assumed.

The prerequisites for this book are those possessed by an advanced undergraduate or a first-year graduate student. The style and approach of the book reflect the fact that the authors are physicists—the level of rigour has, in many cases, been appropriately lowered both to shorten arguments and, we hope, to improve their clarity. Nevertheless we have tried to provide the references necessary for those who wish to read a completely rigorous account.

Applications from condensed matter physics, statistical mechanics and elementary particle theory appear in the book. An obvious ommission here is general relativity—we apologize for this. We originally intended to discuss general relativity. However, both the need to keep the size of the book within reasonable limits and the fact that accounts of the topology and geometry of relativity are already available, for example, in *The Large Scale Structure of Space–Time* by S. Hawking and G. Ellis, made us reluctantly decide to omit this topic.

We would like to warmly thank all the colleagues who encouraged us and criticized the manuscript. In particular we wish to mention David Simms and Richard Ward. Finally we thank Rose Coyne and Breda O'Neill for careful typing of the manuscript.

August, 1982 *Charles Nash and Siddhartha Sen*

To Anita and Edna

Contents

CHAPTER 8. **Morse Theory**

CHAPTER 9. **Defects, Textures and Homotopy Theory**

CHAPTER 10. **Yang–Mills Theories: Instantons and Monopoles**

CHAPTER 1

Basic Notions of Topology and the Value of Topological Reasoning

1.1 INTRODUCTION

Topology can be thought of as a kind of generalization of Euclidean geometry, and also as a natural framework for the study of continuity. Euclidean geometry is generalized by regarding triangles, circles, and squares as being the same basic object. Continuity enters because in saying this one has in mind a continuous deformation of a triangle into a square or a circle, or indeed any arbitrary shape. A disc with a hole in the centre is topologically different from a circle or a square because one cannot create or destroy holes by continuous deformations. Thus using topological methods one does not expect to be able to identify a geometrical figure as being a triangle or a square. However, one does expect to be able to detect the presence of gross features such as holes or the fact that the figure is made up of two disjoint pieces etc. This leads to the important point that topology produces theorems that are usually qualitative in nature—they may assert, for example, the existence or non-existence of an object. They will not in general, provide the means for its construction.

Let us begin by looking at some examples where topology plays a role.

Example 1. Cauchy's residue theorem

Consider the contour integral for a meromorphic function $f(z)$ along the path Γ_1 which starts at a and finishes at b (c.f. Fig. 1.1). Let us write

$$I = \int_{\Gamma_1} f(z)\, dz \tag{1.1}$$

Now deform the path Γ_1 continuously into the path Γ_2 shown in Fig. 1.1. Provided we cross no poles of $f(z)$ in deforming Γ_1 into Γ_2, then Cauchy's

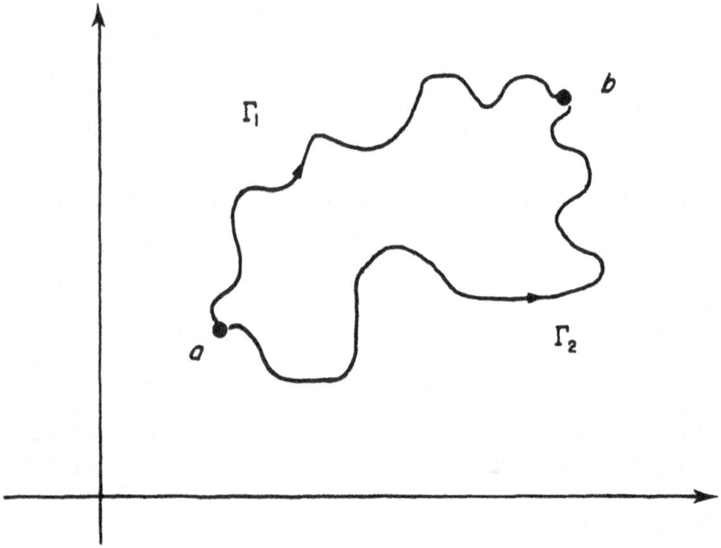

Figure 1.1

theorem for meromorphic functions allows us straightaway to deduce that

$$\int_{\Gamma_1} f(z)\,\mathrm{d}z = \int_{\Gamma_2} f(z)\,\mathrm{d}z \tag{1.2}$$

This is just the statement that

$$\int_C f(z)\,\mathrm{d}z = 0 \tag{1.3}$$

where C is the closed contour made up by joining Γ_1 to Γ_2 and reversing the arrow on Γ_1 so as to give an anticlockwise direction to the contour C. The intuitive content of this result is that to integrate $f(z)$ from a to b in the complex plane is independent of the path joining a to b (under the conditions stated). Even if we relax these conditions and allow that the deformation of Γ_1 into Γ_2 may entail the crossing of some poles of f(z), then we still have complete knowledge of the relationship of the two integrals. It is simply

$$\int_{\Gamma_1} f(z)\,\mathrm{d}z = \int_{\Gamma_2} f(z)\,\mathrm{d}z + 2\pi i \sum res \tag{1.4}$$

where the sum is over the residues, if any, of the poles inside C. This simple example uncovers some topological properties underlying the familiar Cauchy theorem.

Example 2. The fundamental theorem of algebra

In this example we shall use the Cauchy theorem to give a proof of the fundamental theorem of algebra. First of all a simple consequence of Cauchy's theorem is that for a meromorphic function $f(z)$ we have

$$\frac{1}{2\pi i}\int_C \frac{f'(z)}{f(z)}\,\mathrm{d}z = n_0 - n_\mathrm{p} \tag{1.5}$$

where n_0 and n_p are the number of zeroes and the number of poles respectively of $f(z)$ lying inside C. Let $f(z)$ be the polynomial $Q(z)$ of degree q where

$$Q(z) = a_q z^q + a_{q-1} z^{q-1} + \ldots a_1 z + a_0, \qquad a_q \neq 0 \tag{1.6}$$

Then since $Q(z)$ has no poles we have

$$\frac{1}{2\pi i}\int_C \frac{Q'(z)}{Q(z)}\,\mathrm{d}z = n_0 \tag{1.7}$$

It is clear that n_0 is a continuous function of q of the $q+1$ coefficients $a_0, \ldots, a_q$. Let us select the coefficients $a_0, \ldots, a_{q-1}$ and write

$$n_0 \equiv n_0(a_0, \ldots, a_{q-1}) \tag{1.8}$$

However, n_0 also only takes integer values. Now since it is impossible to continuously jump from one integer value to another it follows that n_0 must be invariant under continuous change of the a_i's. This state of affairs permits the following argument: for large $|z|$, say $|z|>R$, $|Q(z)|$, grows as $|a_q||z|^q$ which is a large number. Then if C is a circular contour of radius $R'>R$, C will contain all the zeroes of $Q(z)$. Next we continuously deform a_i by choosing

$$a_i^\varepsilon = \varepsilon a_i, \qquad i = 0, \ldots q-1.$$

Evidently,

$$n_0(a_0^\varepsilon, \ldots a_{q-1}) = n_0(a_0, \ldots a_{q-1}) \tag{1.9}$$

thus when $\varepsilon = 0$ the integral (1.7) becomes trivial to evaluate and we obtain the desired result

$$n_0 = q \tag{1.10}$$

(We cannot, of course, require a_q to tend to zero since this would change the degree of $Q(z)$).

Example 3. Fixed points and their applications

Let f be a function from a set X into itself. The question, does f possess any fixed points?, is one of great interest, and is, in general, tackled using topology. For example if we refer to example 2 above, and choose for f the function

$$f(z) = z + Q(z) \tag{1.11}$$

Then the existence of fixed points, $f(z) = z$, of f amounts to the fundamental theorem of algebra.

Let us have a look at the proof of fixed point theorems for functions of 1 and 2 real variables. For functions of 1 variable from an interval (a, b) into itself the proof is simply the assertion that the graph of $f(x)$ must cross the line $y = x$ somewhere, c.f. Fig. 1.2.

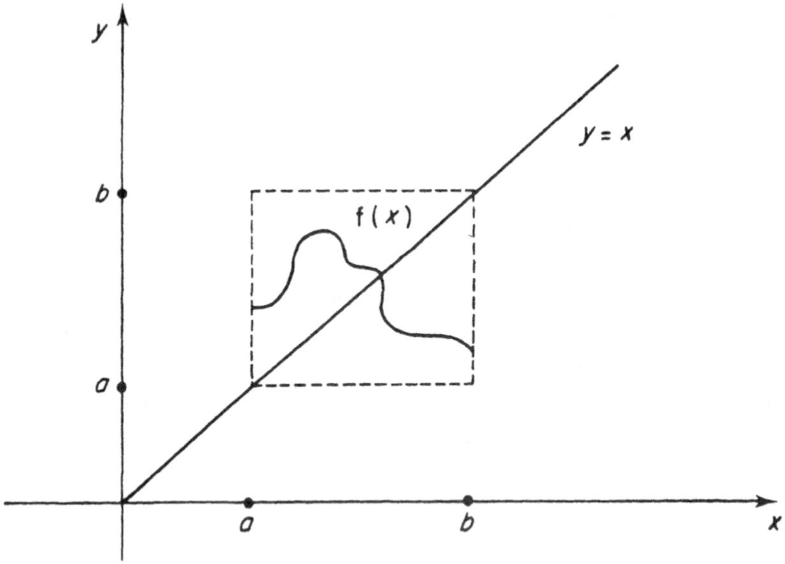

Figure 1.2

For functions of 2 variables let us choose f to be a function from the disc B^2 into itself. B^2 is the disc defined by $x^2 + y^2 \leq a^2$. To prove that $\mathrm{f}(\mathbf{x})$ has a fixed point we start by considering the effect of f on points on the boundary of B^2. Let $\mathbf{x}$ be on the boundary of B^2 and let it be mapped to $\mathrm{f}(\mathbf{x})$, then construct the vector

$$v(\mathbf{x}) = \mathbf{x} - \mathrm{f}(\mathbf{x}) \tag{1.12}$$

shown in Fig. 1.3.

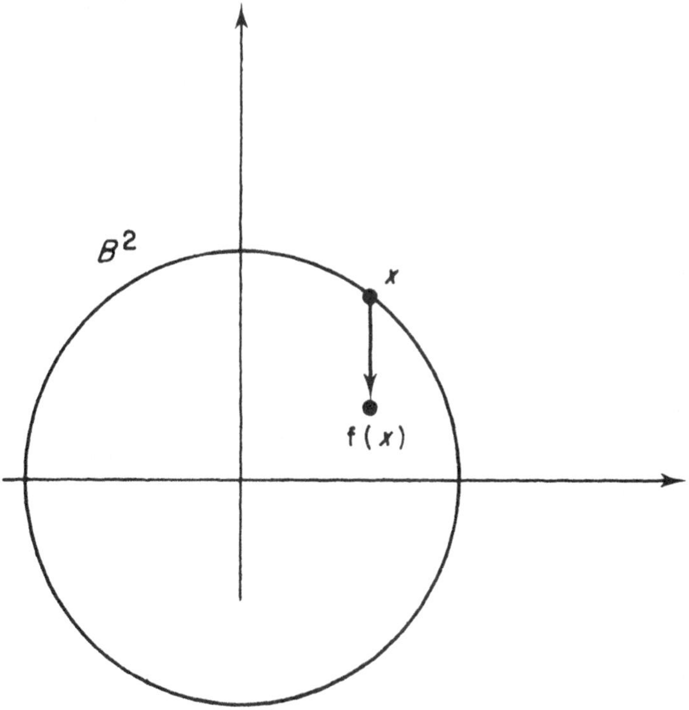

Figure 1.3

We can of course use the Definition (1.12) for all points $\mathbf{x}$ in B^2. This gives us the vector field $v(\mathbf{x})$ on B^2. Returning to the point $\mathbf{x}$ shown in Fig. 1.3, we rotate the point $\mathbf{x}$ through 2π about the centre of B^2. Notice that in doing so $v(\mathbf{x})$ also rotates through 2π once. We then describe this situation by saying that $v(\mathbf{x})$ has index 1. Clearly the index of a vector field, when rotated through 2π, must be an integer. Thus if we continuously deform the boundary of B^2 into the dotted curve shown in Fig. 1.4, then the index of $v(\mathbf{x})$ for this path must still be one—the index is a topological invariant.

We can now prove our result by *reductio ad absurdum*. First of all note that if $v(\mathbf{x})$ has a zero then, since the zero vector has an undefined direction, the index definition breaks down and cannot be defined in the way we gave. Now consider the point $\mathbf{p}$ inside B^2 and assume that $v(\mathbf{p}) \neq \mathbf{0}$ and let

$$|v(\mathbf{p})| = |\mathbf{p} - f(\mathbf{p})| = L \tag{1.13}$$

Then consider a small circle c of radius $\varepsilon < L$ about $\mathbf{p}$ c.f. Fig. 1.5.

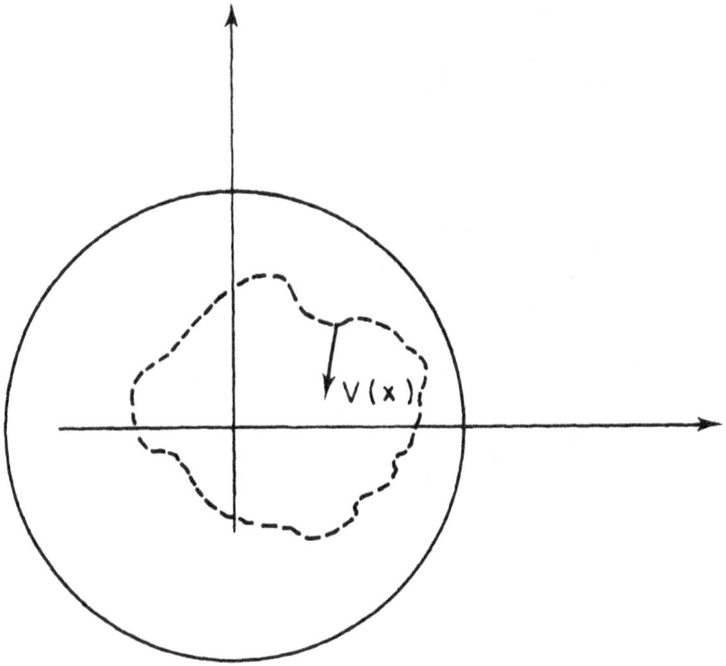

Figure 1.4

By continuity the image of C under f is the dotted circle C' about $f(\mathbf{p})$ of radius $|f'(\mathbf{p})|$. We can choose ε small enough so that $|(\varepsilon + f'(\mathbf{p})\varepsilon)| < L$ so that C and C' do not intersect. Now consider the index of $v(\mathbf{x})$ computed by going round C once. The vector field is graphed in Fig. 1.6 purely assuming $\varepsilon \sim 0$ and continuity of f.

Clearly, since $v(\mathbf{x})$ points from left to right everywhere the index cannot be 1. Thus this assumption of constant unit index must be rejected. This means that $v(\mathbf{x})$ must have a zero somewhere i.e. that $f(\mathbf{x})$ has a fixed point. This may be extended to the case of the solid sphere or ball B^n, $n = 2, 3, \ldots$ etc. This is the celebrated Brouwer fixed point theorem.

The utility of fixed point theorems is enormous. For example if A is a differential operator, say $A = \Delta$ the Laplacian then the existence of solutions to

$$Af = 0 \tag{1.14}$$

is equivalent to the existence of fixed points of the operator $A + I$. Further

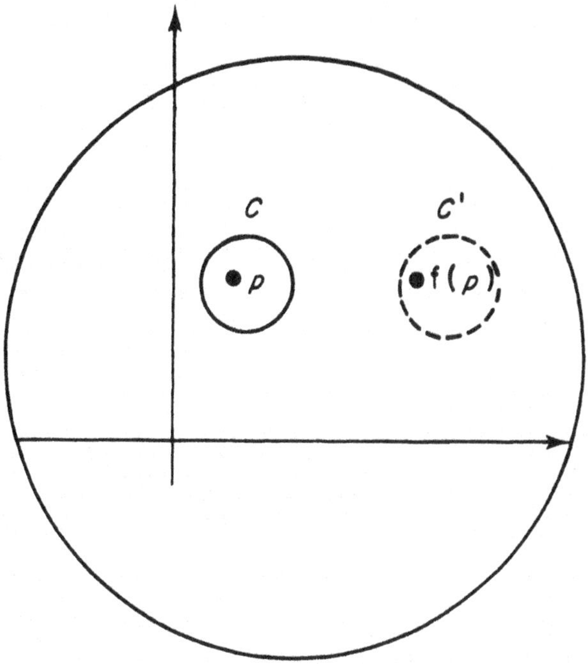

Figure 1.5

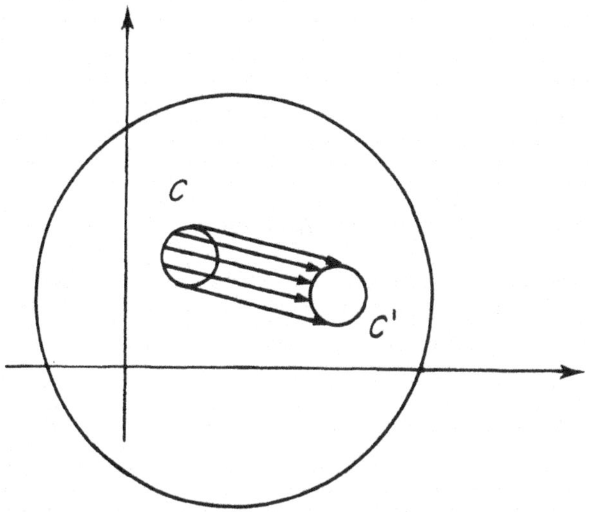

Figure 1.6

important examples are provided by systems of linear differential equations

$$\frac{dx_i}{dt} = f_i(x_1, \ldots, x_n); \qquad i = 1, \ldots, n$$

or

$$\dot{\mathbf{x}} = \mathbf{v} = \mathbf{f} \tag{1.15}$$

Zeroes of the vector field $\mathbf{v}$, commonly called singularities of $\mathbf{v}$, have a special significance. If t is the time, and (1.15) is a set of equations for a dynamical problem, then the singularities of $\mathbf{v}$ are the equilibrium configurations of the dynamical system. In general, if we have a function $f(x_1, \ldots, x_n)$ then the singularities of the vector field $\mathbf{v}$ where

$$\mathbf{v} = \text{grad } f \tag{1.16}$$

are of considerable interest.

The singularities of $\mathbf{v}$ are just the extrema or critical points of f. Their study, using topological methods, provides the foundation for Morse theory, a topic which we shall come to in a later chapter.

1.2 BASIC TOPOLOGICAL NOTATIONS

Having taken a topological view of some mathematical situations, it is now necessary to equip ourselves with some mathematical techniques to progress further. These techniques will be supplied to the reader in the chapters that follow. Some of their names, and perhaps only their names, are already known to the reader, and exposure to them now should breed familiarity for what follows. The names are, homotopy, homology, cohomology and fibre bundles to mention the main ones. The mathematical setting, for the development of these techniques, requires some preliminary definition and illustrative examples. This is where we now begin.

Topological space

The mathematical structure within which we work is called a topological space. There is no reason to restrict ourselves unduly, so we expect this to be a fairly general definition.

Let X be any set and $Y = \{X_\alpha\}$ denote a collection, finite or infinite of subsets of X. Then X and Y form a topological space provided the X_α and Y satisfy:

i. $\phi \in Y, X \in Y$
ii. Any finite or infinite subcollection $\{Z_\alpha\}$ of the X_α has the property that $\bigcup Z_\alpha \in Y$
iii. Any *finite* subcollection $\{Z_{\alpha_1}, \ldots, Z_{\alpha_n}\}$ of the X_α has the property that $\bigcap Z_{\alpha_i} \in Y$ (1.17)

The set X is then called a topological space and the X_α are called open sets. The choice of Y satisfying (1.17) is said to give a topology to X.

Examples

(a) If X is a set and Y is the collection of all subsets of X, i.e. the set 2^X, then (1.17) is clearly satisfied. This is called the discrete topology of X.

(b) If X is any set, then the simple choice $Y = \{\phi, X\}$ satisfies (1.17). This is called the indiscrete or trivial topology.

(c) Let $X = \mathbf{R}$, the real numbers, and let X_α be those subsets of $\mathbf{R}$ which satisfy the following condition. For each $x \in X_\alpha$ there is an open interval (a, b) containing x, i.e. $a < x < b$, and also (a, b) is contained in X_α, i.e. $(a, b) \in X_\alpha$. This gives a topology to $\mathbf{R}$ called the usual topology.

Having given the Definition (1.17) and the examples (a), (b) and (c), we now need to discuss them. Let us begin with the definition of topological space, what is its underlying purpose or motivation? For example the topologically ignorant inquiring mind asks: why do open sets remain open under infinite union but only under finite intersection? Also, what does this abstract Definition (1.17) have to do with the intuitive discussion of continuous deformation which preceded it? In answer to all this it should be gently emphasised that the, admittedly abstract, Definition (1.17) is the end product of many mathematical experiments with various definitions and concepts. These experiments finally gave forth the present definition of a topological space. The purpose of this work was to build a mathematical structure within which continuity, realised by continuous functions, would naturally reside. In other words, the definition of topological space was to be used to give a definition (as general as could be managed, as is always the case in mathematics, especially in this century), of continuous functions. We need therefore, if we are to understand what is going on, to look at this definition.

Definition

A function f mapping from the topological space X to the topological space Y is continuous if the inverse image of an open set in Y is an open set in X. (1.18)

Let us examine this definition using example (c) above of the usual topology on the real numbers **R**. We take X and Y both to be **R** and so we wish to study real valued functions $f(x)$ of a real variable x. Now note that according to example (c) all open intervals (a, b) are automatically open sets. Providing ourselves with a graph of $f(x)$ we can proceed to apply Definition (1.18).

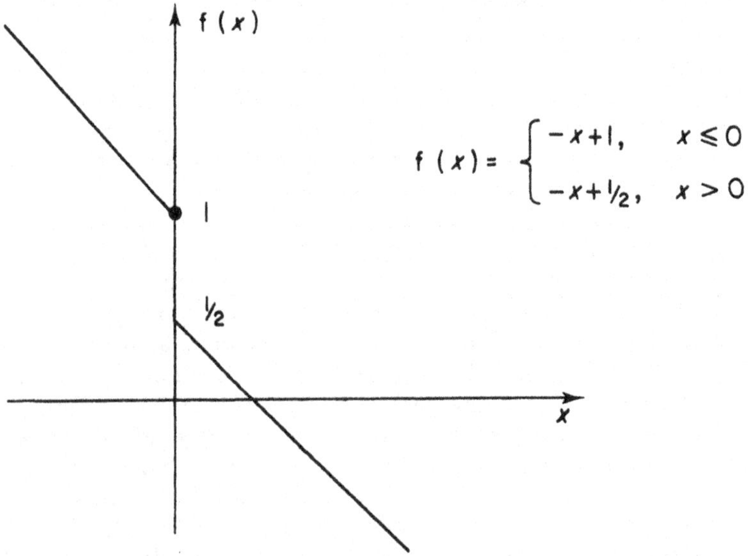

Figure 1.7

Notice that we have given the function $f(x)$ what would normally be called a discontinuity at $x = 0$. To test our Definition (1.18) we take the open set (5, 17) and compute its inverse image $f^{-1}\{(5, 17)\}$. It is immediate to see that

$$f^{-1}\{(5, 17)\} = (-16, -4) \tag{1.19}$$

But $(-16, -4)$ is an open set thus we satisfy (1.18). It is now clear that if we replace (5, 17) by the interval (a, b) and if $1 < a < b$, or, $a < b < \frac{1}{2}$ then $f^{-1}\{(a, b)\}$ will be an open interval which is an open set. This is sufficient to establish the continuity of $f(x)$ according to the usual ε, δ method for values of $x \neq 0$. Clearly at $x = 0$ something different must happen. To see that this is indeed the case we compute the inverse of the open set $(1 - \varepsilon, 1 + \varepsilon)$, $\varepsilon > 0$. Evidently $f^{-1}\{(1 - \varepsilon, 1 + \varepsilon)\}$ is the set of X such that $-\varepsilon < x \leqslant 1$ or

$$f^{-1}\{(1-\varepsilon, 1+\varepsilon)\} = (-\varepsilon, 1] \tag{1.20}$$

But $(-\varepsilon, 1]$ is not an open set. This is because the point 1 cannot be contained in an interval (a, b) which is itself contained in $(-\varepsilon, 1]$. At least for this example then, continuity defined by (1.18) coincides with our usual notion of continuity.

Another point worth examining is the reason why the Definition (1.18) demands that the inverse of an open set remains open under f, rather than the reverse. That is to say would it not be simpler and more natural to define f to be continuous if f maps open sets in X into open sets in Y? We can dispose of this point by a counter example. We display a function f which is clearly discontinuous by reference to our usual intuition and also by reference to Definition (1.18). The function f will, however, have the property that it maps open sets into open sets.

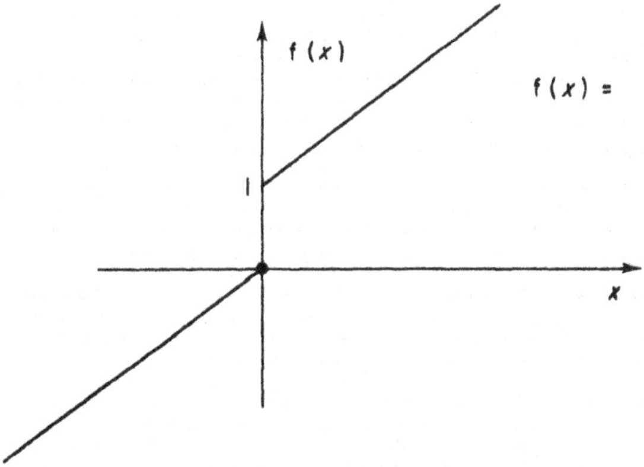

$$f(x) = \begin{cases} x, & x \leqslant 0 \\ 1+x, & x > 0 \end{cases} \tag{1.21}$$

Figure 1.8

We display the actual function $f(x)$ in Fig. 1.8, and we continue to use the usual topology for the real numbers. As can be seen from Fig. 1.8, the topological space X is $\mathbf{R}$, and the union of two disjoint pieces give Y. To show that f is discontinuous is an easy exercise. Nevertheless it is easy to convince oneself by taking examples that f always maps open sets into open sets. This explains why the definition (1.18) needs to be stated in terms of inverses of open sets.

Having attempted to show how the abstract looking objects called open sets can be used to define continuity, we go on to answer another question that we raised earlier. It is the question: why not allow open sets to be open under infinite intersection? The answer is that to allow this in the Definition (1.17) of a topological space would render most topological spaces useless. For example, take the usual topology for **R** obtained as above, now supplement it by the requirement that open sets remain open under infinite intersection. The result of this is that any real number X is now also an open set, and so, because of the union axiom, is any subset of **R**. This is because if $x \in \mathbf{R}$ and we consider the open sets X_α given by

$$\begin{aligned} x_1 &= (x-1, x+1) \\ x_2 &= (x-\tfrac{1}{2}, x+\tfrac{1}{2}) \\ x_3 &= (x-\tfrac{1}{3}, x+\tfrac{1}{3}) \\ &\vdots \end{aligned} \tag{1.22}$$

then the intersection of all the x_α is just the point x, i.e.

$$\bigcap_{\alpha=1}^{\infty} X_\alpha = x \tag{1.23}$$

So the usual topology would, under this stringent requirement concerning intersection, be reduced to the discrete topology on **R**. This would be of no use for studying continuity. To see this note that under the discrete topology since every subset of **R** is open, *all* maps from **R** to **R** are continuous. Clearly we do not want a definition of continuity as broad as this. Now we can begin to see why we only require open sets to remain open under finite intersections. We hope that this examination of some of the motivation behind the Definition (1.17, 18) has helped the reader.

Returning now to examples (a) and (b) above consider the discrete and the indiscrete or trivial topology for X. It is clear that the discrete topology is the largest topology that we can give to X, largest that is in terms of number of open sets. Also the trivial topology is the smallest topology that we can give to X. These two extremes motivate the idea of comparing two topologies on X; this can be done if the open sets defining one of the topologies on X are completely contained in the open sets defining the other topology on X. More precisely let $T_1=\{X_\alpha\}$ and $T_2=\{X'_\alpha\}$ be the open sets defining two topologies T_1 and T_2 on X. Then if $T_1 \supset T_2$ the topology T_1 is said to be larger than the topology T_2, or, alternatively, T_2 is said to be smaller than T_1. In other words, an open set for T_2 is also an open set for T_1 but not the otherway round. Another terminology used describes the above situation by saying that T_2 is coarser than T_1, or equivalently, T_1 is finer than T_2. The origin of this terminology is that we can think of the open sets of T_1 and T_2 as being like a kind of grid on X. This being so, the more open sets we provide, the finer the grid on X. There are more open sets in T_1 than T_2 hence T_1 is finer than T_2.

The situation $T_1 \supset T_2$ is also referred to by saying T_1 is stronger than T_2 or T_2 is weaker than T_1. Thus we have stronger, larger and finer and their opposites: weaker, smaller and coarser. Finally T_1 and T_2 may be such that neither $T_1 \supset T_2$ nor $T_2 \supset T_1$, and then T_1 and T_2 are not comparable.

There now follows a series of short sections where we introduce, and illustrate by examples, the definitions of some standard and regularly occurring properties of topological spaces.

Neighbourhoods and closed sets

Neighbourhoods

Given a topology T on X, then N is a neighbourhood of a point $x \in X$ if N is a subset of X and N contains some open set X_α to which x belongs. Notice that we do not require N itself to be an open set. However, all open sets X_α which contain x are neighbourhoods of x since they are contained in themselves. Thus neighbourhoods are a little bit more general than open sets.

Examples

(i) $\mathbf{R}$: the real line. Take the usual topology on $\mathbf{R}$ then the interval $[-1, 15]$ is a neighbourhood of the point 0. However, $[-1, 15]$ is not an open set because the points -1 and 15 cannot be contained in an open interval (a, b) which also lies within $[-1, 15]$. Clearly $[-1, 15]$ is a neighbourhood of the points $1, 2, 5, 14$ etc. In fact $[-1, 15]$ is a neighbourhood of all x that belong to the open interval $(-1, 15)$.

(ii) $\mathbf{R}^2$: two dimensional Euclidean space. The *usual* topology for $\mathbf{R}^2$ is constructed as follows: consider all rectangles in $\mathbf{R}^2$ of the form $(a, b) \times (c, d)$ where a, b, c and d are all rationals. Then the open sets of the usual topology are given by all such rectangles, and all possible unions of such rectangles. Select now a point (x_1, x_2) in $\mathbf{R}^2$, say $(x_1, x_2) = (1, 2)$. Then the three sets: $(1, 3) \times (-1, 4)$, $[0, 3) \times [-1, 4)$ and the disc $(x-1)^2 + (y-2)^2 \leq \varepsilon^2, \varepsilon > 0$ are all neighbourhoods of the point $(1, 2)$. The first is an open neighbourhood while the latter two are not.

Closed sets

Let T be a topology on X, then any subset U of X is closed if the complement of U in X (which we write as $X - U$), is an open set. Since the complement of the complement of U is U itslef then we could also infer from this that a set is open when its complement is closed. Notice also that this definition applied to the sets X and ϕ shows that they are both open and closed regardless of the details of the topology T.

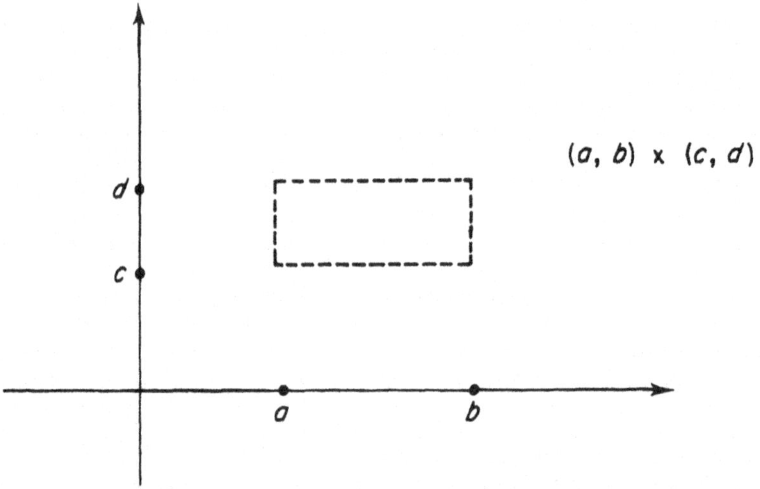

Figure 1.9

Example 1. The interval $[a, b]$. Take the pair of open intervals $(-\infty, a)$ and (b, ∞), their union, $(-\infty, a) \cup (b, \infty)$, is open under the usual topology on $\mathbf{R}$. The complement of $(-\infty, a) \cup (b, \infty)$ is the closed interval $[a, b]$. Thus we see that all closed intervals are closed.

Example 2. The plane $\mathbf{R}^2$ with the usual topology has an obvious closed set: any rectangle of the form $[a, b] \times [c, d]$. In fact by intersecting enough rectangles so as to form polygons with n vertices, and then taking the limit $n \to \infty$ to form a circle $x^2 + y^2 = a^2$, we can see that the disc B^2 defined by $x^2 + y^2 \leq a^2$ is also closed.

Closure of a set

Another closely related notion to that of a closed set is that of the closure of a set. Consider a set U, in general there will be many closed sets which contain U. Denote the family of closed sets with this property by $\{F_\alpha\}$. Then the intersection of all the F_α written $\bigcap F_\alpha$, is called the closure of U and is denoted by $\bar{U}$. In intuitive terms, the intersection of all the F_α is the smallest closed set which contains U.

Example. For the real line $\mathbf{R}$ we can see immediately that if $U = (a, b)$, then $\bar{U} = [a, b]$. Notice that the operation of taking the closure of U rendered the open set (a, b) closed. This will always be so. Further, there is nothing to be gained from taking the closure twice, or $\bar{\bar{U}} = \bar{U}$. If then a set is already closed than it is its own closure, the converse is also true. All these statements became very reasonable and intuitively clear if one works through some simple examples in $\mathbf{R}^n$ such as the ones we have used above.

Boundary and interior

The interior of a set U is the union of all open subsets O_α of U, and is written U^0; evidently $U^0 = \bigcup_\alpha O_\alpha$. In intuitive terms U^0, the interior of U, is the largest open subset of U.

Example 1. Take U to be the closed set B^2 in $\mathbf{R}^2$. Then U^0 is the open set $x^2+y^2<a^2$. On the other hand if we take U to be the open set $x^2+y^2<a^2$, then $U^0=U$. In fact as should be expected by now, $U^0=U$ if and only if U is open.

The boundary of a set U which we shall write $b(U)$ is the complement of the interior of u in the closure of U: $b(U)=\overline{U}-U^0$.

Example 2. If we take the real line $\mathbf{R}$, and U is the set $[a, b)$ then we see that $U^0=(a, b)$ and $\overline{U}=[a, b]$ so that

$$b(U)=\overline{U}-U^0=\{a, b\} \tag{1.24}$$

Note that the sets (a, b), $[a, b]$, $[a, b)$ and $(a, b]$ all have the same boundary namely the pair of points $\{a, b\}$. One can also see from this sort of example that open sets are always disjoint from their boundaries, and that closed sets always contain their boundaries:

$$\begin{aligned} U \cap b(U) &= \phi \Leftrightarrow U \text{ is open} \\ b(U) \subset U &= \Leftrightarrow U \text{ is closed} \end{aligned} \tag{1.25}$$

Example 3. The boundary of B^2 in the usual topology on $\mathbf{R}^2$ is the circle $x^2+y^2=a^2$.

Example 4. With the usual topology on $\mathbf{R}^2$ and the choice: U is the set all points with rational coordinates $(p/q, p'/q')$, we find that $b(U)$ is the whole set $\mathbf{R}^2$. This latter example is an instance of an interesting situation. Rather than the boundary $b(U)$ being disjoint from U, the boundary $b(U)$ contains U. It is also worth noticing that the interior of U is the empty set $\varnothing$. This is because none of the subsets of U are open. If we refer now to the definition of $b(U)$

$$b(U)=\overline{U}-U^0 \tag{1.26}$$

Then we have

$$b(U)=\overline{U}-\varnothing=\overline{U} \tag{1.27}$$

But since

$$b(U)=\mathbf{R}^2$$

then

$$\bar{U} = \mathbf{R}^2$$

In other words the closure of U is also the whole of $\mathbf{R}^2$. When this occurs the set U is said to be *dense* in $\mathbf{R}^2$: more formally, a set U is *dense* in X if the closure $\bar{U}$ or U is X.

Example 5. Take the usual topology on $\mathbf{R}$, then it is clear from the above that the set of all rationals $\mathbf{Q}$ is dense in $\mathbf{R}$. So also though, is the set of all irrationals.

Compactness

Compactness is a very important notion in topology. As with the definition of a topological space its definition needs to be illuminated by some additional discussion. We begin with the definition and follow with the discussion.

A preliminary definition that we need is that of a cover of a set U. Given a family of sets $\{F_\alpha\} = F$ say, then F is a *cover* of U if $\bigcup F_\alpha$ contains U. If all the F_α happen to be open sets then the cover is called an *open cover*. Now consider the set U and all its possible open coverings. The set U is *compact* if for *every* open covering $\{F_\alpha\}$ with $\bigcup F_\alpha \supset U$ there always exists a *finite* subcovering $\{F_1, \ldots, F_n\}$ of U with $F_1 \cup F_2 \ldots \cup F_n \supset U$.

A good way to begin to understand what compactness means is to first examine some sets which are not compact. As usual the simplest place to begin is in Euclidean space $\mathbf{R}^2$. Take for example the infinite strip $|x| \leqslant a$ shown in Fig. 1.10.

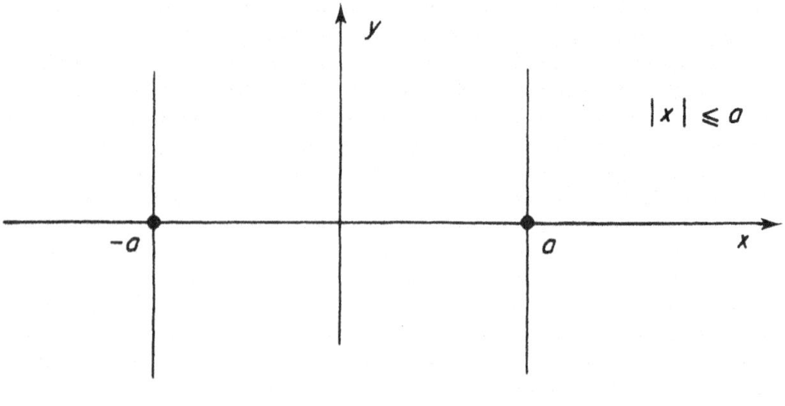

Figure 1.10

Let us denote the strip by X, then an open covering of X is provided by choosing $\{F_\alpha\}$ to be the infinite set of overlapping rectangles where

$$F_\alpha = \begin{cases} |x| < a + \varepsilon, \varepsilon > 0 \\ \alpha/2 < y < \alpha/2 + 1 \end{cases}$$

$$\alpha = \ldots -2, -1, 0, 1, 2, \ldots \tag{1.29}$$

In other words each F_α is open, has width $2a + 2\varepsilon$, and height 1, and all the F_α overlap with their two immediate neighbours. Undoubtedly $\bigcup F_\alpha = X$ so that $\{F_\alpha\}$ is an open cover of X. But there is no finite subcovering of X. Suppose there were a finite subcovering given by the set $\{F_{\alpha_{i_1}}, \ldots, F_{\alpha_{i_n}}\}$. The total area covered by these n rectangles is finite so they cannot cover the strip X which is infinite in area. Since we have found one particular open cover, namely (1.29), which has no finite subcover then X is not compact. In fact all sets in $\mathbf{R}^2$ with infinite area can be seen to be non-compact by choosing for them an open covering $\{F_\alpha\}$ in which all the F_α have finite area—any finite subcovering hence has a finite area and cannot therefore be a cover. We see therefore that for a set X to have a chance of being compact it must at least be finite in area (or volume if we have $\mathbf{R}^n$ in mind). What else is needed before X is guaranteed to be compact? The answer is that if X is a subset of $\mathbf{R}^n$ it will be compact if and only if it is closed and bounded. The boundedness condition is quite close to our experimentally derived notion of finite area or volume; the requirement that X be closed is the additional restriction that we speculated the existence of above. Let us illustrate this by an example. We simply take an open, bounded set in $\mathbf{R}^2$ and show that it is not compact. Let X be the open unit disc B^2: $B^2 = \{(x, y): x^2 + y^2 < 1\}$.

Now choose as an open covering of B^2 a certain family $\{F_\alpha\}$ of concentric open discs, namely

$$F_\alpha = \left\{(x, y): x^2 + y^2 < 1 - \frac{1}{\alpha + 1}\right\}$$

$$\alpha = 1, 2, \ldots \tag{1.30}$$

So F_1 is an open disc of radius $\frac{1}{2}$, F_2 is an open disc of radius $\frac{2}{3}$ etc. Evidently

$$\bigcup_\alpha F_\alpha = X \tag{1.31}$$

Suppose X is compact then we have

$$F_{\alpha_{i_1}} \cup F_{\alpha_{i_2}} \ldots \cup F_{\alpha_{i_n}} = X \tag{1.32}$$

for some $\alpha_{i_1}, \ldots, \alpha_{i_n}$. If (1.32) holds then one of the F_{α_i}, $F_{\alpha'}$ say, has a larger radius than all the others. It will also therefore contain all the other

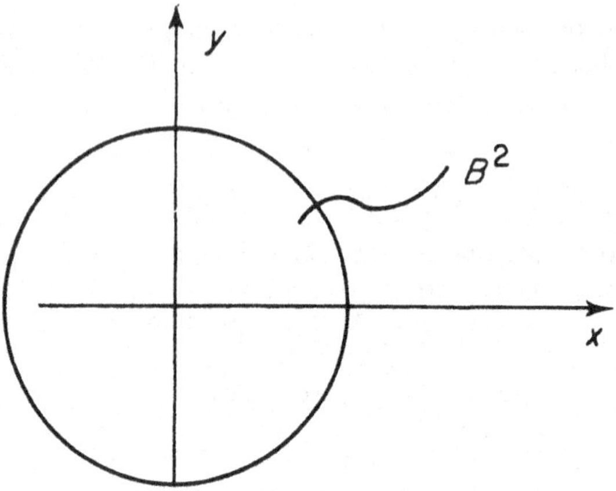

Figure 1.11

F_{α_i}, so (1.32) is really

$$F_{\alpha'} = X \tag{1.33}$$

But $F_{\alpha'}$ has by (1.30) radius r_{α}, where

$$r_{\alpha'} = 1 - \frac{1}{\alpha'+1} \tag{1.34}$$

note though that $r_{\alpha'} < 1$, and so $F_{\alpha'}$ cannot be all of X as asserted in (1.33). Thus X is not compact. It is also clear that it is the open character of X that is forcing it to forego the mantle of compactness. To see this consider the closed disc $Y = \{(x, y){:}\ x^2 + y^2 \leqslant 1\}$. For Y the family $\{F_\alpha\}$ given in (1.30) is not quite open covering because the boundary of Y is not included in the $\{F_\alpha\}$. However a small modification of the $\{F_\alpha\}$ designed to include the boundary, also allows Y to don the mantle of compactness. For example choose for Y the open covering $\{F^\varepsilon_\alpha\}$ where

$$\begin{gathered} F^\varepsilon_\alpha = \left\{(x, y){:}\ x^2 + y^2 < 1 - \frac{1}{\alpha+1} + \varepsilon\right\} \\ \varepsilon \sim 0,\ \varepsilon > 0;, \qquad \alpha = 1, 2, \ldots \end{gathered} \tag{1.35}$$

Certainly we now have

$$\bigcup_\alpha F^\varepsilon_\alpha \supseteq Y \tag{1.36}$$

also there exists a finite subcover of Y namely the family $\{F_1, \ldots, F_n\}$

where n is the first integer for which

$$n > \frac{1}{\varepsilon} - 1 \tag{1.37}$$

The requirement (1.37) is simply the condition that the radius r_n of F_n be greater than one. In fact the single set F_n contains Y and will also serve as a finite subcover for Y. The main point of this lesson concerning the open and closed discs X and Y is that once one makes the transition from the open disc X to the closed disc Y, all open coverings of Y will have finite subcoverings. Finally, to repeat ourselves there is the result that if $X \subset \mathbf{R}^n$ then X is compact if and only if X is closed and bounded. If X is not a subset of $\mathbf{R}^n$ then to decide on the compactness of X we have to apply the more abstract and more general definition in terms of coverings and subcoverings. Our intuitive discussions of compactness, and indeed of our other topological notions, have all drawn on examples taken from $\mathbf{R}^n$. This of course is not completely general. We should not worry however. All these notions had their beginnings in simple spaces like $\mathbf{R}^n$ and as long as we are aware of the above proviso we shall not fall into serious error.

Connectedness

We wish to give a formal definition of what is meant by saying that a set X is connected.

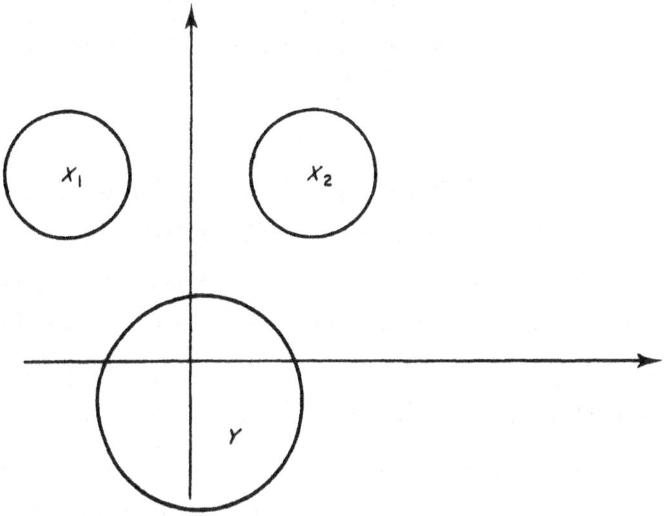

Figure 1.12

For example, in Fig. 1.12, let X be the set $X_1 \cup X_2$ and Y be as shown. It seems reasonable to associate some form of disconnectivity with X (because it is made up of two pieces), but not with Y. We define a set X to be connected if it cannot be written as:

$$X = X_1 \cup X_2 \tag{1.38}$$

where X_1 and X_2 are both open and $X_1 \cap X_2 = \emptyset$. The application of (1.38) to the examples in Fig. 12 is immediate. We have no difficulty in providing examples of connected and disconnected sets. Any discrete set in $\mathbf{R}^n$ containing more than one point is disconnected, the rationals $\mathbf{Q}$ are disconnected, the interval $[a, b]$ is connected and so on. There are several other connectivity properties that topological spaces may possess; spaces can be path connected, arcwise connected and simply connected to mention a few. We shall introduce these definitions when the need for them arises.

1.3 HOMEOMORPHISMS, HOMOTOPY, AND THE IDEA OF TOPOLOGICAL INVARIANTS

Having encountered a good deal of the basic definitions and conceptual material that occur in topology, we can now proceed to another stage in our study of the subject.

The main general idea in topology is to study spaces which can be continuously deformed into one another. This idea is given mathematical substance by the introduction of homeomorphisms. If we take two topological spaces T_1 and T_2 then a map α from T_1 to T_2:

$$\alpha : T_1 \to T_2 \tag{1.39}$$

is called a homeomorphism if it is both continuous and has an inverse which is also continuous. From the equal treatment that this definition accords to α and α^{-1}, we see that if α is a homeomorphism then so is α^{-1}. Indeed if we introduce a third space T_3 then we can immediately verify that if T_1 is homeomorphic to T_2 and T_2 homeomorphic to T_3, then by composing the two homeomorphisms T_1 is homeomorphic to T_3. This means that we immediately are able to divide all topological spaces up into equivalence classes. A pair of spaces T_1 and T_2 belong to the same equivalence class if they are homeomorphic.

The next stage in the topologist's work is to produce enough mathematical criteria to characterise any particular equivalence class. Given any topological space one could, if this were done, place it in its appropriate equivalence class. Except in certain restricted cases, for example two-dimensional closed surfaces, this characterisation is always incomplete and under construction.

However this incompleteness is a feature of most healthy research fields in the pure sciences. The idea behind the characterization is to produce enough topological invariants, i.e. things which do not change under homeomorphisms, to uniquely specify each equivalence class. These topological invariants take many forms: they can be integers such as the dimension n of $\mathbf{R}^n$, they can be certain specific properties of topological spaces such as connectedness or compactness, they can be whole mathematical structures such as homotopy groups, homology and cohomology groups and so on. These latter structures give rise to the subject of algebraic topology. In the search for these invariants two other notions are used, isotopy and homotopy. These notions are inspired, as is the notion of homeomorphism, by the more informal or intuitive notion of deformation. Homotopy is used far more widely than the more difficult and more restrictive notion of isotopy. We shall not therefore deal with isotopy. We turn then to a discussion of homotopy. To follow this discussion we provide a proof of the topological invariance of the dimension n of $\mathbf{R}$, and of the properties of compactness and connectedness.

It is a matter of history that homotopy is now one of the most important and powerful notions in topology. While homeomorphism generates equivalence classes whose members are topological spaces, homotopy generates equivalence classes whose members are continuous maps. Take two continuous maps α_1 and α_2 from a space X to a space Y:

$$\begin{aligned} \alpha_1 &: X \to Y \\ \alpha_2 &: X \to Y \end{aligned} \tag{1.40}$$

(By space we mean of course topological space, note also that we do not require α_1 and α_2 to be invertible as well as continuous.) Then the map α_1 is said to be homotopic to the map α_2 if α_1 can be deformed into α_2, in precise mathematical terms:

$$F: X \times [0, 1] \to Y, \quad F \text{ continuous} \tag{1.41}$$

and F satisfies

$$\begin{aligned} F(x, 0) &= \alpha_1(x) \\ F(x, 1) &= \alpha_2(x) \end{aligned} \tag{1.42}$$

In other words as the real variable t in $F(x, t)$ varies continuously from 0 to 1 in the unit interval $[0, 1]$; the map α_1 is deformed continuously into the map α_2. Homotopy is clearly an equivalence relation and it divides the space of continuous maps from X to Y, which is written $C(X, Y)$, into equivalence classes. Since a homeomorphism is also a continuous map, these equivalence classes are unchanged under homeomorphism of X or Y.

These homotopy equivalence classes are therefore topological invariants of the *pair* of spaces X and Y. How is homotopy to be of use in classifying topological spaces? The trick is to always select for one of X and Y the same topological space, $X = S^n$ the n-sphere is a common choice.

Then allow Y to vary through the family of topological spaces to be studied. So we understand how one space Y' differs from another space Y by comparing them both, in the sense of homotopy, to the same space X which might typically be S^n. We study therefore the equivalence classes under homotopy, written $[S^n, Y]$ of $C(S^n, Y)$ as Y changes from space to space. Suppose now that we have two different equivalence classes E_1 and E_2 in $C(S^n, Y)$. We *cannot* deform *continuously* maps in E_2 into maps in E_1, this leads us to think intuitively that there must be something discrete in the way—some topological obstruction. These obstructions are, as we mentioned above, topological invariants of the pair of spaces Y and S^n, and so can also be called invariants of Y alone. This is how, in general, homotopy invariants arise. The majority of topological invariants are in fact homotopy invariants. Indeed the equivalence classes of $C(S^n, Y)$ can be furnished with a group structure. Then they become the well known homotopy groups $\Pi_n(Y)$. we shall see all this when we return to homotopy groups in a later chapter.

1.4 TOPOLOGICAL INVARIANCE OF COMPACTNESS AND CONNECTEDNESS

We deal first with compactness. Let X be a topological space, f be a homeomorphism to the set Y:

$$f: X \to Y \tag{1.43}$$

and $\{F_\alpha\}$ be an open cover of Y.

Since f is continuous then $f^{-1}(F_\alpha)$ is an open set in X and since f is invertible $\bigcup_\alpha f^{-1}(F_\alpha)$ must be an open cover of X. Now because X is compact the open cover $\{f^{-1}(F_\alpha)\}$ of X always has a finite subcover $f^{-1}(F_{\alpha_{i_1}}), \ldots, f^{-1}(F_{\alpha_{i_n}})$. This is therefore also a finite subcover of Y and so Y must be compact. Now take a connected topological space X and α a homeomorphism to Y. Suppose Y were disconnected, then we would have:

$$\begin{gathered} Y = Y_1 \cup Y_2,\ Y_1 \cap Y_2 = \phi \\ Y_1,\ Y_2 \text{ open} \end{gathered} \tag{1.44}$$

Then because f is continuous the sets $f^{-1}(Y_1)$ and $f^{-1}(Y_2)$ are both open in X, they also satisfy, however

$$f^{-1}(Y_1)\cup f^{-1}(Y_2)=X \tag{1.45}$$

But this would imply that X was not connected which is impossible. Hence Y is connected.

1.5 INVARIANCE OF THE DIMENSION OF $\mathbf{R}^n$

We begin by comparing $\mathbf{R}$ with $\mathbf{R}^2$. Suppose that there is a homeomorphism α from $\mathbf{R}$ to $\mathbf{R}^2$:

$$\alpha:\mathbf{R}\to\mathbf{R}^2 \tag{1.46}$$

If such an α existed then the dimension of $\mathbf{R}$ and $\mathbf{R}^2$ could not be topological invariants. To see that α cannot exist we present the following argument: let $\mathbf{R}^2$ be the (X, Y) plane and $\mathbf{R}$ be its X-axis. Now delete the point $(0, 0)$ from $\mathbf{R}$ and hence the point $X=0$ from $\mathbf{R}$. This transforms $\mathbf{R}$ into a disconnected set but does not make $\mathbf{R}$ disconnected. Thus these two spaces, $\mathbf{R}^2-\{(0, 0)\}$ and $\mathbf{R}-\{0\}$ are *not* homeomorphic. However if α exists then consider the restriction of α to $\mathbf{R}-\{0\}$. This will be a homeomorphism of $\mathbf{R}-\{0\}$ to $\mathbf{R}^2-\alpha\{0\}$, however these two sets are not homeomorphic and hence neither are $\mathbf{R}$ and $\mathbf{R}^2$. The next stage in the argument is to compare $\mathbf{R}^2$ with $\mathbf{R}^3$. $\mathbf{R}^2$ is again the (X, Y) plane while $\mathbf{R}^3$ is the three dimensional X, Y, Z space. As before we suppose α is a homeomorphism from $\mathbf{R}^2$ to $\mathbf{R}^3$ and then we delete the same point from both $\mathbf{R}^2$ and $\mathbf{R}^3$. The restriction of α to $\mathbf{R}^2-\{(0, 0)\}$ should then be a homeomorphism from $\mathbf{R}^2-\{(0, 0)\}$ to $\mathbf{R}^3-\alpha\{0, 0)\}$. We show that this is impossible and hence have our result. Take the two sets $\mathbf{R}^2-\{(0, 0)\}$ and $\mathbf{R}^3-\alpha\{(0, 0)\}$ and consider the circle $x^2+y^2=a^2$. If we consider the limit $a\to 0$, i.e. the circle shrinking smaller and smaller, then we obtain two distinct situations. In $\mathbf{R}^3-\alpha\{(0, 0)\}$ the circle may be shrunk right down to a point since it may if necessary, deformed out of the X, Y plane so as to avoid the point $\alpha\{(0, 0)\}$. But in $\mathbf{R}^2-\{(0, 0)\}$ no deformation can help the shrinking circle to avoid the origin $(0, 0)$, and so it must stop short of shrinking to a point. This would clearly mean a breakdown of the continuity of the supposed homeomorphism α, and so the desired argument is provided. We can continue in this way as the dimension n increases, and by induction we show that $\mathbf{R}^n$ and $\mathbf{R}^m$ are never homeomorphic† unless $n=m$. For example in the general induction

† It is very important that the map α should be a homeomorphism, i.e. continuous and invertible. There are examples of continuous but non-invertible maps from $\mathbf{R}^n$ to $\mathbf{R}^m$, $n\neq m$. There are also examples of invertible but discontinuous maps from $\mathbf{R}^n$ to $\mathbf{R}^m$, $n\neq m$. We see that the topological invariance of dimension really requires α to be a homeomorphism. For an account of this c.f. reference 1.

step, one compares $\mathbf{R}^n$ with $\mathbf{R}^{n+1}$, then one deletes a point p from both spaces to obtain $\mathbf{R}^n - \{p\}$ and $\mathbf{R}^{n+1} - \{p\}$. Then one surrounds the point p with the sphere S^{n-1} and contemplates what happens as the radius of the sphere tends to zero. They may also like to peruse the following more concise description of the method of proof: one considers the equivalence classes of $C(S^{n-1}, \mathbf{R}^n - \{p\})$ and those of $C(S^{n-1}, \mathbf{R}^{n+1} - \{p\})$ and shows that they are not the same. Some technical remarks relevant to our discussion of dimensionality are conveniently supplied to the reader now. The first point is that the circle S^1 in the comparison of $\mathbf{R}$ and $\mathbf{R}^2$ is deviously a prototype of any closed path or loop in a topological space X. When such loops can be continuously contracted to a point, X is called *simply connected.* Secondly, when we restrict the continuous map α to the set $\mathbf{R}^n - \{p\}$, the assumption that α remains continuous under this restriction is tacit. This involves the idea of *relative topology*. The general situation is that one has a subset Y of a topological space X, and one wants to know how, given the open sets $\{X_\alpha\}$ which provide X with its topology, to give Y a topology. Y is given a topology called the *relative topology* of Y to X, the open sets $\{Y_\alpha\}$ of the relative topology are all intersections of the form $X_\alpha \cap Y$ so that $\{Y_\alpha\} = \{X_\alpha \cap Y\}$. Having done this, is the restriction of our particular map α to Y, where $Y = \mathbf{R}^n - \{p\}$, continuous in the relative topology? The answer is that it is because the sets $\mathbf{R}^n - \{p\}$ and $\mathbf{R}^{n+1} - \{p\}$ are both open sets in both their two kinds of possible topology (relative or otherwise). The verification of continuity according to (1.18) is then immediate.

Finally the invariance of the dimension of $\mathbf{R}^n$ also shows that any differentiable manifold has a dimension which is a topological invariant. This is because one can, and does, define the dimension of a manifold to be the dimension of the Euclidean space to which it is locally isomorphic. This brings us to the end of this chapter—the next chapter is an introduction to differentiable manifolds.

REFERENCE

1. HUREWICZ, W. and WALLMAN, H., "Dimension Theory". Princeton University Press, 1948.

CHAPTER 2

Differential Geometry: Manifolds and Differential Forms

We now wish to introduce some of the basic mathematics of differential geometry. We shall start with manifolds and differential forms.

2.1 MANIFOLDS

We expect that the reader has some idea, if only an intuitive one, of what a manifold is. Now some general remarks: if one thinks of topology as the natural mathematical structure within which to study continuity, then the theory of differentiable manifolds is the natural mathematical structure within which to study differentiability. Of course all differentiable functions are also continuous, so to study differentiability is to be less general. However, one might also expect to get more in the way of constructive mathematical results, rather than results of the classification and existence type that are produced by topology. This is indeed so. Topology will still be present though. In fact we shall find that many topological invariants can be expressed in terms of the local geometry of manifolds. For an example of how our distinction between continuous and differentiable operates consider a cube and the sphere S^2 in Fig. 2.1. The map α in Fig. 2.1 represents the continuous deformation of the cube into the sphere. But because of the right-angled bends at the edges of the cube,

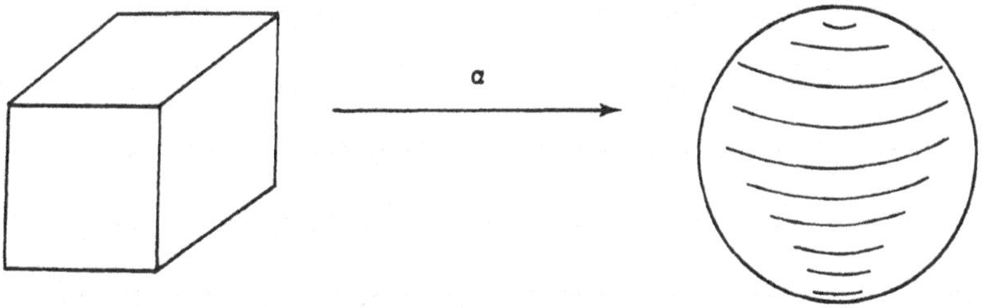

Figure 2.1

we do not expect α to be differentiable along these edges. This leads us to reject a cube as a candidate for a two-dimensional smooth manifold. We do not reject though, a smooth surface such as the sphere S^2 (by smooth we mean arbitrarily differentiable). Nevertheless, considered topologically these two surfaces are equivalent since they are homeomorphic. Before preceeding further we define a manifold.

M is a differentiable manifold if:

i. M is a topological space.

ii. M is provided with a family of pairs $\{(M_\alpha, \phi_\alpha)\}$.

iii. The M_α are a family of open sets which cover M: $\bigcup_\alpha M_\alpha = M$. The ϕ_α are homeomorphisms from M_α to an open subset 0_α of $\mathbf{R}^n$, $\phi_\alpha : M_\alpha \to 0_\alpha$.

iv. Given M_α, M_β such that $M_\alpha \cap M_\beta \neq \phi$, the map $\phi_\beta \circ \phi_\alpha^{-1}$ from the subset $\phi_\alpha(M_\alpha \cap M_\beta)$ of $\mathbf{R}^n$ to the subset $\phi_\beta(M_\alpha \cap M_\beta)$ of $\mathbf{R}^n$ is infinitely differentiable, (written C^∞). (2.1)

The family $\{(M_\alpha, \phi_\alpha)\}$ satisfying (ii), (iii) and (iv) is called an atlas. The individual members (M_α, ϕ_α) of the family are called charts. The content of definition (2.1) of a manifold is made up of two main parts: Firstly, (i), (ii) and (iii) assert that M is a space which is locally Euclidean. That is to say M can be covered with patches M_α which are assigned coordinates in $\mathbf{R}^n$ by the ϕ_α. Within one of these patches M looks like a subset of the Euclidean space $\mathbf{R}^n$. Of course we do not expect, in general, M to be like $\mathbf{R}^n$ globally. This depends on how the patches fit together to form the whole of M. Secondly, (iv) asserts that if two patches overlap, as they will often do, then in the overlap region $M_\alpha \cap M_\beta$, we have two sets of coordinates in $\mathbf{R}^n$ available: $\phi_\alpha(M_\alpha \cap M_\beta)$ and $\phi_\beta(M_\alpha \cap M_\beta)$. It further asserts that if we decide to change from one set of coordinates to the other, i.e. to use the function $\phi_\beta \circ \phi_\alpha^{-1}$, then this can be done in a smooth or C^∞ manner. Thus coordinates have been assigned to the manifold M in such a way that if we move throughout M in whatever fashion, the coordinates that we use vary in a smooth manner. This definition is intended to abstract from common surfaces such as S^2, ellipsoids, surfaces of revolution etc., features which underly the geometry of all of them. The axiomatic format (2.1) of the definition is then possible. It will be useful in all well-known examples of smooth surfaces of whatever dimension, but it also provides a base for a general differential geometry, distinct and free from its rather special origins. Figures 2.2 and 2.3 illustrate the use of a chart ϕ_α and the function $\phi_\alpha \circ \phi_\beta^{-1}$, respectively.

The dimension of the manifold M is defined to be the integer n appearing in the $\mathbf{R}^n$ of its definition. The topological invariance of the dimension of $\mathbf{R}^n$ proved in Chapter 1 automatically establishes the topological invariance of the dimension of M.

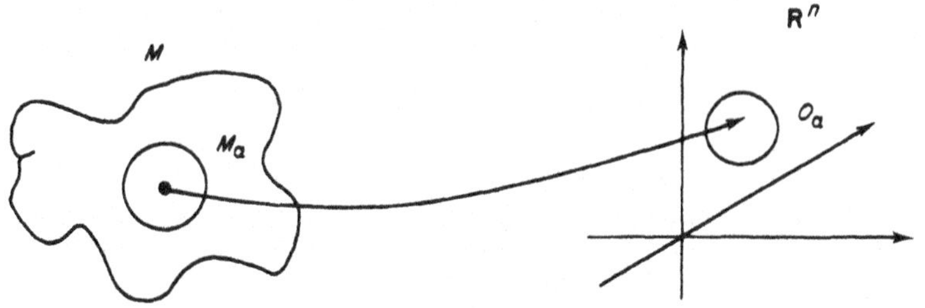

Figure 2.2

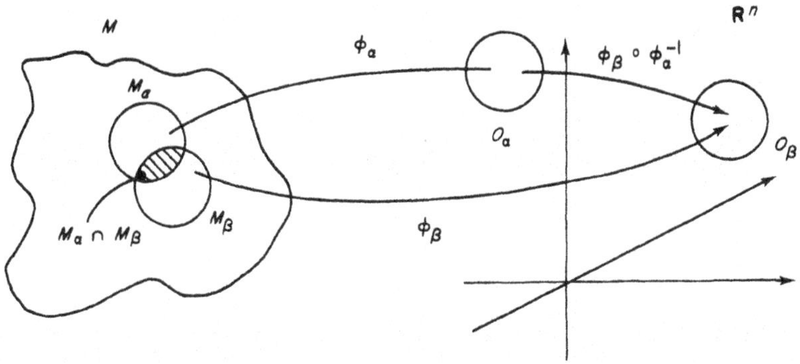

Figure 2.3

Hausdorff and metric spaces

We can conveniently define here the terms Hausdorff space and metric space. Most manifolds M that we shall meet are spaces of this kind. A Hausforff space M is a space for which if any two distinct points x, y, belong to M, then there exists a pair of open sets O_x and O_y, such that $O_x \cap O_y = \phi$ and $x \in O_x$, $y \in O_y$. This is illustrated in Fig. 2.4. Thus no matter how close x and y are, there exist small enough disjoint open sets O_x and O_y which contain x and y respectively.

Example

If we take $M = \mathbf{R}^n$ then M is a Hausdorff space. To see this, let a and b be two distinct points in $\mathbf{R}^n$ and let $|a-b| = d$. Then the two open balls $|x-a| < (d/2) - \varepsilon$ and $|x-b| < (d/2) - \varepsilon$ (where $\varepsilon > 0$ and $\varepsilon < d/2$) are disjoint open sets which contain a and b respectively. We shall assume, unless

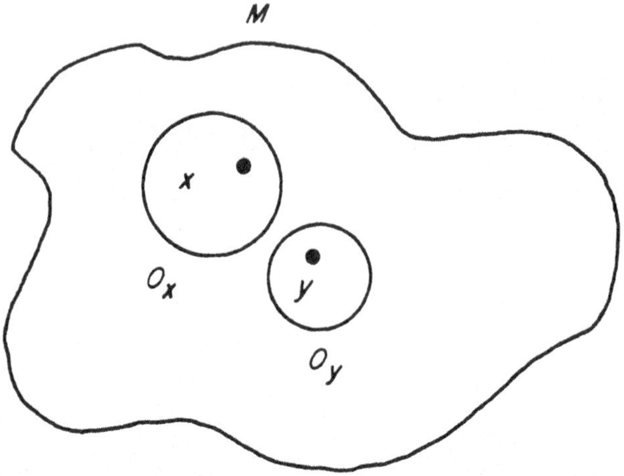

Figure 2.4

there is a specific statement to the contrary, that all our manifolds are Hausdorff.

Metric spaces

A metric space M may be thought of as a particular kind of topological space where the open sets are provided by a distance function $d(x, y)$. The metric $d(x, y)$ is defined so as to correspond to our intuitive notion of the distance between two points:

If $x, y, \in M$, then d satisfies

$$d(x, y) \geqslant 0$$

$$d(x, y) = 0 \Leftrightarrow x = y$$

and if $z \in M$

$$d(x, z) \leqslant d(x, y) + d(y, z) \qquad (2.2)$$

A set M with d satisfying (2.2) is a metric space. It is immediate that M is a topological space. The open sets are all spheres $S_x(\varepsilon)$ and all possible unions of the $S_x(\varepsilon)$ where $S_x(\varepsilon)$ is a sphere centre x and radius ε:

$$S_x(\varepsilon) = \{y : d(x, y) < \varepsilon\} \qquad (2.3)$$

Example 1. If $M = \mathbf{R}^n$ the ordinary Euclidean distance between the points x and y defines a metric. In other words

$$d(x, y) = [(x_1 - x_1) + (x_2 - x_2)^2 \ldots + (x_n - y_n)^2]^{1/2} \qquad (2.4)$$

where $x_1, \ldots, x_n$ and $y_1, \ldots, y_n$ are the Cartesian coordinates of x and y in $\mathbf{R}^n$ defines a metric in $\mathbf{R}^n$.

Example 2. If M is any Riemannian manifold, then M is a metric space. The metric $\mathrm{d}(x, y)$ arises in the following way from the metric tensor g_{ij}: take the points x and $x+\mathrm{d}x$ in M, Then the distance $\mathrm{d}s$ between them is given by $\mathrm{d}s^2 = g_{ij}\,\mathrm{d}x^i\,\mathrm{d}x^j$ in the usual way. Now take two arbitrary points a and b and connect them with a piecewise† once differentiable curve C parameterized by $x(t)$. Let L_{ab} be defined by

$$L_{ab}(C) = \int_a^b g_{ij} \frac{\mathrm{d}x^i}{\mathrm{d}t}\frac{\mathrm{d}x^j}{\mathrm{d}t}\,\mathrm{d}t \tag{2.5}$$

Then $\mathrm{d}(a, b)$ the distance between a and b is defined as the infimum or least upper bound of L_{ab} as L_{ab} varies over all such curves $x(t)$:

$$\mathrm{d}(a, b) = \inf L_{ab}(C) \tag{2.6}$$

Then $\mathrm{d}(a, b)$ defined by (2.6) is a metric and satisfies

$$\begin{aligned} &\mathrm{d}(a, b) \geqslant 0 \\ &\mathrm{d}(a, b) = 0 \Leftrightarrow a = b \\ &\mathrm{d}(a, c) \leqslant \mathrm{d}(a, b) + \mathrm{d}(b, c) \end{aligned} \tag{2.7}$$

in accordance with (2.2). We remind the reader that the metric tensor g_{ij} is positive definite. A space with an indefinite metric g_{ij} such as the Minskowski metric is called a pseudo-Riemannian space.

It is now time to produce some examples of manifolds. To begin with we take examples made simple by choosing their dimension n to be small.

Example 1. Take $n = 1$ and require also that the manifold M be connected. With this restriction there are only two manifolds possible: the real line $\mathbf{R}$ and the circle S^1. The real line $\mathbf{R}$ is of course trivially a manifold. To check that S^1 is a manifold we shall give an atlas for S^1: let S^1 be the circle (x, y): $x^2 + y^2 = 1$ in $\mathbf{R}^2$, then the atlas of S^1 may be given and only contains two charts. Let us write the two charts as (M, f) and (N, g) where M, N, f and g are determined in the following elementary fashion: instead of defining f and g, we define f^{-1} and g^{-1}. We define f^{-1} by saying that f^{-1} is the map from the (open) interval $(0, 2\pi)$ into S^1 under

† The technical term piecewise once differentiable curve C from a to b simply means that rather than $x(t)$ being once differentiable along C, C is partitioned into intervals $[a, a_1]$, $[a_1, a_2] \ldots [a_n, b]$ and $x(t)$ is once differentiable in each interval but with the possibility that the derivatives do not agree at the endpoints.

which $\theta \mapsto (\cos\theta, \sin\theta)$ i.e.

$$f^{-1}:(0, 2\pi) \to S^1$$
$$\theta \mapsto (\cos\theta, \sin\theta) \tag{2.8}$$

The image of $f^{-1}(0, 2\pi)$ in S^1 is the whole of S^1 minus the point $(1, 0)$ i.e. the set $S^1 - \{(1, 0)\}$ *c.f.* Fig. 2.5, the bold dot indicates exclusion of the point $(1, 0)$.

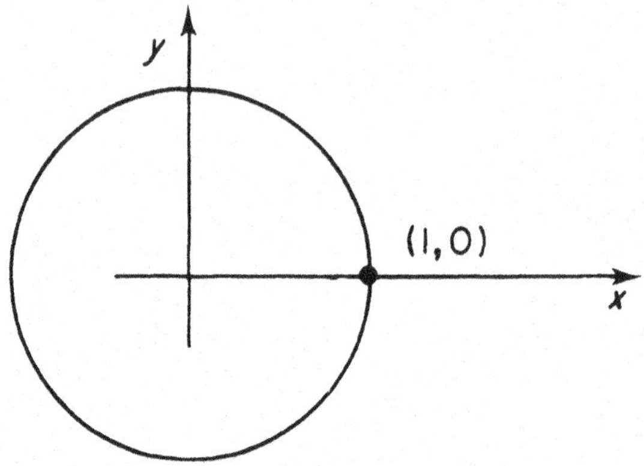

Figure 2.5

Clearly f^{-1} is continuous and invertible and is therefore a homeomorphism. The open set M is then taken to be the set $S^1 - \{(1, 0)\}$. Thus we have defined M and f. To define g we first define g^{-1} and the only way that g^{-1} differes from f^{-1} is that the open set $(0, 2\pi)$ is traded for the open set $(-\pi, \pi)$, The formula for g^{-1} is the same as that for f^{-1}. In other words we have

$$g^{-1}:(-\pi, \pi) \to S^1$$
$$\theta \mapsto (\cos\theta, \sin\theta) \tag{2.9}$$

This time the image of $g^{-1}(0, 2\pi)$ in S^1 is the set $S^1 - \{(-1,0)\}$, c.f. Fig. 2.6.

Again g^{-1} is continuous and invertible and is therefore a homeomorphism. The open set N is obviously taken to be the set $S^1 - \{(-1, 0)\}$ so that now we have defined N and g. Note also that $M \cup N = S^1$ as required by definition (2.1). Thus we have a simple pair of examples namely $\mathbf{R}$ and S^1 of one dimensional manifolds. To obtain higher dimensional examples one can simply take the Cartesian product of the lower dimensional examples. Hence $\mathbf{R} \times \mathbf{R} \ldots \times \mathbf{R} = \mathbf{R}^n$ is an n-dimensional manifold, and $T^n = S^1 \times S^1 \times \ldots S^1$ (n times) is an n-dimensional manifold.

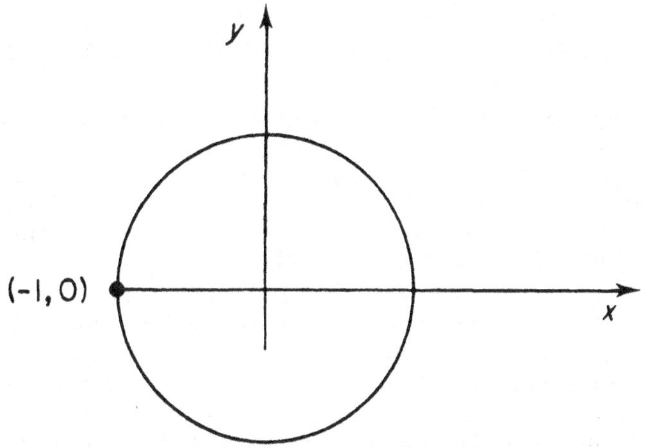

Figure 2.6

Example 2. The manifold T^n above is in fact the n-dimensional torus. If $n = 2$ so that $T^2 = S^1 \times S^1$ then this manifold should correspond to what one usually has in mind when one casually speaks of a torus, c.f. Fig. 2.7.

Now the torus shown in Fig. 2.7 is a subset of $\mathbf{R}^3$. However the torus $T^2 = S^1 \times S^1$ is the Cartesian product of two subsets of $\mathbf{R}^2$ so it is at least a subset of $\mathbf{R}^2 \times \mathbf{R}^2 = \mathbf{R}^4$. Can it in fact differ from the torus shown in Fig. 2.7? The answer is that of course it can, it is only topologically speaking

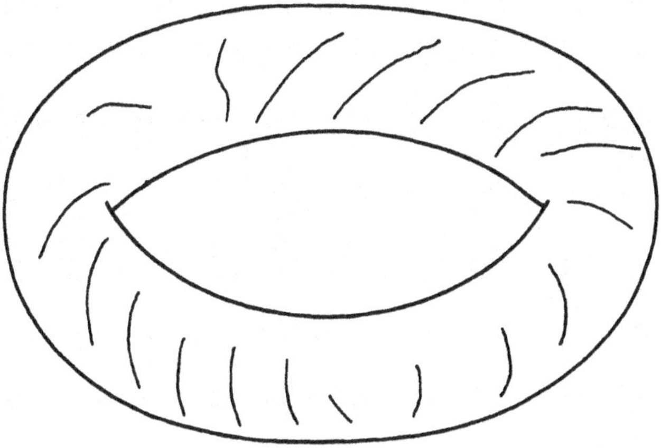

Figure 2.7

that T^2 and the torus of Fig. 2.7 are identified. In mathematical terms there is a homeomorphism from T^2 onto the surface shown in Fig. 2.7. It is worth pursuing this point a little further. The point is that the torus of Fig. 7, which is in $\mathbf{R}^3$, is not flat; but the torus $S^1 \times S^1$ in $\mathbf{R}^4$ can be regarded as being flat. Nevertheless topologically the two tori are the same. All this may seem interesting but a little vague and imprecise. This is indeed so—the precision can only be supplied and the vagueness removed by giving a mathematical notion of flatness—i.e. a definition of curvature. The definition of curvature requires us to specify a Riemannian metric; having done this, one can then answer the question concerning the flatness of T^2. We have not yet dealt with Riemannian metrics and curvature but, as food for thought, we provide a more precise statement of the properties of the torus: the torus of Fig. 2.7 can be endowed with a natural Riemannian metric g_{ij} by taking the Euclidean metric in $\mathbf{R}^3$ and then imposing the restriction that the points in $\mathbf{R}^3$ lie on the torus. With respect to this metric g_{ij} we have a curved torus. On the other hand $T^2 = S^1 \times S^1$ can now be given a metric by taking the distance between two points $t = (x, y)$ and $t' = (x', y')$ in T^2 to be $\mathrm{d}(t, t')$ where

$$\mathrm{d}(t, t') = [(x - x')^2 + (y - y')^2]^{1/2} \tag{2.10}$$

A slight subtlety here is that $x - x'$ and $y - y'$ must be taken to be the minimum distances between x and x' and y and y'. These points have two possible distances between them corresponding to the two possible arcs which connect the points in S^1, c.f. Fig. 2.8.

In any case, it is with respect to this metric that T^2 is flat. Further it is not possible for this flat T^2 to lie in R^3. This is because T^2 is a two-dimensional, compact, connected surface, and in fact, for such a surface

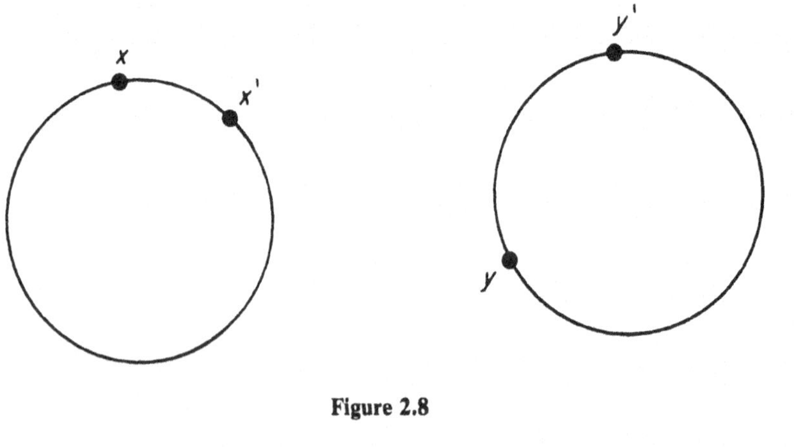

Figure 2.8

to lie in $\mathbf{R}^3$ it must have at least one point of non-zero curvature. We shall deal with these matters when we deal with Riemannian geometry.

Example 3. An example of an n-dimensional manifold which is not a product of n one-dimensional manifolds is provided by the sphere S^n.

It is now clear that examples of manifolds are practically everywhere in mathematics. In fact whenever one deals with differentiable functions of one or more real or complex variables, there is usually a manifold M on which these functions are defined.

2.2 ORIENTABILITY

The property or orientability is an important one for manifolds. It can be defined in a variety of rather different ways. Orientability has its origins in simple observations on the properties of two-dimensional surfaces in $\mathbf{R}^3$: for example, a disc $x^2+y^2 \leqslant a^2$, $z=0$ has a top-side and a bottom-side, or the sphere S^2 has an inside and an outside, as does the torus T^2. Such two-sided surfaces are called orientable since we can use their two sidedness to define orientations or directions in $\mathbf{R}^3$. For example for the disk the unit normal to the surface n has two possible directions: from top to bottom or from bottom to top corresponding to $\mathbf{n} = \pm\mathbf{k}$ c.f. Fig. 2.9.

Similarly the sphere S^2 has two normal directions at each point: inward or outward pointing normals $\mathbf{n} = \pm\mathbf{e}_r$ where $\mathbf{e}_r$ is a unit vector in the normal direction, c.f. Fig. 2.10.

However, if a surface has not got two sides, then one can see difficulties arising in the use of properties of the surface to define an orientation. This is best understood by taking the simplest example of a one-sided surface,

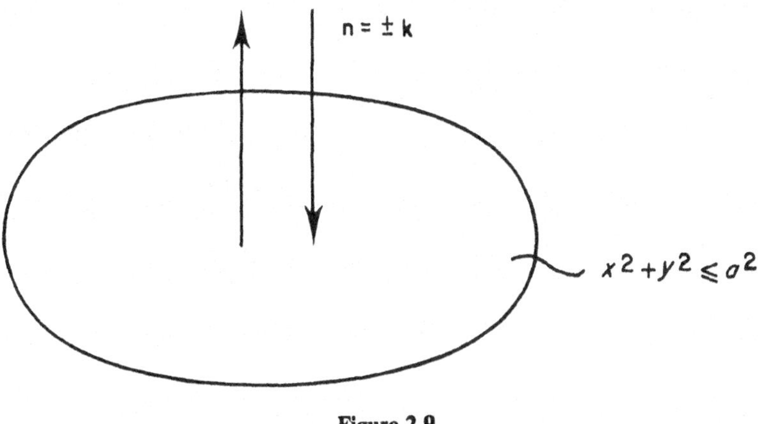

Figure 2.9

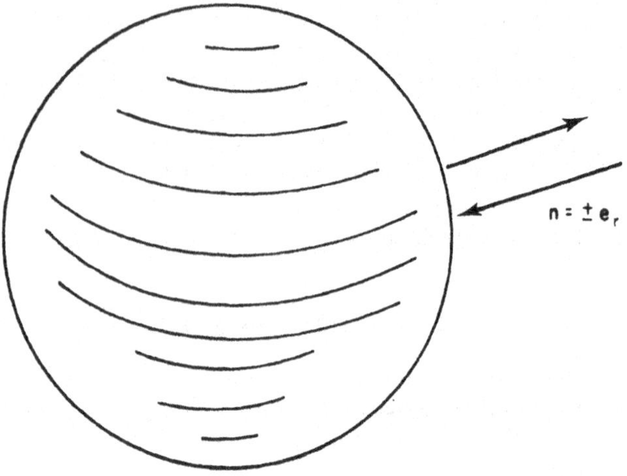

Figure 2.10

this is of course the notorious Möbius strip. The Möbius strip is the two dimensional manifold that is obtained by taking a rectangle and joining two ends having first given one of the ends a twist, c.f. Fig. 2.11.

Now by studying Fig. 2.11 or, better still, by actually constructing a Möbius strip with paper and sellotape, one can verify that it has only one side. Next, one takes a normal $\mathbf{n}$ to the surface at a point p, and moves it round the surface until it returns to the point p. Then one discovers that, because of the necessity of passing through the twist, the normal at p now points in the direction $-\mathbf{n}$ instead of $\mathbf{n}$. This means that the attempt to orient the Möbius strip has to be abandoned—it is a non-orientable surface. In fact, any surface, i.e. any two-dimensional manifold lying in $\mathbf{R}^3$, is orientable if and only if it is two sided.

Having gained some insight into orientability by studying two-dimensional examples, one clearly needs a definition of orientability which can be used in any dimension. We now proceed to such a definition. The key idea is to find a mathematical object which will be able to detect the changes of sign which occur when a manifold is non-orientable. The mathematical object which fulfils this rôle is simply the determinant of a suitable matrix—the Jacobian determinant. Let us see how this arises.

Consider first two different bases $\{\mathbf{e}_1, \ldots, \mathbf{e}_n\}$ and $\{\mathbf{e}'_1, \ldots, \mathbf{e}'_n\}$ of the same vector space V. Let M be the matrix which performs the change of basis:

$$\mathbf{e}'_i = M\mathbf{e}_i, \qquad i = 1, \ldots n. \tag{2.11}$$

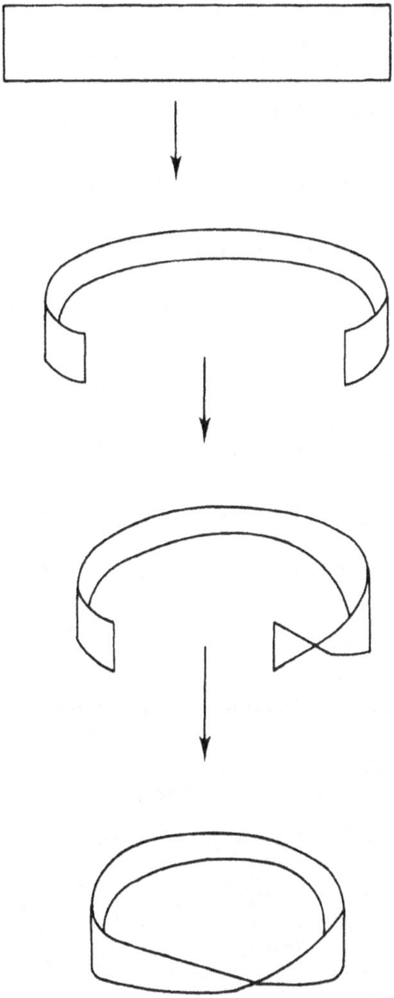

Figure 2.11

Then either det $M > 0$ or det $M < 0$ since M must be invertible. If det $M > 0$ then $\{\mathbf{e}_i\}$ and $\{\mathbf{e}_i'\}$ are said to have the same orientation, if det $M < 0$ the bases are said to have opposite orientations. For example, we can have

$$\{\mathbf{e}_1', \ldots, \mathbf{e}_n'\} = \{\mathbf{e}_1, \mathbf{e}_2, \ldots, \mathbf{e}_{n-1}, -\mathbf{e}_n\} \tag{2.12}$$

so that

$$M = \begin{bmatrix} 1 & & & 0 \\ & \ddots 1 & & \\ & & 1 & \\ 0 & & & -1 \end{bmatrix} \tag{2.13}$$

and

$$\det M = -1 \tag{2.14}$$

or

$$\{\mathbf{e}'_1, \ldots, \mathbf{e}'_n\} = \{\mathbf{e}_1, \mathbf{e}_2, \ldots, \mathbf{e}_{n-2}, -\mathbf{e}_{n-1}, -\mathbf{e}_n\} \tag{2.15}$$

so that

$$M = \begin{bmatrix} 1 & & & 0 \\ & \ddots 1 & & \\ & & -1 & \\ 0 & & & -1 \end{bmatrix} \tag{2.16}$$

and

$$\det M = 1 \tag{2.17}$$

Thus the set of all bases for V is divided into two equivalence classes: each class is transformed into itself by matrices of positive determinant, but one class is transformed into the other by matrices of negative determinant.

Now take an n-dimensional manifold M with atlas the family $\{(M_\alpha, \phi_\alpha)\}$. Suppose that the atlas $\{(M_\alpha, \phi_\alpha)\}$ consists of a single chart—the family $\{(M_\alpha, \phi_\alpha)\} = \{(M, \phi)\}$. Then the coordinates of the manifold M are just given by the function $\phi(x)$, $x \in M$. Since $\phi(x)$ is a point in $\mathbf{R}^n$, it is expressed in terms of some basis, $\{\mathbf{e}_1, \ldots, \mathbf{e}_n\}$ say, and this basis remains fixed for all points x on the manifold. The manifold is orientable because only one basis has to be used for its system of coordinates. However, in general, the family $\{(M_\alpha, \phi_\alpha)\}$ contains many charts and one has to change coordinates in passing from one open set M_α to another open set M_β. In the overlap region $M_\alpha \cap M_\beta$ one can use either the coordinates $\phi_\alpha(M_\alpha)$ or the coordinates $\phi_\beta(M_\beta)$, together with their corresponding bases $\{\mathbf{e}_1, \ldots, \mathbf{e}_n\}$ and $\{\mathbf{e}'_1, \ldots, \mathbf{e}'_n\}$ respectively. If the bases $\{\mathbf{e}_i\}$ and $\{\mathbf{e}'_i\}$ lie in different equivalence classes, then the orientation of M has changed sign. This will be so if the Jacobian matrix $\phi_\beta \circ \phi_\alpha^{-1}$ for changing variables from $\phi_\alpha(M_\alpha)$ to $\phi_\beta(M_\beta)$ has negative determinant. Finally, all this can be put together to give the following definition of orientability:

Given a manifold M with atlas the family $\{(M_\alpha, \phi_\alpha)\}$. M is orientable if $\det(\phi_\beta \circ \phi_\alpha^{-1}) > 0$ for all M_α, M_β such that $M_\alpha \cap M_\beta \neq \emptyset$. (2.18)

We also saw that when a manifold has only one chart it is automatically orientable.

2.3 CALCULUS ON MANIFOLDS

Having devoted a considerable amount of space to giving a set of differentiable coordinates to a manifold, it is reasonable to introduce functions of these coordinates. In fact one can study functions $f(x)$, vectors $v(x)$, and tensors $T(x)$ associated with a manifold M†. The ordinary operations of calculus apply: one can usually differentiate and integrate these mathematical objects. We begin with a discussion of vectors and tensors.

Vectors and tensors

The vectors that we consider are the tangent vectors to the manifold M. Let p be a point on the manifold M and let p trace out the curve $p(t)$.

Suppose that $p(t)$ passes through a chart in which the local coordinates are x^i, $i = 1, \ldots, n$, c.f. Fig. 2.12. In this coordinate patch the coordinates of the curve are

$$x^i(p(t)), \qquad i = 1, \ldots, n \tag{2.19}$$

The velocity vector or tangent vector to this curve is given by

$$\frac{dx^i(p(t))}{dt}, \qquad i = 1, \ldots, n \tag{2.20}$$

It is now established practise in differential geometry to generalize the definition of tangent vector, and to consider a differential operator rather than a vector as being the more fundamental mathematical object. To do this we take a real-valued function $f(p)$ defined on M and consider its rate of change along the curve $p(t)$. The rate of change of $f(p)$ in the direction of $p(t)$ is

$$\frac{d}{dt} f(p(t)) \tag{2.21}$$

† We deal from now on with compact orientable manifolds M unless there is a specific statement to the contrary.

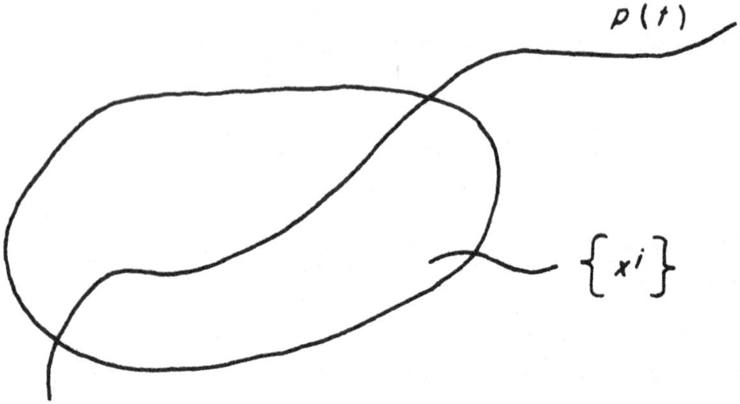

Figure 2.12

Using local coordinates this becomes

$$\frac{\partial f}{\partial x^i}\frac{\mathrm{d}x^i(p(t))}{\mathrm{d}t} \tag{2.22}$$

In other words $\mathrm{d}f(p(t))/\mathrm{d}t$ is obtained by applying the differential operator $\mathbf{X}$ to f where† $\mathbf{X}=a^i(\partial/\partial x^i)$ and $a^i=\mathrm{d}x^i(p(t))/\mathrm{d}t$ i.e.

$$\begin{aligned}\frac{\mathrm{d}f}{\mathrm{d}t}(p(t)) &= \frac{\partial f}{\partial x^i}\frac{\mathrm{d}x^i(p(t))}{\mathrm{d}t}\\ &= \mathbf{X}f\end{aligned} \tag{2.23}$$

It is $\mathbf{X}$ which we now define as the tangent to M at p in the direction given by the curve $p(t)$. If we apply $\mathbf{X}$ to the position coordinates or local coordinates we obtain the traditional velocity vector i.e.

$$\begin{aligned}\mathbf{X}(x^i) &= \frac{\partial x^i}{\partial x^j}\frac{\mathrm{d}x^j}{\mathrm{d}t}\\ &= \frac{\mathrm{d}x^i}{\mathrm{d}t}\end{aligned} \tag{2.24}$$

So a tangent vector is now a differential operator of the form $\mathbf{X}=a^i(\partial/\partial x^i)$. Conversely, every differential operator $\mathbf{0}=O^i(\partial/\partial x^i)$ is the tangent to some

† From now on we use the Einstein summation convention for repeated indices.

curve. To see this, consider the curve passing through the point p at $t = O$ defined by

$$x^i = x^i(p) + O^i t \tag{2.25}$$

Then

$$\left.\frac{dx^i}{dt}\right|_{t=o} = O^i \tag{2.26}$$

Thus $\mathbf{0} = O^i(\partial/\partial x^i)$ is the tangent to this curve at the point p. The space of all possible tangents at the point p is called the tangent space to M at p and is written

$$T_p(M) \tag{2.27}$$

Evidently in terms of local coordinates the differential operators $\partial/\partial x^i$, $i = 1, \ldots, n$ are a basis for $T_p(M)$. Thus $T_p(M)$ has the same dimension as M. These vectors $\mathbf{X}$ that we have defined are called contravariant vectors to correspond with classical nomenclature. To find their companions, the covariant vectors is, in this more general setting, to enter the theory of differential forms. Covariant vectors belong to the vector space dual to the vector space $T_p(M)$ of contravariant vectors. Let us recall, in brief, the important notion of duality. Let V and W be two vector spaces with a linear map α from V to W:

$$\alpha : V \to W$$

The dual spaces of V and W are written V^* and W^* respectively. An element of W^* is any linear map λ say from W to $\mathbf{C}$: (for real vector spaces $\mathbf{C}$ is replaced by $\mathbf{R}$)

$$\lambda : W \to \mathbf{C} \quad \lambda \in W^*$$

Now we may write down the diagram below:

$$V \xrightarrow{\alpha} W \xrightarrow{\lambda} \mathbf{C}$$

Thus α and λ give rise to an element $\gamma \in V^*$ defined by

$$\gamma(v) = \lambda(\alpha(x))$$

In summary, given any element $\lambda \in W^*$, the map α provides us with the map $\gamma \in V^*$ defined above. Thus we have a map $\alpha^* : W^* \to V^*$ defined by

$$\alpha^*(\lambda) = \gamma, \quad \lambda \in W^*, \quad \gamma \in V^*$$

with λ and α defined as above. The important point to remember is that the passage to the duals of V and W changes a map $\alpha : V \to W$ into a map

$\alpha^*: W^* \to V^*$ (duality reverses the direction of maps)

$$V \xrightarrow{\alpha} W$$

$$V^* \xleftarrow{\alpha^*} W^*$$

We now return to the tangent space $T_p(M)$. The dual of $T_p(M)$ is written

$$T_p^*(M) \tag{2.27a}$$

The prototype covariant vector is called a one-form $\boldsymbol{\omega}$. A simple example of a one-form is the differential df of a function f where, in terms of local coordinates,

$$\mathrm{d}f = \frac{\partial f}{\partial x^i}\,\mathrm{d}x^i \tag{2.28}$$

Now an element of the dual of $T_p(M)$ should, by definition, act on an element $\mathbf{X}$ of $T_p(M)$ to give a real number. This action of df on $\mathbf{X}$ is defined by equation (2.29):

$$\langle \mathrm{d}f, \mathbf{X}\rangle = \mathbf{X}f \tag{2.29}$$

So the real number in question is $\mathbf{X}f$ and equation (2.29) is a definition of the notation $\langle \mathrm{d}f, \mathbf{X}\rangle$. An arbitrary one-form $\boldsymbol{\omega}$ is expressed, in terms of local coordinates, by

$$\boldsymbol{\omega} = a_i\,\mathrm{d}x^i \tag{2.30}$$

where the a_i are now not necessarily the partial derivatives of a function f. The differentials $\{\mathrm{d}x^i\}$ form a basis for $T_p^*(M)$. Now if we choose

$$\begin{aligned} f &= x^i \\ \mathbf{X} &= \frac{\partial}{\partial x^j} \end{aligned} \tag{2.31}$$

for some i and j. Then

$$\begin{aligned} \langle \mathrm{d}f, \mathbf{X}\rangle &= \left\langle \mathrm{d}x^i, \frac{\partial}{\partial x^j}\right\rangle \\ &= \frac{\partial x^1}{\partial x^j} = \delta^i_j \end{aligned} \tag{2.32}$$

The basis $\{\mathrm{d}x^i\}$ of one-forms or covariant vectors is then the basis dual to the basis $\{\partial/\partial x^i\}$ of contravariant vectors. Now if we use (2.32) we can evaluate $\langle \boldsymbol{\omega}, \mathbf{X} \rangle$ for arbitrary $\boldsymbol{\omega}$ and $\mathbf{X}$ by simply passing to local coordinates and expanding as $\boldsymbol{\omega}$ and $\mathbf{X}$ as

$$\boldsymbol{\omega} = a_i \, \mathrm{d}x^i$$
$$\mathbf{X} = b^i \frac{\partial}{\partial x^i} \tag{2.33}$$

Tensors of type (a, b) are constructed in the usual way by taking a factors of $T_p(M)$ and b factors of $T_p^*(M)$. The space $\mathbf{T}_b^a$ of tensors at p of contravariant rank a and covariant rank b is defined by

$$\mathbf{T}_b^a = T_p(M) \otimes \ldots \otimes T_p(M) \otimes T_p^*(M) \otimes \ldots \otimes T_p^*(M) \tag{2.34}$$

where there are a factors of $T_p(M)$ and b factors of $T_p^*(M)$. The usual tensor $T_{j_1 \ldots j_b}^{i_1 \ldots i_a}$ of type (a, b) is recovered by expressing an element of $\mathbf{T}_b^a$ in terms of the local coordinate bases so that it becomes

$$T_{j_1 \ldots j_b}^{i_1 \ldots i_a} \, \mathrm{d}x^{j_1} \ldots \mathrm{d}x^{j_b} \frac{\partial}{\partial x^{i_1}} \ldots \frac{\partial}{\partial x^{i_a}}$$

Further, we allow the point p to vary continuously over the whole manifold M. So we have vectors and tensors varying continuously over the whole of M. We are then said to have vector fields and tensor fields on M. Two of the most studied subjects in differential geometry concern continuous vector fields on M, in particular their zeros, and the critical points of differentiable functions $f(x)$ on M. We shall have more to say on these subjects in due course.

We are now ready to take further our discussion of differentiation on a manifold. So far we have encountered two differential operators on M: $(\partial/\partial x^i)$ and d. The operator d is called exterior differentiation. Now while it is a simple matter to take powers of the operator $\partial/\partial x^i$, and thus consider multiple partial differentiations such as $\partial/\partial x^i \partial x^j$ etc; it is not clear whether the operator d can be raised to a power. In fact powers of d may be defined but it is always true that

$$d^2 = 0 \tag{2.35}$$

This property of exterior differentiation will turn out to be very important. To see how this property arises we must consider the space of antisymmetric tensors of type $(0, r)$. Such a tensor is called an r-form and written $\boldsymbol{\omega}$ where

$$\boldsymbol{\omega} = T_{i_1 \ldots i_r} \, \mathrm{d}x^{i_1} \wedge \ldots \wedge \mathrm{d}x^{i_r} \tag{2.36}$$

The notation $\wedge$ in (2.36) denotes the wedge product and is defined by:

$$\mathrm{d}x^{i_1} \wedge \ldots \wedge \mathrm{d}x^{i_p} = 0$$

if any pair of $dx^{i_1}, \ldots, dx^{i_p}$ are equal; $dx^{i_1} \wedge \ldots \wedge dx^{i_p}$ changes sign if any pair of $dx^{i_1}, \ldots, dx^{i_p}$ are interchanged. $dx^{i_1} \wedge \ldots \wedge dx^{i_p}$ is linear in each factor $dx^{i_1}, \ldots, dx^{i_p}$ separately e.g.

$$\begin{aligned}(\alpha\, dx^{i_1} + \beta\, dx^{i_1}) \wedge dx^{i_2} \wedge \ldots \wedge dx^{i_p} \\ = \alpha\, dx^{i_1} \wedge \ldots \wedge dx^{i_p} + \beta\, dx^{i_1} \wedge \ldots \wedge dx^{i_p}\end{aligned} \tag{2.37}$$

The space of r-forms ω is denoted by

$$\Lambda^r T_p^*(M)$$

or

$$\Omega_p^r(M) \tag{2.38}$$

An r-form ω is said to be an element of degree r in the exterior algebra $\Lambda T_p^*(M)$—the word algebra denoting that forms of all degree are members of the exterior algebra. Further, when the point p varies continuously over the whole of M, we say that ω is an r-form on M. For this we use the notation $\Omega^r(M)$. If ω is the r-form (2.36), then we may assume, without less of generality, that the covariant tensor $T_{i_1 \ldots i_r}$ is totally antisymmetric, i.e. it changes sign under an interchange of any pair of indices. An important property of r- forms on M is that if ω is an r-form on M and $r > n$ then $\omega = 0$. This is because when ω is expressed in terms of local coordinates, as in (2.36), then since $r > n$ there is at least one pair of differentials amongst the $dx^{i_1} \ldots dx^{i_r}$ which are equal. Hence, by (2.37), ω is zero. The definition of an r-form is extended to include ordinary functions f on M. Functions on M are called o-forms. So on a manifold M we have a list of only $n+1$ possible kinds of r-form:

$$\dim M = n$$

o-forms	$f(x)$ functions
1-forms	$a_i\, dx^i$ covariant vectors
2–forms	$T_{ij}\, dx^i \wedge dx^j$ antisymmetric covariant tensors of rank 2
⋮	
n-forms	$T_{i_1 \ldots i_n}\, dx^{i_1} \wedge \ldots \wedge dx^{i_n}$ anti-symmetric covariant tensor of rank n (2.39)

If ω is an r-form then the exterior derivative of ω is written $d\omega$ and is defined by the expression in (2.40).

$$\begin{aligned} \omega &= T_{i_1 \ldots i_r}\, dx^{i_1} \wedge \ldots \wedge dx^{i_r} \\ d\omega &= \frac{\partial T_{i_1 \ldots i_r}}{\partial x^{i_{r+1}}} dx^{i_{r+1}} \wedge dx^{i_1} \wedge \ldots dx^{i_r} \end{aligned} \tag{2.40}$$

We may also write $d\omega$ as:

$$d\omega = dT_{i_1 \ldots i_r} \wedge dx^{i_1} \wedge \ldots dx^{i_r}$$

where

$$dT_{i_1 \ldots i_r} = \frac{\partial T_{i_1 \ldots i_r}}{\partial x^{i_{r+1}}} dx^{i_{r+1}} \tag{2.41}$$

In any case one should note that if ω is an r-form then $d\omega$ is an $(r+1)$-form. So d is a map from $\Omega^r(M)$ to $\Omega^{r+1}(M)$:

$$d: \underset{0 \leq r \leq n}{\Omega^r}(M) \to \Omega^{r+1}(M) \tag{2.42}$$

We can now compute d^2 acting on an arbitrary ω and verify that it is zero. For $d^2\omega$ is given, according to (2.40) by the expression in (2.43) below:

$$\begin{aligned} d^2\omega &= \frac{\partial^2 T_{i_1 \ldots i_r}}{\partial x^{i_{r+1}} \partial x^{i_{r+2}}} dx^{i_{r+2}} \wedge dx^{i_{r+1}} \wedge dx^{i_1} \wedge \ldots \wedge dx^{i_r} \\ &= 0 \end{aligned} \tag{2.43}$$

But this is zero by the symmetry of the partial derivatives acting on $T_{i_1 \ldots i_r}$. This remarkable property of the exterior derivative d has a fundamental significance in topology as we shall see when we describe cohomology groups. There is a further useful mathematical device which it is convenient to introduce now. This is simply a map which, given a pair of manifolds M and N, allows one to pass from vectors, tensors, and forms on one manifold, to vectors, tensor, and forms on the other manifold. One first must be given a C^∞ map ϕ between M and N:

$$\begin{aligned} \phi &: M \to N \\ p &\mapsto \phi(p) \end{aligned}$$

The function ϕ provides a way of mapping vectors on M into vectors on N. Let $T_p(M)$ be the tangent space at the point $p \in M$. Then $T_{\phi(p)}(N)$ denotes the image of $T_p(M)$ under ϕ, and $T_{\phi(p)}(N)$ is a tangent space at the point $\phi(p) \in N$. This map between tangent spaces induced by ϕ is written $\phi_*: T_p(M) \to T_{\phi(p)}(N)$. Its precise action on a vector $\mathbf{X} \in T_p(M)$ is such that given f a function on N, so that $f(\phi(p))$ is a function on M, then

$\phi_*\mathbf{X} \in T_{\phi(p)}(N)$ is defined by

$$(\phi_*\mathbf{X})f = \mathbf{X}f(\phi(p))$$

There is a natural extension of the map ϕ_* to contravariant tensors. If $\mathbf{T}^a_o$ is a tensor on M, then we denote the corresponding tensor on N by $\phi_*\mathbf{T}^a_o$. The map ϕ also has consequences for forms. However, because forms are dual to tensors, ϕ transfers forms from N to M rather than from M to N. More precisely if ω is a 1-form on N, $\omega \in \Omega^1_{\phi(p)}(N)$, then $\phi^*\omega$ denotes a 1-form on M defined by the equation

$$\langle \phi^*\omega, \mathbf{X}\rangle = \langle \omega, \phi_*\mathbf{X}\rangle$$

for any vector $\mathbf{X} \in T_p(M)$. Finally, ϕ^* extends to arbitrary r-forms, so that if $\omega \in \Omega^r_{\phi(p)}(N)$ then $\phi^*\omega \in \Omega^r_p(M)$, in the same way ϕ^* transfers covariant tensors $\mathbf{T}^o_b$ on N to covariant tensors $\phi^*\mathbf{T}^o_b$ on M. In addition to partial differentiation and exterior differentiation, a third differential operator is commonly introduced on a manifold M. This is the operator for covariant differentiation. Its introduction, however, requires, in distinction to the former operators, the supplying of additional mathematical structure to the manifold—a connection must be defined on M. Because this is so, we shall postpone our treatment of covariant differentiation until the chapter on fibre bundles, where we shall deal with connections.

It remains for us to describe the integration theory for differential forms. The first requirement for the integration theory on M is for there to be given a measure or volume element. Differential forms do this automatically. To begin consider an n-form ω such that in one particular coordinate patch U, ω is given by

$$\omega = \mathrm{d}x^1 \wedge \mathrm{d}x^2 \wedge \ldots \wedge \mathrm{d}x^n \tag{2.44}$$

We shall see that ω is a candidate for a volume element. Consider a neighbouring and overlapping coordinate patch V, with local coordinates $y^1, \ldots, y^n$. Let us express ω in terms of the coordinates $y^1, \ldots, y^n$ for V. Evidently

$$\omega = \left(\frac{\partial x^1}{\partial y^{i_1}}\mathrm{d}y^{i_1}\right) \wedge \left(\frac{\partial x^2}{\partial y^{i_2}}\mathrm{d}y^{i_2}\right) \wedge \ldots \wedge \left(\frac{\partial x^n}{\partial y^{i_n}}\mathrm{d}y^{i_n}\right) \tag{2.45}$$

But ω is therefore equal to

$$\left(\frac{\partial x^1}{\partial y^{i_1}} \cdots \frac{\partial x^n}{\partial y^{i_n}}\right)\mathrm{d}y^{i_1} \wedge \ldots \wedge \mathrm{d}y^{i_n} = \mathrm{Det}\left(\frac{\partial x^i}{\partial y^j}\right)\mathrm{d}y^1 \wedge \ldots \wedge \mathrm{d}y^n \tag{2.46}$$

by virtue of the antisymmetry properties of the wedge product and the definition of a determinant. But Det $(\partial x^i/\partial y^j)$ is of course the Jacobian determinant for the change of variables from x^i to y^i. Thus we see that the antisymmetry properties of differential forms automatically give ω the transformation properties necessary for a volume element. Now suppose we wish to integrate a function f defined on M. Consider, at first, only one coordinate patch U_α and let the coordinates in U_α be $x^1, \ldots, x^n$. Then the integral on U_α is defined by equation (2.47) below:

$$\int_{U_\alpha} f\omega = \int_{U_\alpha} f(x_1, \ldots, x_n)\, \mathrm{d}x^1\, \mathrm{d}x^2 \ldots \mathrm{d}x^n \tag{2.47}$$

where the right hand side (RHS) of (2.47) is an ordinary multiple integral of f over U_α, and the left hand side (LHS) is the standard differential form notation for this multiple integral. Clearly for any coordinate patch U_α we can now compute $\int_{U_\alpha} f\omega$ the remaining task is to amalgamate these integrals into an integral of f over the whole manifold M This is a purely technical matter and is done by means of what is called a partition of unity. This we now explain.

Take an open covering $\{U_\alpha\}$ of M which is locally finite, locally finite means that each point of M is only covered by a *finite* number of the U_α. Then consider the family of differentiable functions $e_\alpha(x)$ satisfying:

i. $0 < e_\alpha(x) < 1$
ii. $e_\alpha(x) = 0$ if $x \notin U_\alpha$ (2.48)
iii. $e_1(x) + e_2(x) + \ldots = 1$

The e_α are then called a partition of unity subordinate to the cover $\{U_\alpha\}$. Note that the property of local finiteness means that if we fix a point $x \in M$, then the potentially infinite sum in (iii) contains only a *finite* number of non-zero terms. Now we multiply (iii) by the function $f(x)$ and obtain

$$\begin{aligned} f(x) &= \sum_\alpha f(x) e_\alpha(x) \\ &= \sum_\alpha f_\alpha(x) \end{aligned} \tag{2.49}$$

where

$$f_\alpha(x) = f(x) e_\alpha(x) \tag{2.50}$$

The advantage of the decomposition of $f(x)$ into the functions f_α is that because of (2.48) (ii), f_α is zero outside U_α. Hence the integral of each f_α

separately is defined by (2.47). So we now define $\int_M f\omega$ by:

$$\int_M f\omega = \sum_\alpha \int_{U_\alpha} f_\alpha \omega$$
$$= \sum_\alpha \int_{U_\alpha} e_\alpha(x^1_\alpha, \ldots, x^n_\alpha) f(x^1_\alpha, \ldots, x^n_\alpha)\, dx^1_\alpha \ldots dx^n_\alpha \tag{2.51}$$

where the U_α belong to an atlas for M and the local coordinates inside U_α are $x^1_\alpha, \ldots, x^n_\alpha$. It is an easy exercise for the reader to verify that if we choose a different atlas, and a different partition of unity, the integral remains the same†.

So we have seen that an n-form ω provides a volume element suitable for an integration theory on M. If M is a Riemannian manifold then there is a particular useful n-form ω which can be used as a volume element. In local coordinates ω is given by

$$\omega = g^{1/2}\, dx^1 \wedge dx^2 \wedge \ldots dx^n \tag{2.52}$$

with

$$g = [\det g_{ij}] \tag{2.53}$$

The importance of ω is that it is invariant under change of coordinates. In fact if J is the Jacobian determinant for changing coordinates from x^i to $x^{i'}$, then we have

$$g' = \frac{g}{J^2}$$
$$dx^{1'} \wedge dx^{2'} \wedge \ldots dx^{n'} = J\, dx^1 \wedge dx^2 \wedge \ldots dx^n \tag{2.54}$$

Thus

$$\omega' = \omega \tag{2.55}$$

as desired.

We can now proceed to compute some integrals and in doing so we discover an important property of integrals of differential forms. We start with the simplest sort of differential forms possible: o-forms or functions on M. Clearly we cannot integrate o-forms in the way we have just described. However, if f is a function on M then df is a 1-form. Thus if

† The reader may be interested in knowing whether or not there always exists, on M, a partition of unity subordinate to some locally finite cover. We assume that there is and this amounts to assuming that for every atlas $\{U_\alpha, \phi_\alpha\}$ on M contains a *locally finite* atlas $\{V_\alpha, \phi_\alpha\}$. This property is called paracompactness.

we restrict m so as to have dimension 1, e.g. let M be the interval $[a, b]$ on the real line **R**, then we can compute $\int_M$. We have:

$$\int_M \mathrm{d}f = \int_a^b \frac{\partial f}{\partial x}\,\mathrm{d}x = f(b) - f(a) \tag{2.56}$$

Next we increase the dimension of M by 1 and consider the integration of 2-forms. Rather than considering any 2-form, however, we take only 2-forms $\boldsymbol{\theta}$ where $\boldsymbol{\theta} = \mathrm{d}\boldsymbol{\omega}$ and $\boldsymbol{\omega}$ is a one form. Take M then to be a compact surface with a boundary like the disc $x^2 + y^2 \leqslant 1$ for example. We have:

$$\boldsymbol{\theta} = \mathrm{d}\boldsymbol{\omega}, \qquad \boldsymbol{\omega} = a_i\,\mathrm{d}x^i = a_1\,\mathrm{d}x^1 + a_2\,\mathrm{d}x^2$$

$$\int_M \boldsymbol{\theta} = \int_M \mathrm{d}\boldsymbol{\omega} = \int_{\mathrm{disc}} \frac{\partial a_i}{\partial x^j}\,\mathrm{d}x^j \wedge \mathrm{d}x^i \tag{2.57}$$

But

$$\frac{\partial a_i}{\partial x^j}\,\mathrm{d}x^i \wedge \mathrm{d}x^j = \left(\frac{\partial a_1}{\partial x^2} - \frac{\partial a_2}{\partial x^1}\right)\mathrm{d}x^1 \wedge \mathrm{d}x^2 \tag{2.58}$$

So

$$\int_{\mathrm{disc}} \frac{\partial a_i}{\partial x^j}\,\mathrm{d}x^j \wedge \mathrm{d}x^i = \int_{\mathrm{disc}} \operatorname{curl} \mathbf{A} \cdot \mathbf{ds} \tag{2.59}$$

where $\mathbf{A} = (a_1\mathbf{i} + a_2\mathbf{j})$ and $\mathbf{ds}$ is the element of area on the disc. Hence Stokes' theorem transforms the integral (2.59) to an integral around the boundary C of the disc:

$$\int_M \boldsymbol{\theta} = \int_M \mathrm{d}\boldsymbol{\omega} = \int_C \mathbf{A} \cdot \mathbf{dl} \tag{2.60}$$

But the expression $\int_C \mathbf{A} \cdot \mathbf{dl}$ may be written in the notation of differential forms as:

$$\int_{\partial M} \boldsymbol{\omega} \tag{2.61}$$

(∂M is the boundary of M).

Our result can therefore be written

$$\int_M \mathrm{d}\boldsymbol{\omega} = \int_{\partial M} \boldsymbol{\omega} \tag{2.62}$$

Next we increase the dimension of M by 1 to obtain a three-dimensional manifold with boundary. We integrate a 3-form $\boldsymbol{\theta}$, where $\boldsymbol{\theta}$ is given by:

$$\boldsymbol{\theta} = \mathrm{d}\boldsymbol{\omega} \tag{2.63}$$

The 2-form $\boldsymbol{\omega}$ in (2.63) gives rise to a vector $\mathbf{A}$ where

$$\begin{aligned}\boldsymbol{\omega} &= (a_{23}-a_{32})\,\mathrm{d}x^2 \wedge \mathrm{d}x^3+(a_{31}-a_{13})\,\mathrm{d}x^3 \wedge \mathrm{d}x^1+(a_{12}-a_{21})\,\mathrm{d}x^1 \wedge \mathrm{d}x^2 \\ &= a_{ij}\,\mathrm{d}x^i \wedge \mathrm{d}x^j\end{aligned}$$

and

$$\mathbf{A} = (a_{23}-a_{32})\mathbf{i}+(a_{31}-a_{13})\mathbf{j}+(a_{12}-a_{21})\mathbf{k} \tag{2.64}$$

Now note that

$$\mathrm{d}\boldsymbol{\omega} = (\operatorname{div}\mathbf{A})\,\mathrm{d}x^1 \wedge \mathrm{d}x^2 \wedge \mathrm{d}x^3 \tag{2.65}$$

Thus

$$\begin{aligned}\int_M \mathrm{d}\boldsymbol{\omega} &= \int_M (\operatorname{div}\mathbf{A})\mathrm{d}x^1 \wedge \mathrm{d}x^2 \wedge \mathrm{d}x^3 \\ &= \int_{\partial M} \mathbf{A}\cdot\mathbf{ds}\end{aligned} \tag{2.66}$$

by Gauss's theorem. Finally

$$\int_{\partial M} \mathbf{A}\cdot\mathbf{ds} = \int_{\partial M} \boldsymbol{\omega} \tag{2.67}$$

and so we have a result identical in form to that of equation (2.62) namely:

$$\int_M \mathrm{d}\boldsymbol{\omega} = \int_{\partial M} \boldsymbol{\omega} \tag{2.68}$$

As these examples suggest this result generalizes to an n-dimensional compact manifold with boundary. If M is an n-dimensional orientable compact manifold with boundary, and $\boldsymbol{\omega}$ is an $(n-1)$-form, then

$$\int_M \mathrm{d}\boldsymbol{\omega} = \int_{\partial M} \boldsymbol{\omega} \tag{2.69}$$

This general form of the result is called Stokes' theorem. Using differential forms, and the partition of unity, its proof is only a straight forward computation depending essentially on the fundamental theorem of calculus[1]. In fact if $n=1$ then (2.69) is just the fundamental theorem of calculus. In this case $M=[a, b]$ then ∂M, has dimension zero, and is the pair of points $\{a, b\}$. Speaking of manifolds with boundaries, if M is n-dimensional its boundary ∂M is $(n-1)$-dimensional. A discrete set of points has dimension zero. If M is orientable and covered by the family $\{U_\alpha\}$, then in the intersections of U_α with ∂M, the orientation of M induces an orientation on ∂M. Thus ∂M is also orientable.

We close this first discussion of differential geometry with remarks on two subjects: infinite dimensional manifolds and differentiable structures.

2.4 INFINITE DIMENSIONAL MANIFOLDS

All the manifolds that we have discussed, or shall discuss in any detail, have finite dimension. It is worth noting that there are naturally occurring examples of differentiable manifolds of infinite dimension. These examples occur in the modern treatment of the calculus of variations—the Morse theory. The points in the manifold are simply the geodesics of some Riemannian manifold like a torus or a sphere. These manifolds are of infinite dimension, and are subspaces of the space of all possible paths, (i.e. not just the geodesics), which is also of infinite dimension. In this respect they are rather like the Riemannian manifolds of dimension n which are subspaces of some $\mathbf{R}^N$ for large enough N. We shall not make any extensive use of this infinite dimensional aspect of Morse theory in out treatment of the subject.

2.5 DIFFERENTIABLE STRUCTURE

In providing a differentiable manifold M with an atlas we can also be described as giving M a differentiable structure. The presence of this differentiable structure on manifolds leads one to the notion of a diffeomorphism which is, as one might expect from the name, a differentiable homeomorphism. Take two manifolds M and N and take an invertible map f from M to N

$$\begin{aligned} f&:M\to N \\ f^{-1}&:N\to M \end{aligned} \tag{2.70}$$

Using the charts on M and N we can differentiate f and f^{-1}. A C^∞ function f which has a C^∞ inverse is called a diffeomorphism. Because diffeomorphisms must be continuous, all diffeomorphisms are homeomorphisms. It then is natural to ask, are all homeomorphic manifolds M and N also diffeomorphic? The answer is no they are not. Thus homeomorphic manifolds are divided up into equivalence classes under diffeomorphisms. The fact that the answer to the above question is no, means that there is more than one equivalence class under the equivalence relation of diffeomorphism. Manifolds belonging to different equivalence classes have therefore *different* differentiable structures on them. If we take the case of spheres S^n, then there is only one equivalence class for $n \leqslant 6$. But for $n = 7$ there are 28 equivalence classes, and as n increases the number of equivalence classes runs into several millions. We cannot say more on this subject here, however it is of interest to the physicist, who is always differentiating functions of various kinds, to realise that there are quite distinct differentiable structures, and hence methods of differentiating, for some manifolds. Which ones are the natural ones for physical problems? We do not know the answer to this latter question and find this an appropriate point to close this chapter.

REFERENCE

1. SPIVAK, M., "Differential Geometry". Vol. 1, Publish or Perish Inc, 1970.

CHAPTER 3

The Fundamental Group

3.1 INTRODUCTION

In this chapter we introduce the fundamental or first homotopy group associated with a topological space X. We shall first establish some of the properties of the fundamental group and then show how, for triangulable spaces, this group may be determined in a routine manner. In a later chapter we shall see that the fundamental group is also useful in physics. For instance, it turns up in gauge theories and in the classification of defects in solid state physics.

Let us start by considering a simple example. Figure 3.1 shows two rectangular regions X_1 and X_2 of $\mathbf{R}^2$. X_1 contains a hole, shown as a shaded area while X_2 does not. It is intuitively clear that any loop in X_2 can be shrunk to a point. It is also clear that this is not so in X_1. Thus the space X_2 has only one kind of loop: a loop which can be shrunk to a point.

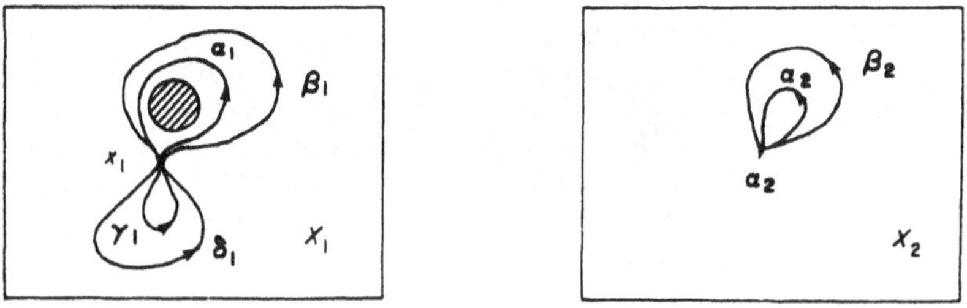

Figure 3.1

However there are two kinds of loops in X_1: loops like γ_1 and δ_1 which can be shrunk to a point, and loops like α_1 and β_1 which contain the hole—these latter cannot be shrunk to a point. Continuing in this manner we observe that an equivalence between loops can be introduced. Two loops are equivalent, or homotopic, if one can be obtained from the other by a process of continuous deformation. For instance α_2 and β_2 are homotopic, written $\alpha_2 \simeq \beta_2$, in X_2. Also $\alpha_1 \simeq \beta_1$ and $\gamma_1 \simeq \delta_1$ in X_1. But α_1 is not homotopic to γ_1. Thus we see that although there are an infinite number of loops in X_1 and X_2, if homotopic loops are regarded as being equivalent then there is just one homotopy class of loops in X_2. In X_1 there is one homotopy class h_n associated with each integer n. A loop belongs to h_n if it encircles the hole n times, $n > 0$ is taken to mean clockwise encirclement, $n < 0$ is taken to mean anti-clockwise encirclement and $n = 0$ no encirclement. Working with homotopy classes of loops rather than the loops themselves thus allows us to distinguish the space X_1 from the space X_2. This example suggests that the study of equivalence classes of loops or closed paths in any topological space X might be a way of determining the 'holes' in the space. We shall do this and we shall also see that such loops can be furnished with a group structure. This is the fundamental group $\pi_1(X)$. To see how this group structure arises we return to our simple example. Observe that the two loops γ and β starting from the same point x_1 say, can be combined to produce a third loop $\tilde{\gamma}$ also starting from x_1. We now describe $\tilde{\gamma}$ which we also write as $\gamma * \beta$. $\tilde{\gamma}$ as the loop which starts from x_1 and first goes round the loop γ and then goes round the loop β. Similarly by $\tilde{\gamma} = \beta * \gamma$ we mean the loop that goes round β first and γ second. The inverse of a loop β, β^{-1} starts from x_1 and goes round the loop β in the opposite sense to β. Finally the identity loop is defined to be the loop which stays at x_1 all the time.

Let us consider a simple example. Consider the loop $\varepsilon = \beta * \beta^{-1}$ based at x_1. If the loops themselves were to form a group then ε would have to be the identity loop. But it is not. However ε is homotopic to the identity loop. This is demonstrated by the series of pictures in Fig. 3.2. We are thus led to expect that the elements of the fundamental group should be homotopy classes of loops rather than the loops themselves. This also agrees with our earlier observation that information such as the presence of a hole in X_1 is contained in the equivalence classes of loops in X_1. Let us now make precise the intuitive ideas discussed above. We have the following definitions:

Definition

A path $\alpha(t)$ in X from x_0 to x_1 is a continuous map from $[0, 1]$ into X with $\alpha(0) = x_0$, $\alpha(1) = x_1$, written as:

$$\alpha : [0, 1] \to X \tag{3.1}$$

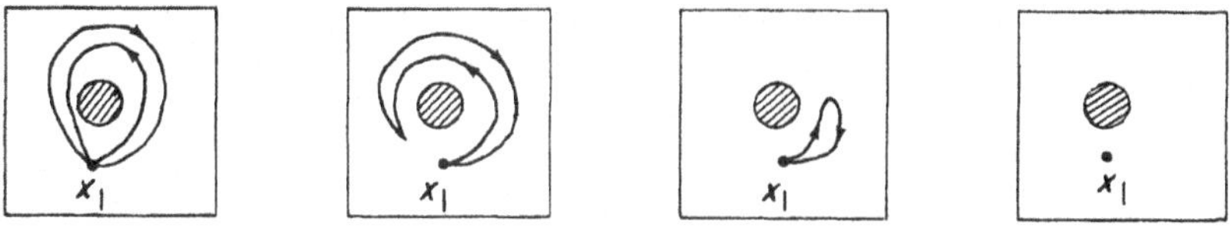

Figure 3.2

Definition

A topological space X is called arc-wise connected or path connected if there always exists a path α between any pair of points x_0 and x_1 in X. (3.2)

In connection with definition (3.2) note that in Chapter 1 a connected topological space was defined. The notion of path connected just defined is stronger than the notion of connectedness. In other words a space can be connected but not path connected, but a path connected space is always connected. For most reasonable topological spaces however the two notions coincide.

Example

In $\mathbf{R}^2$ consider $X=\{(0, x_2)|-1<x_2<1\}$, $Y=\{(x_1, \sin \pi/x_1)|0<x_1<1\}$. It is possible to prove that $X \cup Y$ is connected but not path connected. A picture of the spaces X and Y helps to make the result plausible, c.f. Fig. 3.3.

Definition

A closed path or loop in X at the point x_0 is a path $\alpha(t)$ for which $\alpha(0)=\alpha(1)=x_0$. Where

$$\alpha:[0, 1]\rightarrow X \tag{3.3}$$

Definition

The product γ of two loops based at $x_0 \in X$ is written as $\gamma=\alpha * \beta$ and is defined by:

$$\gamma(t)=\begin{cases}\alpha(2t) & 0\leqslant t\leqslant \frac{1}{2}\\ \beta(2t-1) & \frac{1}{2}\leqslant t\leqslant 1\end{cases} \tag{3.4}$$

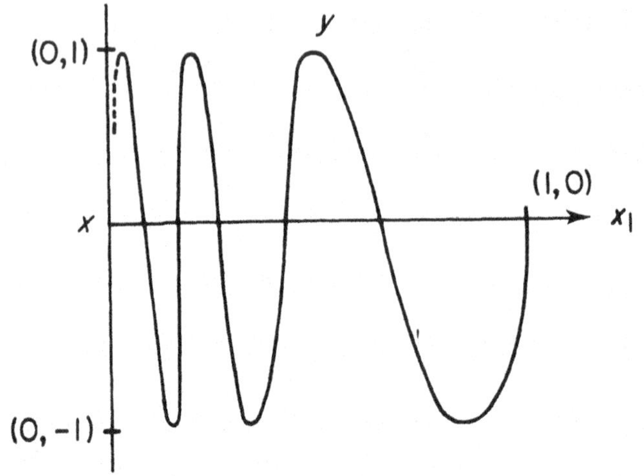

Figure 3.3

This definition agrees with our earlier intuitive discussion on how two loops were to be combined. The precise details of the definition become clear once it is realised that α, β, γ must all be loops and hence must continuously map [0, 1] into X. It is straightforward to check that these conditions are satisfied by definition (3.4).

Definition

The inverse loop $\alpha^{-1}(t)$ based at $x_0 \in X$ is defined by

$$\alpha^{-1}(t) = \alpha(1-t), \qquad 0 \leqslant t \leqslant 1 \tag{3.5}$$

Definition

The constant loop c based at x_0 is the map:

$$c:[0, 1] \to X$$

with

$$c(t) = x_0, \qquad 0 \leqslant t \leqslant 1 \tag{3.6}$$

The homotopy of loops can be defined using essentially the definition (1.41, 1.42) of homotopy of maps given in Chapter 1.

Definition

Two loops α and β based at x_0 are homotopic, $\alpha \simeq \beta$, if there exists a continuous map H from $[0, 1]\times[0, 1]$ to X

$$H:[0, 1]\times[0, 1]\to X$$

and H satisfies:

$$\begin{aligned} H(t, 0) &= \alpha(t), \qquad 0 \leq t \leq 1 \\ H(t, 1) &= \beta(t), \qquad 0 \leq t \leq 1 \\ H(0, s) = H(1, s) &= x_0, \qquad 0 \leq s \leq 1 \end{aligned} \tag{3.7}$$

The map H is called a homotopy between α and β.

A few remarks might be useful. Note that $H(t, s)$ for any fixed value of s as t varies is a loop in X based at x_0 which interpolates in a continuous manner between the loops α and β. A more suggestive notation for $H(t, s)$ would thus be $\alpha_s(t)$. It is sometimes useful to draw a picture of the source of the homotopy $H(t, s)$ as a square box (Fig. 3.4). The top of the box $(s = 1)$ represents β, the bottom $(s = 0)$ α, while the both sides $(t = 0; t = 1)$ are maps to x_0. To get some practice in the use of this definition we will prove in detail one part of the following trivial Lemma.

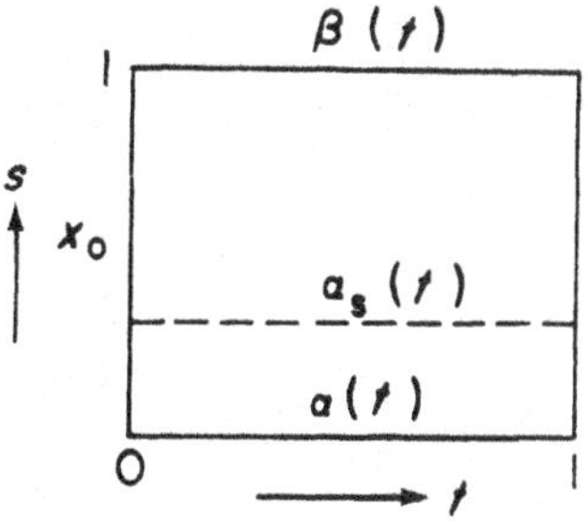

Figure 3.4

Lemma

If $\alpha_0, \beta_0, \gamma_0, \ldots, \alpha_1, \beta_1, \gamma_1 \ldots$ represent loops based at $x_0 \in X$ then

i. $\alpha_0 \simeq \alpha_0$

ii. $\alpha_0 \simeq \beta_0$, implies $\beta_0 \simeq \alpha_0$

iii. $\alpha_0 \simeq \beta_0$, $\beta_0 \simeq \gamma_0$ implies $\alpha_0 \simeq \gamma_0$ (3.8)

iv. $\alpha_0 \simeq \alpha_1$, implies $\alpha_0^{-1} \simeq \alpha_1^{-1}$

v. $\alpha_0 \simeq \alpha_1$, $\beta_0 \simeq \beta_1$, implies $\alpha_0 * \beta_0 \simeq \alpha_1 * \beta_1$

Proof

We will prove (iii). Using the pictorial representation of homotopies, our problem is the following: We are given (Fig. 3.5):

Figure 3.5

We want to show that a homotopy H, drawn below (Fig. 3.6), can be constructed from F and G.

Thus the required homotopy H between α_0 and γ_0 is given by:

$$H(t,s)=\begin{cases}F(t,2s) & 0\leq s\leq \frac{1}{2}\\ G(t,2s-1) & \frac{1}{2}\leq s\leq 1\end{cases}$$

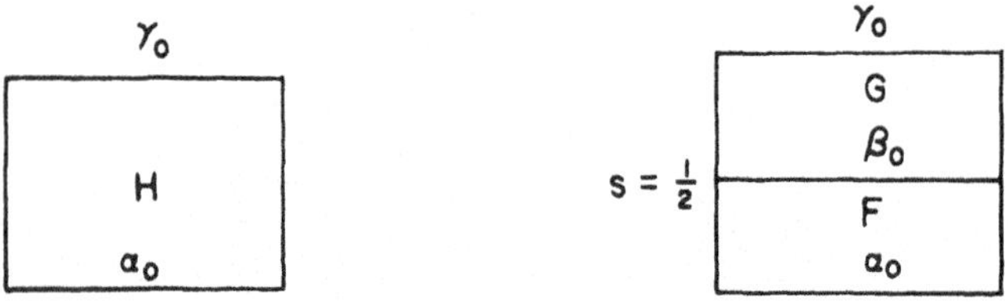

Figure 3.6

A few remarks are in order. The Lemma tells us that homotopy is an equivalence relation. This means that the space of loops can be partitioned into disjoint equivalence classes, members in a given class being homotopic to each other. Let us denote by $[\alpha]$ the equivalence class of loops homotopic to the loop α.

3.2 DEFINITION OF THE FUNDAMENTAL GROUP

Let us denote by $\pi_1(X, x_0)$ the set of homotopy classes of loops based at x_0. If $[\alpha]$ and $[\beta]$ belong to $\pi_1(X, x_0)$ then the product $[\alpha]\circ[\beta]$ can be defined as

$$[\alpha]\circ[\beta]=[\alpha * \beta] \tag{3.9}$$

Thus the product of two homotopy classes is defined to be the class determined by their representative elements. Lemma (3.8) shows that this product rule is well defined. We can now state the following theorem.

Theorem

$\pi_1(X, x_0)$ with the product law just defined is a group. The unit element of this group is the homotopy class of the constant loop at x_0. The group $\pi_1(X, x_0)$ is called the first homotopy group or the fundamental group of the path connected topological space X based at x_0. (3.10)

Proof

To prove the theorem we have to check that the product law defined satisfies all the group axioms. This is fairly straightforward so that we will only go through the argument for the associativity law. To verify that the associativity law is obeyed we have to show that

$$(\alpha * \beta) * \gamma \simeq \alpha * (\beta * \gamma) \tag{3.11}$$

as this implies from (3.9) that

$$([\alpha]\circ[\beta])\circ[\gamma]=[\alpha]\circ([\beta]\circ[\gamma]) \tag{3.12}$$

To establish (3.11) we have to find a homotopy which, pictorially, has $(\alpha * \beta) * \gamma$ on the bottom and $\alpha * (\beta * \gamma)$ on the top. We note that

$$\alpha * (\beta * \gamma)(t)=\begin{cases}\alpha(2t) & 0\leq t\leq \frac{1}{2}\\ \beta * \gamma(2t-1) & \frac{1}{2}\leq t\leq 1\end{cases} \tag{3.13}$$

$$=\begin{cases}\alpha(2t) & 0\leq t\leq \frac{1}{2}\\ \beta(4t-2) & \frac{1}{2}\leq t\leq \frac{3}{4}\\ \gamma(4t-3) & \frac{3}{4}\leq t\leq 1\end{cases} \tag{3.14}$$

Similarly

$$(\alpha * \beta) * \gamma=\begin{cases}\alpha(4t) & 0\leq t\leq \frac{1}{4}\\ \beta(4t-1) & \frac{1}{4}\leq t\leq \frac{1}{2}\\ \gamma(2t-1) & \frac{1}{2}\leq t\leq 1\end{cases} \tag{3.15}$$

A picture of the required homotopy H can now be drawn and is shown in Fig. 3.7.

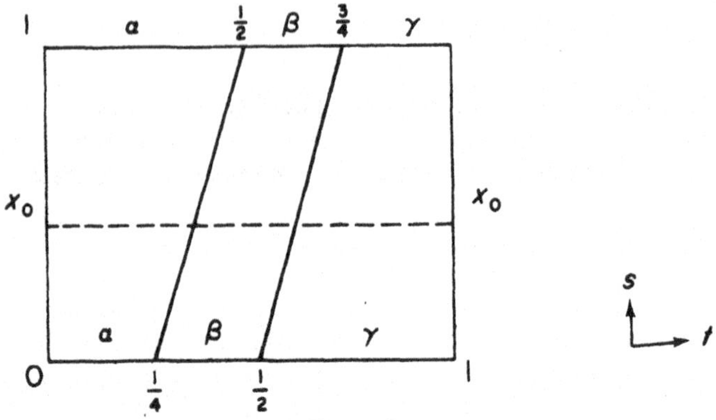

Figure 3.7

As s changes from 0 to 1 the loop $(\alpha * \beta) * \gamma$ changes to the loop $\alpha * (\beta * \gamma)$. By assuming that this change is linear i.e. by joining the points $(t = \frac{1}{4}, s = 0)$, $(t = \frac{1}{2}, s = 1)$ and $(t = \frac{1}{2}, s = 0)$, $(t = \frac{3}{4}, s = 1)$ by straight lines an explicit expression for $H(t, s)$ can be obtained from the geometry of Fig. 3.7 it is:

$$H(t,s) = \begin{cases} \alpha\left(\dfrac{4t}{s+1}\right) & 0 \leqslant t \leqslant \dfrac{s+1}{4} \\ \beta(4t - s - 1) & \dfrac{(s+1)}{4} \leqslant t \leqslant \dfrac{(s+2)}{4} \\ \gamma\left(\dfrac{4t-2-s}{2-s}\right) & \dfrac{s+2}{4} \leqslant t \leqslant 1 \end{cases} \tag{3.16}$$

Thus (3.11) is established.

It would appear from Theorem (3.10) that $\pi_1(X, x_0)$ depends on the base point x_0 chosen for the loops. This would be something of a disadvantage if it were generally true. Happily it is not true if the space X is path connected, as we have assumed it to be, then $\pi_1(X, x_0)$ is isomorphic to $\pi_1(X, x_1)$ for two different base points.

Theorem

If X is a path connected topological space and $x_0, x_1 \in X$ then the groups $\pi_1(X, x_0)$ and $\pi_1(X, x_1)$ are isomorphic. (3.17)

Proof

This theorem can be proved by reference to Fig. 3.8.

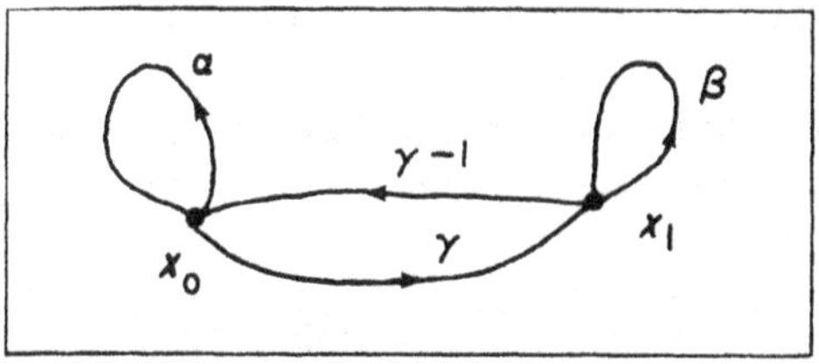

Figure 3.8

Let α be a loop based at x_0, β be a loop based at x_1 and let γ be a path (not a loop) joining x_0 to x_1. From Fig. 3.8 it is clear that with the help of the paths γ, γ^{-1} the loop α based at x_0 can be converted into a loop based at x_1 and similarly the loop β based at x_1 can be converted into a loop based at x_0. This is essentially all that is involved in the proof of the theorem. Using γ, and γ^{-1} we have the homomorphism

$$\sigma_\gamma : \pi_1(X, x_0) \to \pi_1(X, x_1); \, x_0, x_1 \in X \tag{3.18}$$

given by:

$$\sigma_\gamma : ([\alpha], x_0) \to ([\gamma^{-1} * \alpha * \gamma], x_1)$$

and

$$\sigma_{\gamma^{-1}} : \pi_1(X, x_1) \to \pi_1(X, x_0) \tag{3.19}$$

given by:

$$\sigma_{\gamma^{-1}} : ([\beta], x_1) \to ([\gamma * \beta * \gamma^{-1}], x_0)$$

In order to show (3.18, 3.19) make sense we have to define the product rule for two paths since our previous discussions were concerned with loops only.

Definition

A path γ can be combined with a path σ to give a new path $\gamma * \sigma$ only if $\gamma(1) = \sigma(0)$ (i.e. the beginning of σ is the end of γ)

$$\text{Then } \gamma * \sigma(t) = \begin{cases} \gamma(2t) & 0 \leqslant t \leqslant \frac{1}{2} \\ \sigma(2t-1) & \frac{1}{2} \leqslant t \leqslant 1 \end{cases} \tag{3.20}$$

Using this definition it is possible to show that the maps σ_γ, $\sigma_{\gamma^{-1}}$ are well defined. From (3.18, 3.19) it follows that $\sigma_{\gamma^{-1}} \circ \sigma_\gamma$ and $\sigma_\gamma \circ \sigma_{\gamma^{-1}}$ are both identity homomorphisms, i.e. isomorphisms. This means that $\sigma_{\gamma^{-1}}$ and σ_γ

are themselves isomorphisms, not just homomorphisms, which establishes the theorem.

Thus $\pi_1(X, x_0)$ for a path connected topological space depends, up to isomorphism, on the space X, and not on the base point x_0 selected. On the other hand it is also clear from the proof we have just given that there is no canonical isomorphism between $\pi_1(X, x_0)$ and $\pi_1(X, x_1)$. The isomorphism can change if we make a different choice for γ. Is the group $\pi_1(X, x_0)$ a topological invariant? An easy corollary of the next theorem shows that it is.

Theorem

If X and Y are two path connected topological spaces of the same homotopy type then $\pi_1(X, x_0)$ is isomorphic to $\pi_1(Y, y_0)$.

$$(x_0 \in X,\ y_0 \in Y) \tag{3.21}$$

We, have to explain what homotopy type means. We recall that if X and Y are homeomorphic spaces then there must exist continuous maps $f \cdot X \to Y$; $g : Y \to X$ such that $f \circ g = 1_y$, the identity map on Y and $g \circ f = 1_x$ the identity map on X. We now have:

Definition

Two spaces X and Y are of the same homotopy type if we have continuous maps f and $g : F : Y \to X$; $g : Y \to X$ but now $f \circ g$ need no longer be equal to the identity map 1_y but only be homotopic to it, i.e. we require:

$$f \circ g \simeq 1_y,\ g \circ f \simeq 1_x \tag{3.22}$$

Since 1_x, 1_y are clearly homotopic to themselves it follows that two homeomorphic spaces are of the same homotopy type. Thus:

Corollary

If X and Y are homeomorphic path-connected topological spaces then $\pi_1(X, x_0)$ is isomorphic to $\pi_1(Y, y_0)$

$$(x_0 \in X,\ y_0 \in Y) \tag{3.23}$$

Corollary (3.23) establishes the fundamental group as a topological invariant of a space.

Before proving Theorem (3.21) let us dispose of a technical difficulty. It might be useful to recall that the definition of homotopy used in

Chapter 1 was more general than the one we have used for loops in this chapter. The essential difference lay in our insistence that we would only allow continuous deformation which do not disturb the base point of the loops. There was good reason for this. Without this restriction all loops would be homotopic to the constant loop.

The homotopies allowed in Definition (3.22) need not have this property. The base point can be disturbed by these homotopies but not in an arbitrary way. The complications due to this are handled by the following Lemma:

Lemma

Let $F: X\times I \to Y$ be a homotopy between two maps $f_i:(X, x_0)\to(Y, y_i)$ $(i=0, 1)$ and γ a path connecting y_0 to y_1 in Y.

Then $\sigma_\gamma * f_0^* = f_1^*$ where $f_i^*: \pi_1(X, x_0)\to\pi_1(Y, y_i)$, a group homomorphism induced by:

$$f_i^*:([\alpha], x_0)\to([f(\alpha)], f(x_0)), \qquad (i=1, 2)$$

and (3.24)

$$\sigma_\gamma:([\alpha], x_0)\to([\gamma^{-1}*\alpha*\gamma], x_1), \qquad [\alpha]\in\pi_1(X, x_0)$$

Proof

We are interested in the way the homomorphisms f_0^*, f_1^* are related due to the fact that the maps f_0, and f_1 are homotopic. The homomorphisms arise out of the action of f_0 and f_1 on loops in X based at x_0. We are thus led to consider:

$$G:[0, 1]\times[0, 1]\to Y \tag{3.25}$$

defined by:

$$G(t, s)=F(\alpha(t), s) \tag{3.26}$$

where $\alpha(t)$ represents a loop in X based at x_0. A picture of G is drawn below: (Fig. 3.9)

We now make the observation that (3.25) can be regarded as a map from a space $R=[0, 1]\times[0, 1]$ to Y. Any loop in R is contractible. Since G is a continuous map the image under G of any loop in R must also be contractible i.e. homotopic to the constant loop in Y. The constant loop in a space, we recall, represents the identity element of the fundamental group of the space. Thus we have (reading Fig. 3.9 as a mapping of a loop from R to Y) the result:

$$[f_0(\alpha)*\gamma*f_1(\alpha)^{-1}*\gamma^{-1}]=\text{identity}$$

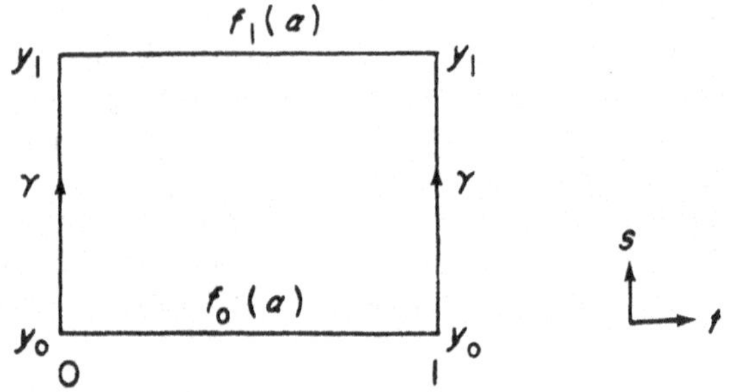

Figure 3.9

or

$$[f_1(\alpha)]=[\gamma^{-1}*f_0(\alpha)*\gamma] \tag{3.27}$$

i.e.

$$f_1^*=\sigma_\gamma\circ f_0^*$$

which establishes the lemma.

With the help of this lemma proving Theorem (3.21) is easy. Since X and Y are of the same homotopy type we must have maps $f:X\to Y$, $g:Y\to X$ and homotopies F, H such that

$$f\circ g\overset{F}{\simeq}1_Y$$

$$f\circ f\overset{H}{\simeq}1_X$$

From the lemma it follows that

$$\begin{aligned}\sigma_\gamma\circ 1_y^*&=(f\circ g)^*=f^*\circ g^*\\ \sigma_{\gamma-1}\circ 1_x^*&=(g\circ f)^*=g^*\circ f^*\end{aligned} \tag{3.28}$$

We know from Theorem (3.17) that σ_γ, $\sigma_{\gamma^{-1}}$ are isomorphisms. The relations (3.28) then implies that $f^*\circ g^*$ and $g^*\circ f^*$ are also isomorphisms. Hence f^* and g^* are isomorphisms which establishes the theorem.

One can readily provide an example of a pair of spaces X and Y which are of the same homotopy type but are not homeomorphic. This means that if two spaces have isomorphic fundamental groups they need not be homeomorphic. Now for our example. Take X to be the closed curve C and Y to be C with a line segment, PQ, length L say, added to C of Fig. 3.10.

Let $f:X\to Y$ be the map $f(x)=x$ and $g:Y\to X$ be the map:

$$g(y)=\begin{cases} y & \text{if } y\in C\\ P & \text{if } y\in PQ\end{cases} \tag{3.29}$$

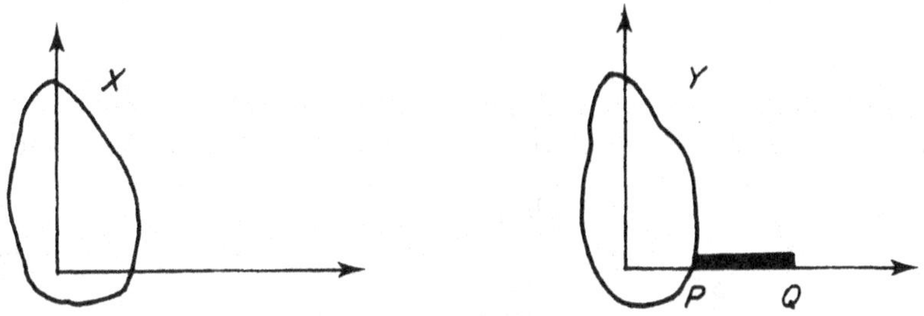

Figure 3.10

where P is the point where the line segment joins the curve C. Now X and Y are not homeomorphic, this is because removal of a point from X does not render X disconnected, but removal of a point from the line segment PQ of Y does render Y disconnected. The spaces X and Y are, however, of the same homotopy type, i.e. f and g satisfy:

$$\begin{aligned} g \circ f &= I_x, \quad \text{the identity on } X \\ f \circ g &\simeq I_Y, \quad \text{the identity in } Y \end{aligned} \tag{3.30}$$

To see that $g \circ f = I_x$ is obvious. To show that $f \circ g \simeq I_y$ just think of a continuous deformation which contracts the line segment down to the point P.

This example leads directly to the notions of retract and deformation retracts. We shall explain what these notions are as our last piece of theoretical discussion before showing how to actually calculate $\pi_1(X, x_0)$. Intuitively if $A \subset X$ is a deformation retract of X it means that X can be continuously deformed to A without moving points of A at any stage.

Definition

A subset A of a topological space X is called a retract of X if there exists a continuous map, called a retraction:

$$r: X \to A$$

such that $r(a) = a$, for any $a \in A$. (3.31)

Definition

A subset A of a topological space X is a deformation retract of X if there is a retraction $r: X \to A$ and a homotopy $H: X \times [0, 1] \to X$ such that

$$\begin{aligned} &H(x, 0) = x \\ &H(x, 1) = r(x) \\ &H(a, t) = a,\ a \in A,\ t \in [0, 1] \end{aligned} \tag{3.32}$$

In our simple example if we consider the space Y which consisted of the curve C with the line segment PQ adjoined then C is a deformation retract of Y. In terms of the fundamental group the interest in deformation retracts lies in the following theorem.

Theorem

If X is a path connected topological space and A a deformation retract of X, then $\pi_1(X, a)$ is isomorphic to $\pi_1(A, a)$, $a \in A$. (3.33)

The proof follows immediately from the fact that the deformation retractedness property of A implies that A and X are of the same homotopy type.

The usefulness of the theorem just established should be apparent. If we can show that a space A is a deformation retract of a space X and we know how to calculate the fundamental group of one of the spaces we have, in effect, determined the fundamental group of the other space.

Example 1. Let $\mathbf{R}^n$ represent n-dimensional Euclidean space, i.e. $\mathbf{R}^n = \{x \mid x = (x_1, \ldots, x_n)\}$. Let Y denote the space consisting of the single point $Y = \{0\}$; we claim that Y is a deformation retract of $\mathbf{R}^n$.

Define

$$H: \mathbf{R}^n \times [0, 1] \to \mathbf{R}^n \tag{3.34}$$

by the formula

$$H(x, t) = tx,\ t \in [0, 1],\ x \in \mathbf{R}^n$$

Observe that this establishes the claim. We also have the result: $\pi_1(\mathbf{R}^n, 0)$ is isomorphic onto $\pi_1(\{0\}, 0) =$ identity element.

Example 2. The unit $(n-1)$-sphere is a deformation retract of $D^n - \{0\}$, the closed unit disc with the origin removed. To prove this define

$$H: (D^n - \{0\} \times [0, 1]) \to (D^n - \{0\})$$

by the formula:

$$H(x,t)=(1-t)x+t\frac{x}{|x|}$$
$$t\in[0,1],\quad x\in\mathbf{R}^n,\quad |x|=\surd(x\cdot x) \tag{3.35}$$

A picture explains what is involved for $n=2$ (Fig. 3.11).

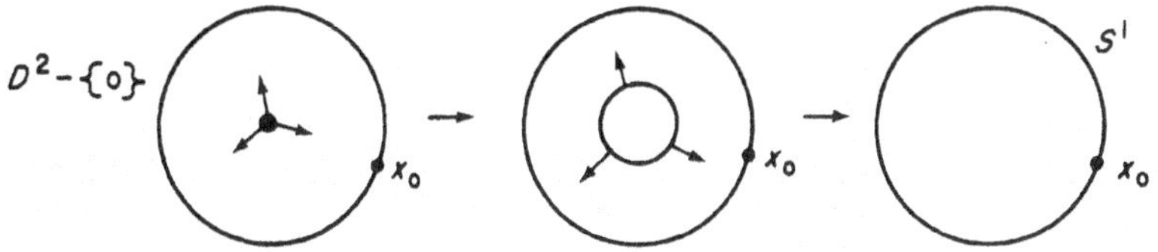

Figure 3.11

We also have the result that $\pi_1(D^2-\{0\}, x_0)$ is isomorphic to $\pi_1(S^1, x_0)$. Where $x_0\in S^1$. From our intuitive discussion on loops at the beginning of this chapter we expect $\pi_1(S^1, x_0)$ to be isomorphic to the group of integers under addition.

Example 3. Two circles with one point in common (i.e. Fig. 3.12) is a deformation retract of $D^2-\{p\}-\{q\}$, the closed unit disc in two dimensions with the points $\{p\}$ and $\{q\}$ removed. Instead of writing down a formula for the homotopy as in the previous examples we simply draw pictures to establish the result (Fig. 3.12).

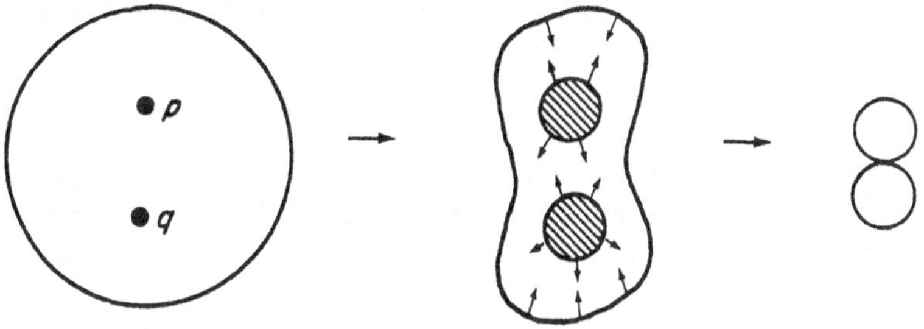

Figure 3.12

Again we have $\pi_1(D^2-\{p\}-\{q\}, x_0)$ isomorphic to $\pi_1(8, x_0)$ where $x_0\in 8$. Later on we will learn how to calculate $\pi_1(8, x_0)$; for the moment we draw pictures to suggest that the group $\pi_1(D^2-\{p\}-\{q\}, x_0)$ is not abelian. The relevant pictures are given in Fig. 3.13.

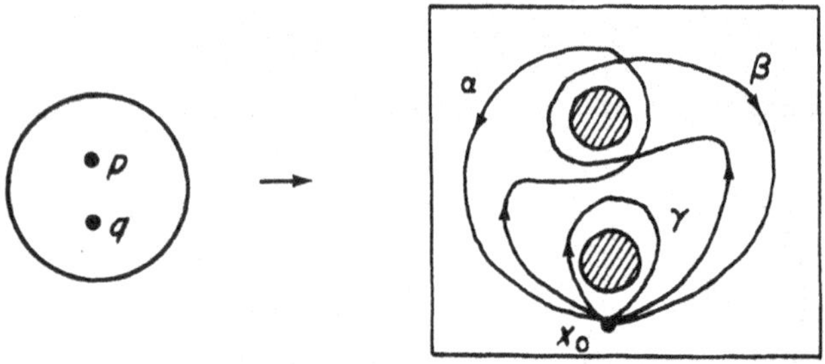

Figure 3.13

It is reasonable, from the picture, to conclude that the loop α is not homotopic to the loop β if x_0 is kept fixed. We write this as:

$$\alpha \not\simeq \beta \tag{3.36}$$

On the other hand with the help of the path γ (which encircles the lower circular hole) we can convert the loop α from being a loop which encircles the upper circular hole while staying to the left of the lower hole to a loop which encircles the upper circular hole while staying to the right of the lower hole. Hence we expect:

$$\gamma * \alpha * \gamma^{-1} \simeq \beta \not\simeq \alpha \tag{3.37}$$

Hence:

$$[\gamma]\circ[\alpha] \neq [\alpha]\circ[\gamma]$$

i.e. the group $\pi_1(D-\{p\}-\{q\}, x_0)$ is not abelian. Until now we have not discussed methods of determining the fundamental group for a given topological space. We remedy this by first introducing a special kind of topological space called a polyhedron and then proving a theorem which allows the determination of the fundamental group for a polyhedron in a routine manner.

A polyhedron can be regarded as a subspace of some Euclidean $\mathbf{R}^n$ which is obtained by properly gluing together certain elementary spaces called simplexes. A m-simplex (written σ^m) is a generalized triangle in m-dimensions. Thus σ^2 is a triangle and σ^3 a tetrahedron. The simplexes are glued together in such a way that two simplex, if they meet, have a common vertex or edge. Let us make these ideas precise.

3.3 SIMPLEXES AND THE CALCULATING THEOREM

Definition

Let $x_1, x_2, \ldots, x_{m+1}$ be distinct points in $\mathbf{R}^n$. The sets of points $x_1, \ldots, x_{M+1}$ are independent if the M vectors

$$x_2 - x_1, x_3 - x_1, \ldots, x_{m+1} - x_1 \tag{3.38}$$

are linearly independent vectors.

Definition

An M-simplex, σ^M is the set of points x in $\mathbf{R}^n$ given by:

$$\sigma^M = \left\{ x = \sum_{i=1}^{M+1} \lambda_i x_i \middle| \lambda_i \geqslant 0, \sum_{i=1}^{M+1} \lambda_i = 1 \right\}$$

where $x_1, \ldots, x_{M+1}$ are independent.

We often write $\sigma^M = [x_1, \ldots, x_{M+1}]$ and call $x_1, \ldots, x_{M+1}$ the vertices of the M-dimensional simplex (or M-simplex) σ^M. The λ_i are called the barycentric co-ordinates of the simplex. (3.39)

We note that the point x contained in M simplex, σ^M, corresponding to a given set of barycentric coordinates $\lambda_1, \lambda_2, \ldots, \lambda_{M+1}$ can be regarded as the centre of mass of the system with masses $\lambda_1, \lambda_2, \ldots \lambda_{M+1}$ placed at the vertices $x_1, x_2, \ldots x_{M+1}$ respectively. This physical analogy leads us to expect that if all the λ_i are non-zero then the corresponding set of points x represent the interior of σ^M while if any $\lambda_{j=0}$ then the set of points x represents a 'face' of σ^M opposite to the vertex x_j. Thus we have.

Definition

The set $\{\lambda_1 x_1 + \cdots + \lambda_{M+1} x_{M+1} | \lambda_j = 0|\}$ is called the jth face of the simplex σ^M. It lies opposite the jth vertex x_j. (3.40)

Example

Consider the two-simplex σ^2. If $x \in \sigma^2$, then $x = \lambda_1 x_1 + \lambda_2 x_2 + \lambda_3 x_3$; x_1, x_2, x_3 being independent points, λ_1, λ_2, $\lambda_3 \geqslant 0$ and $\lambda_1 + \lambda_2 + \lambda_3 = 1$. Thus geometrically if λ_1, λ_2, λ_3 are all non-zero x is a point inside the triangle formed with x_1, x_2, x_3 as vertices. Now notice that if $x \in \sigma^2$ and λ_1 is 0 then: $x = \lambda_2 x_2 + \lambda_3 x_3$, with $\lambda_2 + \lambda_3 = 1$, λ_2, $\lambda_3 \geqslant 0$. This represents the set of points on the line joining x_2 and x_3, i.e. the face $x_2 x_3$ of σ^2 which lies opposite to x_1. We next define what we mean by a polyhedron.

Definition

A simplicial complex K is a finite collection of simplexes in some $\mathbf{R}^n$ satisfying:

i. If $\sigma^p \in K$, then all faces of σ^p belong to K
ii. If $\sigma^p, \sigma^q \in K$ then either $\sigma^p \cap \sigma^q = \emptyset$ or $\sigma^p \cap \sigma^q$ is a common face of σ^p and σ^q.

The dimension of K, $\dim K$, is defined to be the maximum of the dimensions of the simplexes of K. (3.41a)

Definition

A simplicial complex K is path connected if for every pair of vertices u and v of K there is a sequence $v_0, v_1, \ldots, v_n$ of vertices in K such that $v_0 = u$ and $v_n = v$ and $v_i v_{i+1}$ is a 1-simplex of K for all $i = 0, 1, \ldots, n-1$. (3.41b)

Definition

The union of the members of K with the Euclidean subspace topology is called the polyhedron associated with K. A polyhedron is path connected if the simplicial complex with which it is associated is path connected. (3.42)

Theorem (the calculating theorem)

Let K be a path connected polyhedron with a_0 a vertex of K. Let L be a one-dimensional sub-polyhedron of K which is contractible (i.e. has the same homotopy type as a point) and contains all the vertices of K. Let G be the group generated by the symbols g_{ij} one for each ordered 1-simplex $\{a_i, a_j\}$ of K, subject to the relations $g_{ij}g_{jk}g_{ik}^{-1} = 1$ one for each ordered 2-simplex $\{a_i, a_j, a_k\}$ of $K - L$. If $\{a_i, a_j\} \in L$ then $g_{ij} = 1$. Then G is isomorphic to the fundamental group $\pi_1(K, a_0)$. (3.43)

The statement of the theorem is involved and needs explanation. We start by explaining what is meant by an ordered simplex. If all the vertices of K are ordered in the form $a_0 < a_1 \ldots < a_{M+1}$ then each simplex σ^N of K can be written as $\{a_{i_0}, a_{i_1}, \ldots a_{i_n}\}$ where $i_0 < i_1 \ldots < i_n$. These are the ordered simplexes of K. Next we observe that the statement of the theorem assumes that for any path-connected polyhedron K there is always a one-dimensional subpolyhedron L which is contractible and contains all the vertices of K. This is something we will, in fact, prove. Note also that the 1- and 2-simplexes of K alone completely determine G. Finally the theorem gives an algorithm for determining the fundamental group of K in terms of a set of 'generators' and relations. Let us explain what this means. Our discussion will be sketchy and we refer the reader to the books on group theory listed at the end of this chapter for details. If H is a subset of a group G then H is said to generate G if every element of G can be written as a product of positive and negative powers of elements of H. If the set H contains elements such that the product of these elements is equal to the identity then such a product will be called a relation between the elements of the generating set H. If a group G is completely determined up to isomorphism by the set of generators H and the set of relations $\{r_i\}$ then the set of relations is said to be complete. Thus, we have:

Definition

A presentation of a group G is a pair $(H, \{r_i\})$ consisting of a set of generators for G and a complete set of relations between these generators. The presentation is said to be finite if both H and $\{r_j\}$ are finite sets, and the group G is said to be finitely presented if it has at least one finite presentation. (3.44)

A given group can have many different presentations which may look quite different. Conversely it is often difficult—if not impossible—to tell if two groups are isomorphic from their presentations. Thus although we can determine $\pi_1(K, a_0)$ using Theorem (3.43) the 'presentation' method of specifying $\pi_1(K, a_0)$ is not theoretically very satisfying.

Now for the proof of the Theorem (3.43). Instead of giving a complete proof we will only explain the essential ideas involved and then refer the reader to references [1 or 2] listed at the end of this Chapter for the details.

Theorem (3.43) relates $\pi_1(K, a_0)$ to a group G generated by the 1-simplexes (edges) of K. Since the basic elements of $\pi_1(K, a_0)$ are arbitrary loops in K based at a_0 while the basic elements of G are not loops but edges in K, as a first step in the proof of the theorem we try and replace loops α in K by homotopically equivalent edge loops e_α in K. Figure 3.14 explains what is involved when $K = \sigma^2$.

To carry out this step involves properly defining edge loops and homotopy classes of edge loops and then proving that for a polyhedron it is always possible to replace a loop α in K by a homotopically equivalent edge loop l_α. The proofs are straightforward but lengthy. Once this is done we next have to establish a correspondence between edges and homotopy classes

Figure 3.14

of edge loops to prove the theorem. One way of setting up such a correspondence is to use the fact that in K there is always a maximal contractible one-dimensional subpolyhedron L containing all the vertices of K. That there is such an L is easy to prove. Suppose this was not the case. Then since K is path connected there must be an edge $\{a, b\}$ in K such that $a \in L, b \notin L$. Then $L \cup \{a, b\} \cup \{b\}$ is a one-dimensional subpolyhedron bigger than L which is contractible contrary to hypothesis. Now for the correspondence between edges and edge loops. With each edge $\{x, y\} \in L$ assign the constant edge loop based at a_0. If the edge $\{a, b\} \notin L$ then it can be made to correspond to the loop obtained by first travelling from a_0 to a along edges belonging to L then going from a to b along $\{a, b\}$ and finally returning from b to a_0 again along edges belonging to L. This is always possible because L contains all the vertices of K as we have just shown.

The construction outlined explains how a correspondence between elements of G and elements of $\pi_1(K, a_0)$ can be established—using it and a similar inverse correspondence Theorem (3.43) can be proved. Instead of giving further details of the proof we turn to applications of the Theorem. In order to use Theorem (3.43) to determine the fundamental group of a given topological space X we have to find a polyhedron which is homeomorphic to it. It is convenient to introduce:

3.4 TRIANGULATION OF A SPACE WITH EXAMPLES

Definition

A topological space X which is homeomorphic to a polyhedron K is said to be triangulable and the polyhedron K (which is not unique) is called a triangulation of X. (3.45)

All our applications will involve only one- and two-dimensional systems. For a two dimensional space a triangulation actually does correspond to representing the space by gluing various triangles together, making sure that any two distinct triangles either are disjoint, have a single vertex, or an entire edge in common as required by Definition (3.42).

Example 1.

$$\pi_1(S^1, x_0)$$

S^1 is the circumference of a circle of unit radius in two dimensional Euclidean space and $x_0 \in S^1$.

We first triangulate S^1. Since S^1 is a one-dimensional space this means finding a collection of suitably joined one simplexes which is homeomorphic to S^1. Pictorially we might try to open up the circle S^1 to get Fig. 3.15a.

This is not a permitted triangulation because a 1-simplex must have two distinct vertices. We remedy this in Fig. 3.15b but this is still not permitted because the two simplexes in Fig. 3.15b are supposed to be distinct but have identical vertices. This leads us to Fig. 3.15c which is a proper triangulation of S^1. The polyhedra K_0 associated with S^1 obtained in Fig. 3.15c, totally ordered can be written as:

$$K_0 = \{1\} \cup \{2\} \cup \{3\} \cup \{1, 2\} \cup \{1, 3\} \cup \{2, 3\}$$

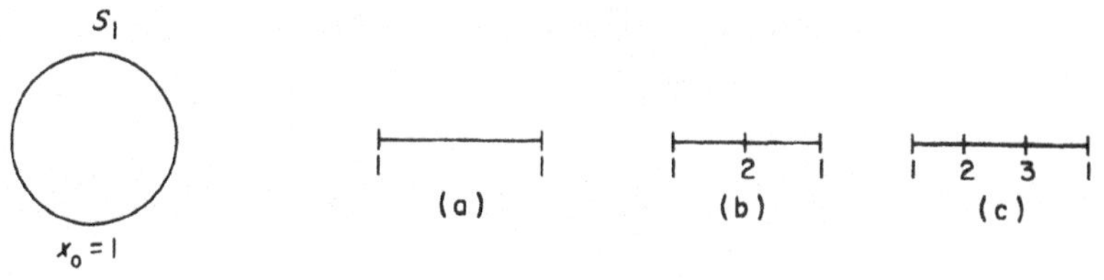

Figure 3.15

i.e. K is the union of 3-0-simplexes and 3-1-simplexes. L_0, the contractible subpolyhedra contained in K in this case is given by:

$$L_0 = \{1, 3\} \cup \{2, 3\}$$

Thus G the group generated by the symbols g_{ij}, one for each ordered 1-simplex of K_0, is generated in this case by one element: $g_{12} = g$, since $g_{13} = g_{23} = 1$ being elements of L_0. Thus $\pi_1(K_0, x_0)$ is isomorphic to the group generated by one element g which is isomorphic to $\mathbf{Z}$ {group of integers under addition} we write $\pi_1(S', x_0) \simeq \mathbf{Z}$.

Example 2.

$$\pi_1(D, x_0), \quad \text{where } D = \{(x_1, x_2) | x_1^2 + x_2^2 \leq 1\}$$

The triangulation of D is shown in (Fig. 3.16).

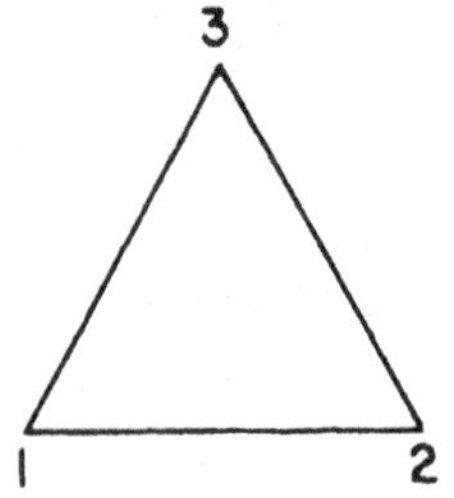

Figure 3.16

In this case the polyhedra associated with D is $K = K_0 \cup \{1, 2, 3\}$, K_0 represents the triangulation of S^1 discussed in Example 1. L, the contractible subpolyhedra contained in K which contains all the vertices of K is the same as L_0 i.e.

$$L = \{1, 3\} \cup \{2, 3\}$$

K as in the previous example contains only one element $g_{12} \neq 1$, however this time there is a 2-simplex present, namely $\{1, 2, 3\}$. Thus there is the relation:

$$g_{12} \cdot g_{23} \cdot g_{13}^{-1} = \{1\}, \text{ the unit element of } G$$

Since g_{23}, g_{13} are elements of L they can be regarded as unit elements. Thus $g_{12} = \{1\}$. Hence $\pi_1(K, x_0) \simeq \{1\}$, the group generated by the identity. Then $\pi_1(D, x_0) \simeq \{1\}$.

Example 3. $\pi(S^2, x_0)$, S^2 being the surface of a sphere of unit radius in three-dimensional Euclidean space.

We triangulate S^2 on the boundary of a tetrahedron (Fig. 3.17). Thus

$$K = \{1\} \cup \{2\} \cup \{3\} \cup \{4\}$$

$$\{1, 2\} \cup \{2, 3\} \cup \{1, 3\} \cup \{1, 4\} \cup \{3, 4\} \cup \{2, 4\}$$

$$\{1, 2, 4\} \cup \{2, 3, 4\} \cup \{1, 3, 4\} \cup \{1, 2, 3\}$$

while $L = \{12\} \cup \{23\} \cup \{34\}$

The generators g_{12}, g_{23}, g_{34} are all equal to $\{1\}$, the identity element of G since they are elements of L. The remaining generators g_{14}, g_{24}, g_{13} must satisfy 4 relations, one from each of the 2-simplexes of K. Thus $g_{12} \cdot g_{24} \cdot g_{14}^{-1} = \{1\}$ etc., these imply: $g_{14} = g_{24} = g_{13} = \{1\}$. Hence $\pi_1(S^2, x_0) \simeq \{1\}$.

Example 4. The Möbius band. A triangulation starts with the observation that the Möbius band is obtained by taking a rectangle and joining one of the edges to the edge parallel to it after a twist. We can represent this by the following picture (Fig. 3.18).

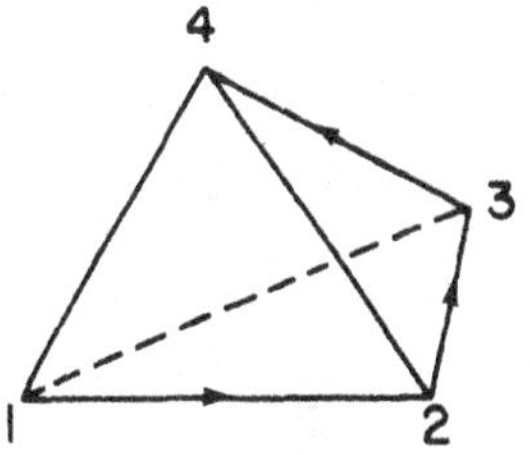

Figure 3.17

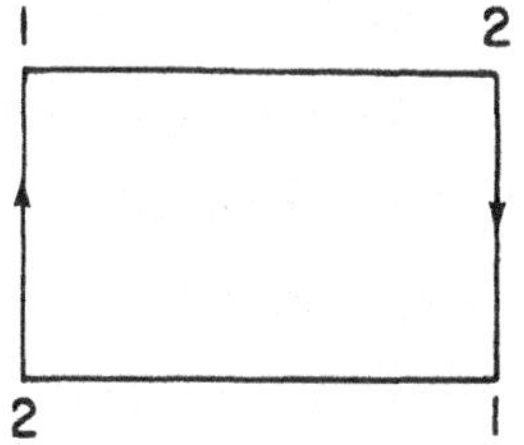

Figure 3.18

The vertices are labelled indicating the way the edges are to be identified. A possible triangulation of the band is given below (Fig. 3.19).

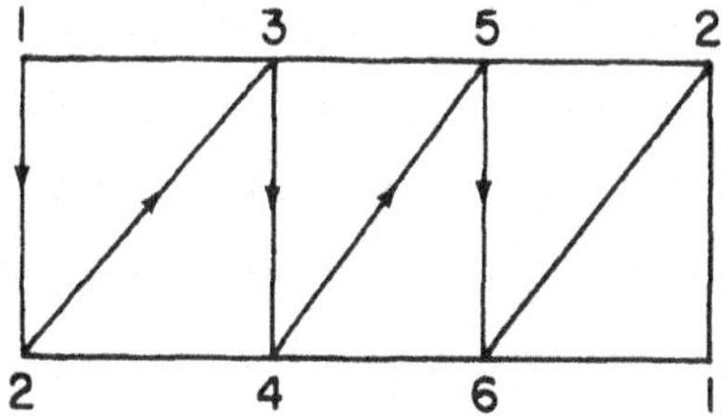

Figure 3.19

Thus the polyhedra K corresponding to the Möbius band is given by:

$$K = \{1\}, \{2\}, \{3\}, \{4\}, \{5\}, \{6\}$$

$$\{1, 2\}, \{1, 3\}, \{1, 6\}, \{2, 3\}, \{2, 4\}, \{2, 5\}, \{2, 6\}$$

$$\{3, 4\}, \{3, 5\}, \{4, 5\}, \{4, 6\}, \{5, 6\}$$

$$\{1, 2, 3\}, \{1, 2, 6\}, \{2, 3, 4\}, \{2, 5, 6\}, \{3, 4, 5\}, \{4, 5, 6\}$$

A collection of six 0-simplexes, twelve 1-simplexes and six 2-simplexes. The one-dimensional contractible polyhedron L which contains all the vertices of K can be selected to be:

$$L = \{1, 2\} \cup \{2, 3\} \cup \{3, 4\} \cup \{4, 5\} \cup \{5, 6\}$$

Thus the generators $g_{ij} = \{i, j\} = 1$ when the simplex $\{i, j\} \in L$. The remaining generators, are g_{13}, g_{24}, g_{35}, g_{25}, g_{46}, g_{16} and g_{26}. These generators must satisfy 6 relations, one from each 2-simplex, in K namely

$$g_{12} \cdot g_{23} = g_{13}$$

$$g_{24} \cdot g_{43} = g_{23}$$

$$g_{35} \cdot g_{54} = g_{34}$$

$$g_{46} \cdot g_{65} = g_{45}$$

$$g_{25} \cdot g_{56} = g_{26}$$

$$g_{12} \cdot g_{26} = g_{16}$$

Since $g_{12} = g_{23} = g_{34} = g_{45} = g_{56} = 1$ (the corresponding one simplexes are elements of L). We find that $g_{13} = 1$, $g_{24} = 1$, $g_{35} = 1$, $g_{46} = 1$ and $g_{25} = g_{26} = g_{16} = g$, say.

Thus π_1 (Möbius band, x_0) $\simeq$ group generated by one generator $\simeq \mathbf{Z}$.

Notice, however that the path along the boundary of the Möbius band corresponds to the element ± 2, unlike the circle where the corresponding path corresponds to ± 1.

Example 5. The torus T. A triangulation of the torus is given below (Fig. 3.20).

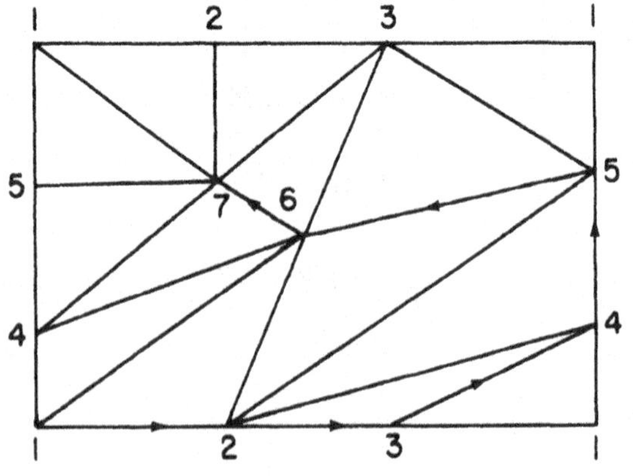

Figure 3.20

The triangulation K consists of seven 0-simplexes, fourteen 1-simplexes and fourteen 2-simplexes. The contractible 1-dimensional sub-polyhedron of K which contains all the vertices of K is indicated in the figure by arrows. Thus the generators $g_{12}=g_{23}=g_{34}=g_{45}=g_{56}=g_{67}=1$ since the corresponding 1-simplexes are elements of L. The remaining 15 generators (total number of 1-simplexes minus the number of 1-simplexes belonging to L) have to satisfy 14 relations, one from each of the fourteen 2-simplexes of the triangulation. These give:

$$g_{13}=g_{14}=g_{64}=g_{47}=g_{75}=g, \text{ say}$$

$$g_{17}=g_{27}=g_{37}=g_{36}=g_{35}=k, \text{ say}$$

$$g_{16}=g_{26}=g_{25}=g_{24}=1$$

and $kg=gk$, i.e. the two generators g and k commute. Then π_1 (torus) $\simeq$ group generated by two commuting generators $\simeq \mathbf{Z}\oplus\mathbf{Z}$.

Example 6. The projective plane. A triangulation is given below (Fig. 3.21).

We recall that the projective plane can be regarded as the space obtained by taking a finite disc and identifying each pair of diametrically opposite points.

In this case the triangulation consists of six 0-simplexes, fifteen 1-simplexes and ten 2-simplexes. The contractible 1-dimensional subpolyhedra L which contains all the vertices of K can be chosen in the manner indicated by arrows on Fig. 3.21. Thus $g_{12}=g_{23}=g_{35}=g_{54}=g_{46}=1$ since

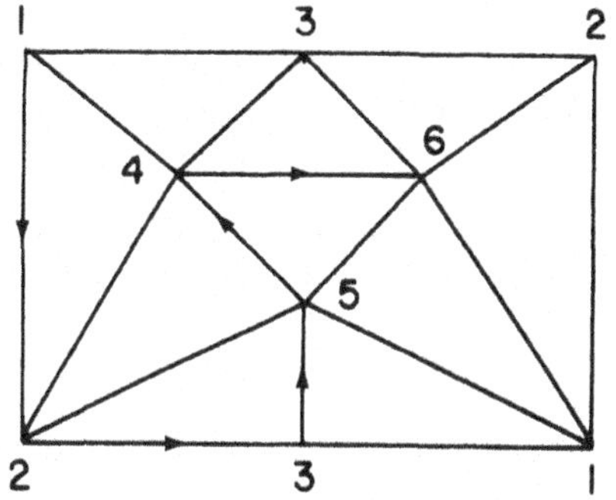

Figure 3.21

the corresponding 1-simplexes belong to L. The remaining 10 generators (10 = total number of 1-simplexes of the triangulation minus the number of 1-simplexes belonging to L) must satisfy 10 relations, one from each of the ten 2-simplexes in the triangulation. These give:

$$g_{25} = g_{24} = g_{14} = g_{56} = 1$$

$$g_{13} = g_{61} = g_{15} = g_{62} = g_{63} = g_{43} = g, \text{ say}$$

and $g^2 = 1$.

Thus

$$\begin{aligned}\pi_1\,(\text{projective plane}) &\simeq \text{group generated with one element of the relation } g^2 = 1 \\ &\simeq \text{group of integers under addition modulo } 2 \simeq \mathbf{Z}/2\end{aligned}$$

Example 7. The Klein bottle. The Klein bottle is obtained from a cylinder by identifying the two circular ends with the orientation of the two circles reversed. A triangulation is given below (Fig. 3.22).

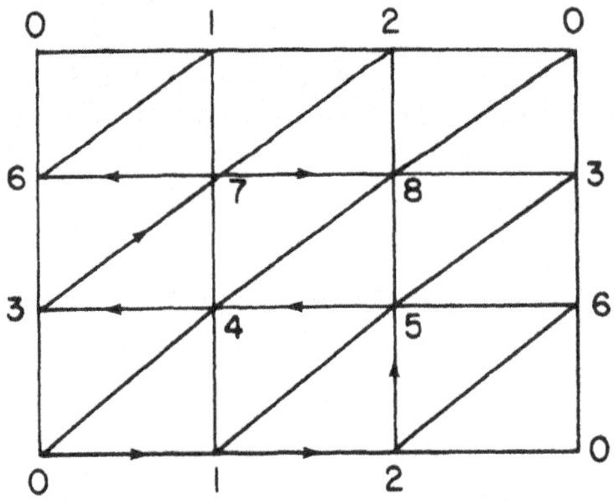

Figure 3.22

The triangulation K consists of eight 0-simplexes, twenty seven 1-simplexes, and eighteen 2-simplexes. The contractible 1-dimensional subpolyhedra contained in K which contains all the vertices of K is indicated by arrows on Fig. 3.22. Thus $g_{01} = g_{12} = g_{25} = g_{45} = g_{34} = g_{37} = g_{78} = g_{67} = 1$, since the corresponding 1-simplexes are elements of L. The remaining 19 generators $(27 - 8)$ have to satisfy 18 relations, one relation from every

2-simplex present in the triangulation. We find that

$$g_{26} = g_{56} = g_{83} = g_{80} = g_{53} = g$$

$$g_{17} = g_{27} = g_{28} = g_{06} = g_{16} = k$$

$$g_{67} = g_{36} = g_{15} = g_{14} = g_{04} = g_{03} = g_{47} = g_{48} = g_{58} = 1$$

$$\text{and } g\,k\,g\,k^{-1} = 1$$

Thus

$$\pi_1 \text{ (Klein bottle)} \simeq \text{group generated by two generators } g, k \text{ with one relation } g\,k\,g\,k^{-1} = 1$$

After Theorem (3.43) was stated we noted that only the 1-simplex and 2-simplex structure of the polyhedra K were needed to determine $\pi_1(K, a_0)$. Using geometrical language this could be taken to mean that $\pi_1(K, a_0)$ can only spot the two dimensional holes present in a space. For instance we found that $\pi_1(S^2, x_0) = \{1\}$ although it is clear that the space S^2 encloses a three-dimensional hole. In Chapter 5 we will discuss higher dimensional analogues of the fundamental group called higher homotopy groups $\pi_n(X)$ which can detect the higher dimensional holes present in a space. Unfortunately there is no analogue of Theorem (3.43) for these higher homotopy groups and their determination is, in general, very difficult. In the next chapter an alternate method of spotting holes of arbitrary finite dimension for triangulable spaces will be described. These simplicial homology groups $H_n(X)$ are easier to determine but contain in them less information regarding the underlying topological space than the corresponding homotopy groups, $\pi_n(X)$.

We end this chapter by establishing another theorem useful for calculations.

3.5 FUNDAMENTAL GROUP OF A PRODUCT X×Y

Theorem

The fundamental group of the product of two topological spaces X and Y is isomorphic to the direct product of their fundamental groups; in symbols:

$$\pi_1(X \times Y, x_0 \times y_0) \simeq \pi_1(X, x_0) \oplus \pi_1(Y, y_0), \qquad x_0 \in X,\ y_0 \in Y \quad (3.46)$$

We recall that if G_1 and G_2 are two groups then their direct product denoted by $G_1 \oplus G_2$ is the set of all ordered pairs (g_1, g_2), $g_1 \in G_1$, $g_2 \in G_2$

with multiplication (addition for abelian groups) defined according to the rule:

$$(g_1, g_2) \times (g_1', g_2') = (g_1 g_1', g_2 g_2') \tag{3.47}$$

The proof of the theorem is straightforward and follows from the remark that a loop in $(X \times Y, x_0 \times y_0)$ is exactly a pair of loops (X, x_0) and (Y, y_0).

Example 1. The torus T is homeomorphic to the product $S^1 \times S^1$. Hence (using an additive notation for the abelian groups involved):

$$\begin{aligned}\pi_1(T, t_0) &\simeq \pi_1(S^1, x_0) \oplus \pi_1(S^1, y_0), \; t_0 = (x_0 \times y_0)\\ &\simeq \mathbf{Z} \oplus \mathbf{Z}\end{aligned}$$

Example 2. A closed cylinder C is the product of a circle S^1 and a closed interval $[0, 1]$. Thus

$$\begin{aligned}\pi_1(C, c_0) &\simeq \pi_1(S^1, x_0) \oplus \pi_1([0, 1], t_0), \; c_0 = x_0 \times t_0\\ &\simeq \mathbf{Z} \oplus \{0\}, \text{ since } [0, 1] \text{ is contractible}\end{aligned}$$

REFERENCES

1. Hu, S. T., "Homotopy Theory". Academic Press, 1959.
2. Maunder, C. R. F., "Algebraic Topology". Van Nostrand Reinhold Co, 1972.
3. Massey, W. S., "Algebraic Topology: An Introduction". Springer, 1967.
4. Kurosh, A. G., "The Theory of Groups". Chelsea, 1955.
5. Rotman, J. J., "The Theory of Groups". Allyn and Bacon, 1965.

CHAPTER 4

The Homology Groups

4.1 INTRODUCTION

In this chapter we give an account of the homology groups $H_p(x)$, $p = 0, 1, 2, \ldots$, associated with a topological space X. Rather than consider an arbitrary space X we will suppose that the space X is triangulable, i.e. homeomorphic to some polyhedron K. The homology groups $H_p(K)$ can then be defined in terms of the simplexes of K and are, for this reason, called the simplicial homology groups of the polyhedron K. At the end cf the chapter we will briefly describe one way of defining homology groups for an arbitrary, not necessarily triangulable, space X.

In our discussions we will also accept without proof the following important theorems:

Theorem

If X and Y are two topological spaces of the same homotopy type then $H_p(X)$ is isomorphic to $H_p(Y)$ for all p.
We write:

$$H_p(X) \equiv H_p(Y) \tag{4.1a}$$

An immediate implication of this theorem is that the homology groups are topological invariants. This is because homeomorphic spaces are necessarily of the same homotopy type.

Theorem

If K_1 and K_2 are two triangulations of the same topological space K then

$$H_p(K_1) = H_p(K_2), \forall p \tag{4.1b}$$

A proof of these two theorems can be found in references [1] or [3] listed at the end of the Chapter.

We now proceed to explain the geometrical ideas underlying the simplicial homology groups. We do this by examining once again the two spaces X_1 and X_2 introduced at the beginning of Chapter 3. X_1 and X_2 were rectangular regions of $\mathbf{R}^2$. X_1 contained a hole while X_2 did not. X_2 is thus homeomorphic to the 2-simplex σ^2 while, in view of Theorem (4.1a), X_1 can be replaced by its deformation retract, the edges of σ^2. Figure 4.1 explains what is involved.

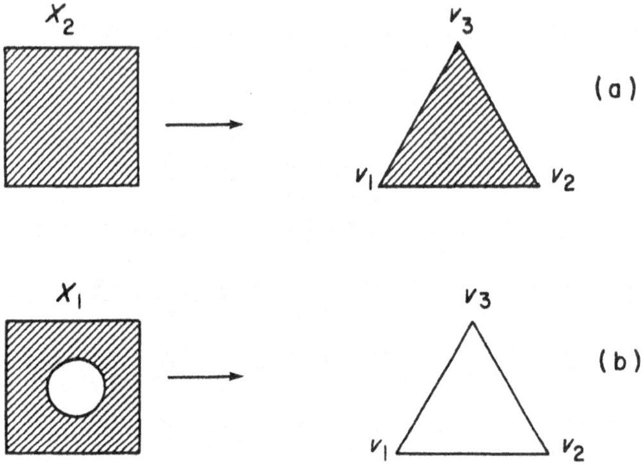

Figure 4.1

The shaded region represents the surface, the unshaded region the hole. In Chapter 3 the topological difference between X_1 and X_2 was studied using homotopy classes of loops. Now we proceed differently. Observe that the boundary of Fig. 4.1(a) is the boundary of a connected region. But Fig. 4.1(b) consists only of the edges of the triangular region v_1, v_2, v_3 and is not the boundary of any region. This simple observation suggests a method of spotting holes in a space. A closed two-dimensional region has a hole if its boundary, some closed curve, is not the boundary of a connected region. The usefulness of this procedure is two-fold. Firstly, the idea can be generalized to higher dimensions. Secondly, for a triangulable space, a simple algebraic definition of the boundary of the region can be given. Let us explain how this is done by considering some examples. We begin with the simplest case, namely, when our space K is the 0-simplex $\sigma^0=[v]$. Then K does not have any boundary and we require:

$$\partial[v]=0$$

where $\partial[v]$ represents the boundary of the 0-simplex $[v]$. What this equation means is the following: a correspondence between a positive or negative integer and $[v]$ is set up. In terms of such a correspondence $\partial[v]$ is mapped to zero.

Next suppose K is the 1-simplex $\sigma^1=[v_1, v_2]$, together with its faces. Geometrically the boundary of σ^1 consists of the two end points of σ^1, namely $[v_1]$ and $[v_2]$. We write this formally as:

$$\partial[v_1, v_2]=[v_1]+[v_2] \tag{4.2}$$

where, as before, the formal sum can be understood in terms of a correspondence set up between the simplexes $[v_1, v_2]$, $[v_1]$ and $[v_2]$ and the positive and negative integers. Proceeding in this manner the boundary of the 2-simplex $\sigma^2=[v_1, v_2, v_3]$ can be written as:

$$\partial[v_1, v_2, v_3]=[v_1, v_2]+[v_2, v_3]+[v_1, v_3] \tag{4.3}$$

and the boundary of a K-simplex $\sigma^K=[v_0, \ldots, v_K]$ as:

$$\partial[v_0, \ldots, v_K]=\sum_{j=1}^{K}[v_0, \ldots, \hat{v}_j, \ldots, v_K] \tag{4.4}$$

where $[v_0, \ldots, \hat{v}_j, \ldots, v_K]$ represents the $(K-1)$-simplex σ^{K-1} obtained from the K-simplex σ^K by omitting the vertex v_j. Geometrically this is reasonable since such a $(K-1)$-simplex is a face of the K-simplex σ^K.

Does the boundary operator ∂ defined by equation (4.4) give the boundary of a polyedron? Let us consider the simple polyhedron shown in Fig. 4.2.

Figure 4.2

From the figure it is clear that

$$\partial K=[v_1]+[v_3] \tag{4.5}$$

It is also clear that K is obtained by joining the two 1-simplexes $[v_1, v_2]$ and $[v_2, v_3]$. It is tempting to write:

$$K=[v_1, v_2]+[v_2, v_3] \tag{4.6a}$$

where the formal sum is understood in terms of a correspondence between the simplexes of K and the integers. In order to use equation (4.4) to determine the boundary of K we have to assume that the boundary operator

∂ acts linearly on the simplexes of K, i.e.

$$\begin{aligned}\partial K &= \partial[v_1, v_2]+\partial[v_2, v_3]\\ &=[v_1]+[v_2]+[v_2]+[v_3]\end{aligned} \tag{4.6b}$$

which is different from (4.5). There are two ways of saving the situation. One is to formally set $[v_2]+[v_2]=0$. The other way is to modify the definition of ∂ so that the two factors of $[v_2]$ in (4.6b) appear with opposite signs and hence cancel. Since ∂ was defined in terms of simplexes, this suggests that each simplex should be given a sign. How should this be done? Consider the K-simplex $\sigma^K=[v_0, v_1, \ldots, v_k]$, we can clearly write σ^K in a variety of different ways merely by permuting the vertices $v_0, v_1, \ldots, v_K$. Is there any way of separating all of these different possible representations into two classes? There is: namely any given representation of $\sigma^K=[v_{i_0}, v_{i_1}, \ldots, v_{i_K}]$, where $0<i_1<\ldots<i_K$ be obtained from the 'standard' form $[v_0, v_1, \ldots, v_{i_K}]$ by an even or odd number of permutations. We could thus assign a positive sign to a member of the even permutation class and a negative sign to a member of the odd permutation class. What we are really doing is orienting the simplexes in the sense of Chapter 2. To see this let us consider the case of $\sigma^2=[v_1, v_2]$. According to the convention we have just described: $+\sigma=[v_1, v_2]$, and $-\sigma^2=[v_2, v_1]$. On the other the basis vectors v_1, v_2 can be transformed to v_2, v_1 by means of the matrix $\left[\begin{smallmatrix}0 & 1\\ 1 & 0\end{smallmatrix}\right]$ and the determinant of this matrix is equal to minus one. The orientations of $[v_1, v_2]$ and $[v_2, v_1]$ are thus different.

Let us next see how the boundary operator ∂ should be modified in order to make it consistant with the sign convention introduced. We still expect $\partial\sigma^K$ to be a formal linear combination of the faces σ^{K-1} of σ^K but now each of the simplexes has to be given a sign. Consider the case of σ^2. We write:

$$\partial\sigma^2=\partial[v_1, v_2]=[\hat{v}_1, v_2]+[v_1, \hat{v}_2] \tag{4.7}$$

where, as before, $\hat{v}_i$ indicates that the corresponding vertex v_i is to be removed. Writing: $[\hat{v}_1, v_2]=[v_2]$ leads us to expect:

$$[v_1, \hat{v}_2]=(-)[\hat{v}_2, v_1]=(-)[v_1]$$

Thus we get:

$$\partial[v_1, v_2]=[v_2]-[v_1] \tag{4.8}$$

In terms of the modified boundary operator we have:

$$\begin{aligned}\partial K &= \partial[v_1, v_2]+\partial[v_2, v_3]\\ &=([v_2]-[v_1])+([v_3]-[v_2])\\ &=[v_3]-[v_1], \text{ as required}\end{aligned}$$

We can now return to our original problem of distinguishing the space X_1 from the space X_2 in an algebraic manner using the boundary operator ∂. X_1 is a polyhedron made up of the following simplexes:

$$X_1 = \{[v_1, v_2]; [v_2, v_3]; [v_1, v_3]\} \text{ and their faces}$$

while

$$X_2 = \{[v_1, v_2, v_3]; \text{ and all its faces}\}$$

We observe that in general if K, is a closed region then ∂K, the boundary of K, is expected to be a closed surface. ∂K should not itself have any boundary, i.e. $\partial(\partial K) = 0$. Thus once we have defined a boundary operator ∂ it is straightforward to spot a boundary surface b. We just verify that $\partial b = 0$. It is also possible to tell if the boundary b is the boundary of a hole or of a connected region. This is because if b were the boundary of some connected region then we expect to find a K such that $b = \partial K$. Let us now apply these ideas to the spaces X_1 and X_2. We note that

$$b = [v_2, v_3] - [v_1, v_3] - [v_1, v_2]$$

in X_1 is a boundary since $\partial b = 0$. On the other hand there are no higher dimensional simplexes in X_1, i.e. $b \neq \partial K$. Thus X_1 contains a hole. For X_2, b is also a boundary but it is precisely the boundary of the 2-simplex $\sigma^2 = [v_1, v_2, v_3]$. Thus X_2 does not contain a hole. Using the geniune boundary b of X_1 an abelian group can be generated simply by noting that if b is a boundary then so is $\pm Nb$, where N is an integer. This abelian group is the homology group $H_1(X_1)$. Our discussion suggests that $H_1(X_1)$ is isomorphic to $\mathbf{Z}$, the group of positive and negative integers under addition.

Let us now make precise the intuitive ideas discussed.

4.2 ORIENTED SIMPLEXES AND THE DEFINITION OF THE HOMOLOGY GROUPS

Definition

An oriented p-simplex $p > 1$ is obtained from a p-simplex $\sigma^p = [v_0, \ldots, v_p]$ by choosing an ordering for its vertices. The equivalence class of even permutations of the chosen ordering determines the positively oriented simplex $+\sigma^p$, while the equivalence class of odd permutations determines the negatively oriented simplex $(-)\sigma^p$. A simplicial complex whose simplexes have been assigned an orientation is called an oriented simplicial complex. (4.9)

Example

For the 2-simplex, if we choose $+\sigma^2 = [v_0, v_1, v_2]$, then

$$+\sigma^2 = [v_0, v_1, v_2] = [v_1, v_2, v_0] = [v_2, v_0, v_1]$$

while

$$-\sigma^2 = [v_0, v_2, v_1] = [v_2, v_1, v_0] = [v_1, v_0, v_2]$$

Our next step is to associate with each p-simplex σ_i^p $(p = 0, 1, \ldots, n)$, of a simplicial complex K, an abelian group $C_p(K)$ called the chain group. Once this is done the geometric notion of a boundary which we discussed can be changed into an algebraic statement involving the chain group.

Definition

Let K be a n-dimensional simplicial complex containing l_p, p-simplexes. The p chain of K, $C_p(K)$ is the free abelian group generated by the orientated p-simplexes of K. What this means is the following: an arbitrary element $c_p \in C_p(K)$ can be written as the formal sum:

$$c_p = \sum_{l=1}^{l_p} f_i\sigma_i^p, f_i \in \mathbf{Z}$$

where

$$\sigma_i^p + (-\sigma_i^p) = 0, \forall i, p$$

and

$$\sum_{i=1}^{l_p} f_i\sigma_i^p + \sum_{i=1}^{l_p} g_i\sigma_i^p = \sum_{i=1}^{l_p} (f_i + g_i)\sigma_i^p (f_i, g_i \in \mathbf{Z})$$

The statement that K is n-dimensional simply means that $P = 0, 1, \ldots, n$. It is often convenient to define $C_p(K) = \{0\}$ for $P > n$ (4.10)

The boundary operator ∂ can now be properly defined.

Definition

The boundary operator ∂_p is the map:

$$\partial_p : C_p(K) \to C_{p-1}(K)$$

with the following properties

i. It is linear: $\partial_p \{\sum_i f_i\sigma_i^p\} = \sum_i f_i \partial\sigma_i^p$

ii. For an oriented p-simplex

$$\sigma^p = [v_0, \ldots, v_p]$$

$$\partial[v_0, \ldots, v_p] = \sum_{j=0}^{p} (-1)^j [v_0, \ldots, \hat{v}_j, \ldots, v_p]$$

where $[v_0, \ldots, \hat{v}_j, \ldots, v_p]$ represents the $(p-1)$-simplex σ^{p-1} obtained from the p-simplex σ^p by omitting the vertex v_j.

iii. The boundary of every zero chain is defined to be zero. (4.11)

It is straightforward to check that ∂_p is a homomorphism from $C_p(K)$ to $C_{p-1}(K)$. Very often we will omit the subscript p of the boundary operator ∂_p and write simply ∂. We next check that the boundary of a polyhedron K does not itself have a boundary, i.e. we have:

Theorem

$$\partial_{p-1} \circ \partial_p = 0 \tag{4.12}$$

Proof

Because of the linearity of ∂p it is sufficient to show that

$$\partial_{p-1} \circ \partial_p \sigma^p = 0$$

Now

$$\begin{aligned}
\partial_{p-1} \circ \partial_p \sigma^p &= \partial_{p-1} \circ \partial_p [v_0, \ldots, v_p] \\
&= \partial_{p-1} \left\{ \sum_{j=0}^{p} (-1)^j [v_0, \ldots, \hat{v}_j, \ldots, v_p] \right\} \\
&= \sum_{j=0}^{p} (-1)^j \partial_{p-1} [v_0, \ldots, \hat{v}_j, \ldots, v_p] \\
&= \sum_{j=0}^{p} (-1)^j \left\{ \sum_{i=0}^{j-1} (-)^i [v_0, \ldots, \hat{v}_i, \ldots, \hat{v}_j, \ldots, v_p] \right. \\
&\quad \left. + \sum_{i=j+1}^{p} (-1)^{i-1} [v_0, \ldots, \hat{v}_j, \ldots, \hat{v}_i, \ldots, v_p] \right\} \\
&= \sum_{i<j} (-)^{i+j} [v_0, \ldots, \hat{v}_i, \ldots, \hat{v}_j, \ldots, v_p] \\
&\quad + \sum_{i>j} (-1)^{i+j-1} [v_0, \ldots, \hat{v}_j, \ldots, \hat{v}_i, \ldots, v_p] \\
&= \sum_{i<j} [(-1)^{i+j} + (-1)^{i+j-1}] [v_0, \ldots, \hat{v}_i, \ldots, \hat{v}_j, \ldots, v_p] \\
&= 0
\end{aligned}$$

which establishes the result.

We now proceed to first identify all the p-dimensional boundaries present (cycles) and then to identify the ones that are boundaries of connected regions in terms of the chain group.

Definition

$z_p \in C_p(K)$ is called a p-dimensional cycle or p-cycle if $\partial z_p = 0$. The family of p-cycles is thus the kernel of the homomorphism: $\partial : C_p \to C_{p-1}$ and is a subgroup of $C_p(K)$. This subgroup is called the p-dimensional cycle group of K and is denoted by $Z_p(K)$. (4.13)

Definition

$b_p \in C_p(K)$ is called a p-dimensional boundary or p-boundary if there is a $(p+1)$ chain C_{p+1} such that $\partial C_{p+1} = b_p$. The family of p-boundaries is thus the homomorphic image $\partial C_{p+1}(K)$ and is a subgroup of $C_p(K)$. This subgroup is called the p-dimensional boundary group of K and is denoted by $B_p(K)$ (4.14)

Note, because of Theorem (4.12) any element b_p of $B_p(K)$ has the property that $\partial b_p = 0$. Thus $B_p(K)$ is a subgroup of $Z_p(K)$. In order to spot the $(p+1)$-dimensional holes we have thus to weed out the elements belonging to $B_p(K)$ contained in $Z_p(K)$. This is achieved by introducing the homology group $H_p(K)$.

Definition

The p-dimensional homology group of K denoted by $H_p(K)$ is the quotient group:

$$H_p(K) = Z_p(K)/B_p(K) \tag{4.15}$$

An element h_p of $H_p(K)$ is thus an equivalent class $[z_p]$, defined by the relation $\overset{1}{z}_p$ is equivalent to $\overset{2}{z}_p$ if $\overset{1}{z}_p - \overset{2}{z}_p \in B_p(K)$. This equivalence relation is called homology and if $\overset{1}{z}_p - \overset{2}{z}_p$ is in $B_p(K)$ then $\overset{1}{z}_p$ and $\overset{2}{z}_p$ are said to be homologous. The fact that $H_p(K)$ is a group is easy to verify and follows from the fact that the $Z_p(K)$ and $B_p(K)$ are both abelian groups.

From the way it is defined it might seem quite remarkable that the groups $H_p(K)$ do not depend on the triangulation of K (Theorem 4.1b). This is because the groups $C_p(K)$, $Z_p(K)$ and $B_p(K)$ certainly do depend on the triangulation of K. On the other hand from the geometrical ideas which motivate the definition of $H_p(K)$ Theorem (4.1b) is understandable. Different triangulations of K are certainly expected to lead to different cycle groups and boundary groups since the number of boundaries introduced depends on triangulation. But $H_p(K)$, which depends on the number of $(p+1)$-dimensional holes present in the space, should not.

Using Theorem (4.1a) and the definition of the homology group $H_p(K)$ just introduced we can prove our first general Theorem.

Theorem

If K is a contractible space i.e. has the homotopy type of a single point then

$$H_p(K) = \begin{cases} \{0\}, \ p \neq 0 \\ \mathbf{Z}, \ p = 0 \end{cases} \tag{4.16}$$

Proof

From Theorem (4.1b) it follows that if K and the point v_0 have the same homotopy type then

$$H_p(K) = H_p([v_0]), \forall p$$

Thus, as far as the homology groups are concerned, the complex corresponding to K is the 0-simplex $\sigma^0 = [v^0]$. The dimension of K is thus zero, $C_p(K) \equiv \{0\}$ for $p > 0$ by definition and hence $H_p(K) = \{0\}$ for $p \neq 0$. For $p = 0$ we note that $Z_0(K) = C_0(K)$ since all elements C_0 of $C_0(K)$ have property $\partial C_0 = 0$. On the other hand, as there are no higher dimensional simplexes present, $B_0(K) \equiv \{0\}$. So that

$$\begin{aligned} H_0(K) &= Z_0(K) = \{z_0 | z_0 = f[v_0], f \in \mathbf{Z}\} \\ &= \mathbf{Z} \end{aligned}$$

which completes the proof.

A few examples illustrating how the homology groups $H_p(K)$ are calculated might be useful.

Example 1. Let $K = \sigma^2 = [v_0, v_1, v_2]$. We will calculate $H_k(K)$ for $k = 0, 1, 2, \ldots$. We first note that dim $K = 2$. Thus, by definition $C_p(K) \equiv \{0\}$ for $p > 2$, hence $H_k(K) \equiv \{0\}$ for $k > 2$. Thus we only have to calculate $H_0(K)$, $H_1(K)$, and $H_2(K)$.

Calculation of $H_0(K)$. Since $H_0(K) \equiv Z_0(K)/B_0(K)$, we have to determine $Z_0(K)$ and $B_0(K)$ we recall that $z_0 \in Z_0(K)$ if $z_0 \in C_0(K)$ and $\partial z_0 = 0$ From definition. (4.11) it follows that $C_0(K) = Z_0(K)$ since all zero chains have zero boundaries. Also any element of $C_0(K)$ can be written as:

$$a_0[v_0]+b_0[v_1]+c_0[v_2]$$

where $a_0, b_0, c_0 \in \mathbf{Z}$; $Z_0(K)$ thus has three independent generators so that:

$$Z_0(K)=\mathbf{Z}\oplus\mathbf{Z}\oplus\mathbf{Z}=\mathbf{Z}^3$$

Next we study $B_0(K)$ we recall that $b_0 \in B_0(K)$ if $b_0 \in C_0(K)$ and $b_0=\partial C_1$ where $C_1 \in C_1(K)$. Any element of $C_1(K)$ can be written as:

$$C_1=a_1[v_0, v_1]+b_1[v_0, v_2]+c_1[v_1, v_2]$$

Thus

$$\begin{aligned}\partial C_1&=a_1([v_1]-[v_0])+b_1([v_2]-[v_0])+C_1([v_2]-[v_1])\\&=(a_1-c_1)[v_1]-(a_1+b_1)[v_0]+(b_1+c_1)[v_2]\end{aligned}$$

Hence any element b_0 of $B_0(K)$ can be written as:

$$b_0=a_0[v_0]+b_0[v_1]+c_0[v_2]$$

with $a_0+b_0+c_0=0$. This means that $B_0(K)$ has two independent generators. Hence $B_0(K)=\mathbf{Z}\oplus\mathbf{Z}$. Finally if $h_0 \in H_0(K)$ we can write h_0 as the coset:

$$\begin{aligned}h_0&=z_0+B_0(K),\\&=a_0[v_0]+b_0[v_1]+c_0[v_2]+\{-a_0[v_0]-b_0[v_1]+(a_0+b_0)[v_2]\}\\&=(a_0+b_0+c_0)[v_2]=d_0[v_2],\ d_0 \in Z\end{aligned}$$

Thus $H_0(K)$ has only one independent generator and $H_0(K)=\mathbf{Z}$

Calculation of $H_1(K)$. Since $H_1(K)\equiv Z_1(K)/B_1(K)$, we have to determine $Z_1(K)$ and $B_1(K)$. Now $z_1 \in Z_1(K)$ if $z_1 \in C_1(K)$ and $\partial z_1=0$. From our previous calculation we know that if $z_1 \in C_1(K)$ then $z_1=a_1[v_0, v_1]+b_1[v_0, v_2]+c_1[v_1, v_2]$ and $\partial z_1=(a_1-c_1)[v_1]-(a_1+b_1)[v_0]+(b_1+c_1)[v_2]$. Hence the requirement $\partial z_1=0$ means that

$$z_1=a_1[v_0, v_1]-a_1[v_0, v_2]+a_1[v_1, v_2]$$

i.e.

$$Z_1(K)=\mathbf{Z}$$

Next we have to study $B_1(K)$. We note that $b_1 \in B_1(K)$ if $b_1 \in C_1(K)$ and $b_1=\partial C_2$ where $C_2 \in C_2(K)$. Any element of $C_2(K)$ can be written as:

$$C_2=a_2[v_0, v_1, v_2]$$

Then

$$\partial C_2=a_2\{[v_1, v_2]-[v_0, v_2]+[v_0, v_1]\}$$

Hence any element b_1 of $B_1(K)$ can be written as:

$$b_1 = a_2\{[v_1, v_2]-[v_0, v_2]+[v_0, v_1]\}$$

This means that $B_1(K)=\mathbf{Z}$. Finally if $h_1 \in H_1(K)$ we write $h_1 = z_1 + B_1(K)$, and note that $z_1 \in B_1(K)$ so that $H_1(K)=\{0\}$.

Calculation of $H_2(K)$. Since $H_2(K) \equiv Z_2(K)/B_2(K)$, we have to determine $Z_2(K)$ and $B_2(K)$. As before $z_2 \in Z_2(K)$ if $z_2 \in C_2(K)$ and $\partial z_2 = 0$. From our previous calculation it follows that $Z_2(K)=\{0\}$, since $\partial z_2 = 0$, implies $a_2 = 0$. Again $b_2 \in B_2(K)$ if $b_2 \in C_2(K)$ and $b_2 = \partial C_3$. Since there are no 3-simplexes in K, $b_2 \equiv 0$. So that $H_2(K)=\{0\}$.

Thus when $K = \sigma^2 = [v_0, v_1, v_2]$

$$H_0(K)=\mathbf{Z}, H_k(K)=\{0\}, k \neq 0. \tag{4.17}$$

Example 2

$$K = S^1 = \partial\sigma^2$$

Now dim $K = 1$, so that $C_k(S^1) \equiv \{0\}$ for $k > 1$. Hence $H_K(S^1) \equiv \{0\}$, for $k > 1$. Thus we only have to calculate $H_0(S^1)$ and $H_1(S^1)$.

Calculation of $H_0(S^1)$. Since $H_0(S^1) \equiv Z_0(S^1)/B_0(S^1)$ we have to determine $Z_0(S^1)$ and $B_0(S^1)$. Again $z_0 \in Z_0(S^1)$ if $z_0 \in C_0(S^1)$ and $\partial z_0 = 0$. From Example 1 it follows that $H_0(S^1)=\mathbf{Z}$ since the 0-simplex and 1-simplex structure of σ^2 and $\partial\sigma^2$ are the same.

Calculation of $H_1(S^1)$. Since $H_1(S^1) \equiv Z_1(S^1)/B_1(S^1)$ we have to determine $Z_1(S^1)$ and $B_1(S^1)$. From Example 1 it follows that $Z_1(S^1)=\mathbf{Z}$ since $Z_1(S^1)$ only involves the 1-simplexes. However $B_1(S^1) \equiv \{0\}$, since there are no 2-simplexes in S^1. So that $H_1(S^1)=\mathbf{Z}$.

Thus $H_0(S^1)=\mathbf{Z}$, $H_1(S^1)=\mathbf{Z}$

$$H_k(S^1)=\{0\}, k>1 \tag{4.18}$$

In all of these examples one notes that $H_0(K)=\mathbf{Z}$. The reason for this is the following Theorem which, again, we will not prove:

Theorem

If K is a connected polyhedron then

$$H_0(K)=\mathbf{Z} \tag{4.19}$$

A proof of the theorem can be found in references [1] or [3] listed at the end of the Chapter.

From the fact that H_0 is the quotient group of two abelian groups Z_p and B_p a few remarks regarding the general structure of the homology groups

$H_p(K)$ can be made. To do this a few results from the theory of abelian groups are needed which we state without proof.

4.3 ABELIAN GROUPS

Definition

Let G be an abelian group. A set $\{g_i\}$ of elements G is called a set of generators of G if every element $g \in G$ can be expressed in the form of a finite sum

$$\sum_{i=1}^{K} n_i g_i \text{ where } n_i \in \mathbf{Z}. \tag{4.20}$$

Definition

The set $\{g_i\}$ of Definition (4.20) freely generates G if for each $g \in G$ the expression

$$g = \sum_{i=1}^{K} n_i g_i$$

is unique i.e. the elements $\{g_i\}$ are linearly independent over $\mathbf{Z}$. An abelian group G which is freely generated by a set of generators is called a free abelian group and a free generating set is called a basis. (4.21)

Definition

An abelian group is said to be finitely generated if it has a set generators consisting of a finite number of elements. (4.22)

It is a theorem in linear algebra that the number of elements in a basis of a free finitely generated abelian group is independent of the choice of the basis. This number is called the *rank* of the group. The two theorems on finitely generated abelian groups that we need can now be stated.

Theorem

Let F be a free finitely generated abelian group and let R be a subgroup of F. Then R is free and finitely generated. (4.23)

Theorem

Let A be a finitely generated (not free) abelian group generated by n generators, say. Then

$$A \simeq F/R = G + Z_{h_1} + \ldots + Z_{h_m}$$

where F and R are free finitely generated abelian groups with $R \subset F$, G is a free abelian group of rank $(n-m)$ and Z_{h_i} is cyclic of order h_i. The rank $(n-m)$ of G and the numbers $h_1, h_2, \ldots, h_m$ are uniquely determined by A. Very often we write

$$A \simeq G \oplus T, \text{ where } T \simeq Z_{h_1} \oplus Z_{h_2} \ldots Z_{h_m} \tag{4.24}$$

and call T the torsion subgroup of A.

Theorems (4.19, 4.20) immediately imply that the boundary groups $B_p(K)$ and cycle groups $Z_p(K)$ are free finitely generated abelian groups while

$$H_p(K) \simeq Z_p^!(K)/B_p(K) \simeq G_p \oplus T_p \tag{4.25}$$

where G_p is a free finitely generated abelian group and T the torsion subgroup of $H_p(K)$. This result is interesting. We introduced the homology group $H_p(K)$ as an algebraic object which could spot the $(P+1)$ dimensional holes present in K. This feature of $H_p(K)$ is reflected in its free finitely generated abelian group part G_p. The rank $R_p(K)$ of $G_p(K)$ counts the number of such holes and is called the pth Betti number of K. But as we now discover $H_p(K)$ contains in it even more information. For even if $G = \{0\}$, $H_p(K) \neq \{0\}$, if $T_p \neq 0$. The group $T_p(K)$ contains information about the manner in which the space K is twisted as the following example illustrated.

Example

The projective plane P. We recall that the projective plane is obtained from a finite disc by identifying each pair of diametrically opposite points. A triangulation of P was given in Example 6, Chapter 3. For convenience the triangulation is given below (Fig. 4.3)

We note that $\dim P = 2$, so that $C_p(P) = \{0\}$, for $p > 2$. Hence $H_p(P) \equiv \{0\}$ for $p > 2$ and we only have to calculate $H_0(P)$, $H_1(P)$ and $H_2(P)$.

Calculation of $H_2(P)$. Since $H_2(P) \equiv Z_2(P)/B_2(P)$ we have to determine $Z_2(P)$ and $B_2(P)$.

We recall that if $b_2 \in B_2(P)$, then $b_2 \in C_2(P)$ and $b_2 = \partial C_3$, where $C_3 \in C_3(P)$. Since P does not contain any 3-simplexes, $b_2 \equiv 0$ and $B_2(P) \equiv \{0\}$.

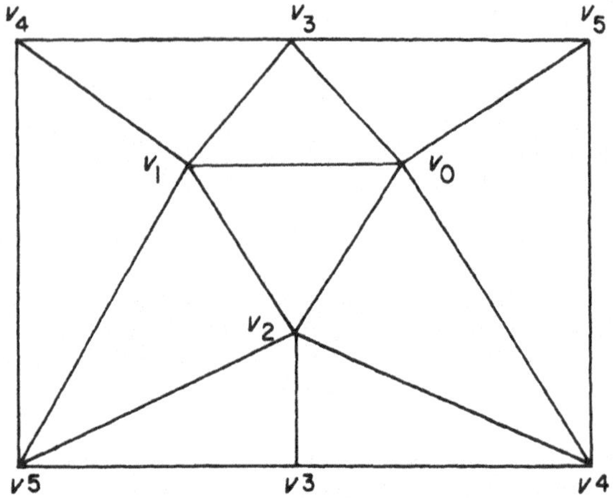

Figure 4.3

Again if $z_2 \in Z_2(P)$, then $z_2 \in C_2(P)$ and $\partial z_2 = 0$. There are ten 2-simplexes in P thus

$$\begin{aligned} z_2 = {} & a_1[v_0, v_1, v_2] + a_2[v_0, v_1, v_3] + a_3[v_0, v_3, v_5] + a_6[v_1, v_3, v_4] \\ & + a_4[v_0, v_4, v_5] + a_5[v_0, v_2, v_4] + a_7[v_1, v_4, v_5] + a_8[v_1, v_2, v_5] \\ & + a_9[v_2, v_3, v_5] + a_{10}[v_2, v_3, v_4] \end{aligned} \tag{4.26}$$

and

$$\begin{aligned} \partial z_2 = {} & (a_1 + a_8)[v_1, v_2] - (a_1 - a_5)[v_0, v_2] + (a_1 + a_2)[v_0, v_1] \\ & + (a_2 + a_6)[v_1, v_3] - (a_2 - a_3)[v_0, v_3] + (a_3 + a_9)[v_3, v_5] \\ & - (a_3 + a_4)[v_0, v_5] + (a_4 + a_7)[v_4, v_5] + (a_4 - a_5)[v_0, v_4] \\ & + (a_5 - a_{10})[v_2, v_4] + (a_6 + a_{10})[v_3, v_4] - (a_6 - a_7)[v_1, v_4] \\ & - (a_7 + a_8)[v_1, v_5] + (a_8 - a_9)[v_2, v_5] + (a_9 + a_{10})[v_2, v_3] \end{aligned} \tag{4.27}$$

In order to see the geometrical significance of this expression it is convenient to write it as:

$$\begin{aligned} \partial z_2 = {} & a_1\{[v_0, v_1] + [v_1, v_2] + [v_2, v_0]\} \\ & + a_2\{[v_1, v_3] + [v_3, v_0] + [v_0, v_1]\} \\ & + a_3\{[v_0, v_3] + [v_3, v_5] + [v_5, v_0]\} \\ & + a_4\{[v_5, v_0] + [v_0, v_4] + [v_4, v_5]\} \end{aligned}$$

$$+a_5\{[v_0, v_2]+[v_2, v_4]+[v_4, v_0]\}$$
$$+a_6\{[v_1, v_3]+[v_3, v_4]+[v_4, v_1]\}$$
$$+a_7\{[v_4, v_5]+[v_1, v_4]+[v_5, v_1]\}$$
$$+a_8\{[v_1, v_2]+[v_2, v_5]+[v_5, v_1]\}$$
$$+a_9\{[v_3, v_5]+[v_5, v_2]+[v_2, v_3]\}$$
$$+a_{10}\{[v_2, v_3]+[v_3, v_4]+[v_4, v_2]\} \tag{4.28}$$

Now one can explicitly see that each term in (4.28) traces out a closed cycle through the complex. From (4.27) it follows that $\partial z_2=0$ implies $a_1=a_5$, $a_2=a_3$, $a_4=a_5$, $a_5=a_{10}$, $a_6=a_7$, $a_8=a_9$, $a_1+a_8=0$, $a_2+a_6=0$, $a_2+a_4=0$, $a_2+a_8=0$, $a_4+a_6=0$, $a_6+a_8=0$ and $a_4+a_8=0$ which in turn means that

$$a_1=a_2=a_3=a_4=a_5=a_6=a_7=a_8=a_9=a_{10}=0$$

Thus $Z_2(P)=\{0\}$. Hence $H_2(P)=\{0\}$.

Calculation of $H_1(P)$. Since $H_1(P)\equiv Z_1(P)/B_1(P)$ we have to determine $Z_1(P)$ and $B_1(P)$.

We note that $b_1\in B_1(P)$ means that $b_1\in C_1(P)$ and $b_1=\partial C_2$, where $C_2\in C_2(P)$. From the calculation of $H_2(P)$ we know the form of b_1. It is given by equation (4.27) when C_2 is taken to be equation (4.26). Thus $B_1(P)=\mathbf{Z}^{10}$, the direct sum of 10 copies of the abelian group $\mathbf{Z}$. Next we consider $z_1\in Z_1(P)$. Then $z_1\in C_1(P)$ and $\partial z_1=0$. There are 15 one-simplexes in P so that:

$$\begin{aligned} z_1 = {} & d_1[v_0, v_1]+d_2[v_0, v_2]+d_3[v_0, v_3]+d_4[v_0, v_4] \\ & +d_5[v_0, v_5]+d_6[v_1, v_2]+d_7[v_1, v_3]+d_8[v_1, v_4] \\ & +d_9[v_1, v_5]+d_{10}[v_2, v_3]+d_{11}[v_2, v_4]+d_{12}[v_2, v_5] \\ & +d_{13}[v_3, v_4]+d_{14}[v_3, v_5]+d_{15}[v_4, v_5] \end{aligned} \tag{4.29}$$

and

$$\begin{aligned} 0=\partial z_1 = {} & -[v_0]\{d_1+d_2+d_3+d_4+d_5\}+[v_1](d_1-d_6-d_7-d_8-d_9\} \\ & +[v_2]\{d_2+d_6-d_{10}-d_{11}-d_{12}\}+[v_3]\{d_3+d_7+d_{10}-d_{13}-d_{14}\} \\ & +[v_4]\{d_4+d_8+d_{11}+d_{13}-d_{15}\}+[v_5]\{d_5+d_9+d_{12}+d_{14}+d_{15}\} \end{aligned} \tag{4.30}$$

There are thus six constraint equations that the fifteen integer co-efficients $d_1, d_2, \ldots, d_{15}$ must satisfy. However only 5 of them are linearly independent. Thus 10 of the 15 co-efficients in (4.29) can be freely chosen, so that

$$Z_1(P)=\mathbf{Z}^{10}$$

Finally we turn to the determination of $H_1(P)$. Since $B_1(P)$, $Z_1(P)$ are both $\mathbf{Z}^{10}$ we examine if $h_1 \in H_1(P)$ also imply $h_1 \in B_1(P)$ that would, of course, mean that $H_1(P) \equiv \{0\}$. Comparing (4.25) with (4.29) we see that for this to be the case we must have:

$$\begin{array}{llll} d_1 = a_1 + a_2 & d_5 = -a_3 - a_4 & d_9 = -a_7 - a_8 & d_{13} = a_6 + a_{10} \\ d_2 = a_5 - a_1 & d_6 = a_1 + a_8 & d_{10} = a_9 + a_{10} & d_{14} = a_3 + a_9 \\ d_3 = a_3 - a_2 & d_7 = a_2 + a_6 & d_{11} = a_5 - a_{10} & d_{15} = a_4 + a_7 \\ d_4 = a_4 - a_5 & d_8 = a_7 - a_6 & d_{12} = a_8 - a_9 & \end{array} \tag{4.31}$$

It is straightforward to check that these equations are consistant with the constraint equations implied by equation (4.30) for $d_1, d_2, \ldots, d_{15}$. Since $a_1, a_2, \ldots, a_{10}$ are non-zero positive or negative integers, equation (4.31) implies that whenever $d_1, d_2, \ldots, d_{15}$ are even every such element of $H_1(P)$ belongs to $B_1(P)$. When $d_1, d_2, \ldots, d_{15}$ are not all even, such a cycle need not be a boundary. An example of such a cycle is obtained by dividing the set of equation (4.27) by 2 and setting $a_1 = a_2 = \ldots = a_{10} = +1$.

But if we consider any arbitrary cycle z_1 then $2z_1$ will be a cycle with even coefficients and thus belong to $B_1(P)$. Thus

$$H_1(P) = \mathbf{Z}/2, \text{ the group of integers modulo 2}$$

Calculation of $H_0(P)$. Since $H_0(P) \equiv Z_0(P)/B_0(P)$ we proceed to determine $Z_0(P)$ and $B_0(P)$.

We have already noted several times that $Z_0(P) = C_0(P)$. Since P contains 6 zero-simplexes, $Z_0(P) = \mathbf{Z}^6$. While if $b_0 \in B_0(P)$, then $b_0 \in C_0(P)$ and $b_0 = \partial C_1$ where $C_1 \in C_1(P)$. Equation (4.26) gives the general structure of b_0. We note that there are only five independent coefficients in (4.26). Thus $B_0(P) = \mathbf{Z}^5$ and $H_0(P) = \mathbf{Z}$.
Thus

$$\begin{aligned} H_k(P) &= \{0\}, k > 2 \\ H_2(P) &= \{0\} \\ H_1(P) &= \mathbf{Z}/2 \\ H_0(P) &= \mathbf{Z} \end{aligned} \tag{4.32}$$

We note that $H_1(P)$ does not contain any freely generated abelian group but only a cyclic group of order 2. This group reflects the twisted nature of the space P.

4.4 RELATIVE HOMOLOGY GROUPS

From the examples it should be apparent that the determination of the homology groups $H_p(K)$ of a polyhedron K starting from the definition of H_p although theoretically feasable is a laborious process. Very often a good deal of the labour can be avoided by considering a subpolyhedron $L \subset K$ and relating $H_p(K)$ to $H_p(L)$. This becomes very fruitful if the notion of the relative homology group $H_p(K, L)$ is introduced. The homology groups $H_p(K)$, $H_p(L)$, and $H_p(K; L)$ then form what is called an exact sequence. This is an algebraic structure which, in certain cases, can actually determine the groups $H_p(K)$ themselves. Our aim now is to establish this important result. First we have to define the relative homology groups $H_p(K; L)$. The intuitive idea is that anything in K belonging to the subpolyhedron L is regarded as belonging to the identity element of $H_k(K; L)$. More precisely let K be a complex and $L \subset K$ be a subcomplex. Let the corresponding chain groups be $C_p(K)$ and $C_p(L)$ respectively then we have:

Definition

The p-dimensional chain group of K modulo L or the relative p-chain group with integer co-efficients is the quotient group

$$C_p(K; L) = C_p(K)|C_p(L), p > 0$$

Thus each member of $C_p(K, L)$ is a coset

$$c_p + C_p(L), \text{ where } c_p \in C_p(K). \tag{4.33}$$

Definition

For $p > 1$, the relative boundary operator $\bar{\partial}_p$ is the map:

$$\bar{\partial}_p : C_p(K; L) \to C_{p-1}(K; L)$$

defined by:

$$\bar{\partial}_p(c_p + C_p(L)) = \partial_p c_p + C_{p-1}(L)$$

where $c_p + C_p(L) \in C_p(K; L)$ and $\partial_p c_p$ denotes the usual boundary of the p-chain C_p. It is easy to check that the relative boundary operator is a homomorphism. (4.34)

Definition

The group of relative p-dimensional cycles on K modulo L, denoted by $Z_p(K;L)$ is the kernel of the relative boundary operator. For $p=0, Z_0(K;L)=C_0(K;L)$ (4.35)

Definition

The group of relative p-dimensional boundaries on K modulo L, denoted by $B_p(K;L)$ is the image of $C_{p+1}(K;L)$ under the relative boundary homomorphism. (4.36)

The relative homology group can now be defined. We have:

Definition

The relative p-dimensional homology group of K modulo L, denoted by $H_p(K;L)$, is the quotient group:

$$H_p(K;L)=Z_p(K;L)/B_p(K;L), p>0$$

The members of $H_p(K;L)$ are $z_p+C_p(L)$. Note that it is required that ∂z_p be a $(p-1)$ dimensional chain on L, not that z_p be an actual cycle. (4.37)

An example illustrating the definitions might be useful.

Example

Let K be the 2-skeleton of $\sigma^2 \equiv [v^0, v^1, v^2]$, i.e. K contains all the simplexes of σ^2 and its faces. Let L be the subcomplex $[v^0, v^1]$, $[v^0]$ and $[v^1]$. We will determine $H_p(K;L)$, $\forall p$

A. $H_0(K;L)$. Since $H_0(K;L)\equiv Z_0(K;L)/B_0(K;L)$ we must determine $Z_0(K;L)$ and $B_0(K;L)$.

We start with $Z_0(K;L)$. If $z_0 \in Z_0(K;L)$ then by def. $z_0 \in C_0(K;L)$ since $C_0(K;L)=Z_0(K;L)$. $z_0 \in C_0(K;L)$ means:

$$z_0=a_0[v^2]+C_0(L)$$

where $C_0(L)=\{b_0[v^1]+g_0[v^0]|, b_0, g_0 \text{ integer}\}$. Thus $Z_0(K;L)=\mathbf{Z}$

Next consider $B_0(K;L)$. For b_0 to be an element of $B_0(K;L)$ we must have:

$$b_0 \in C_0(K;L) \text{ and } b_0=\bar{\partial}C_1,\ C_1 \in C_1(K;L)$$

$$b_0 \in C_0(K,L) \Rightarrow b_0=h_0[v^2]+C_0(L)$$

For $b_0 = \bar{\partial} C_1$ we note that

$$C_1 \in C_1(K;L) \Rightarrow C_1 = a_1[v^1, v^2] + b_1[v^0, v^2] + C_1(L)$$

where $C_1(L) \equiv \{h_1[v^0, v^1], h_1 = \text{integer}\}$

$$\begin{aligned} \bar{\partial} G &= (a_1 + b_1)[v^2] - a_1[v_1] - b_1[v^0] + C_1(L) \\ &= (a_1 + b_1)[v^2] + C_1(L) \end{aligned}$$

Thus any element of $C_0(K;L)$ is the relative boundary of some element C_1 of $C_1(K;L)$. Thus

$$B_0(K;L) = \mathbf{Z} \quad \text{and} \quad H_0(K;L) \equiv \{0\}$$

B. $H_1(K;L) \equiv z_1(K;L)/B_1(K;L)$

$$z_1 \in Z_1(K;L) \Rightarrow z_1 \in C_1(K;L) \quad \text{and} \quad \bar{\partial} z_1 = 0$$

$$z_1 \in C_1(K;L) \Rightarrow z_1 = a_1[v^1, v^2] + b_1[v^0, v^2] + C_1(L)$$

$$0 = \bar{\partial} z_1 = (a_1 + b_1)[v^2] + C_1(L) = 0$$

Therefore

$$a_1 + b_1 = 0, \qquad a_1 = -b_1 = g,$$

say

$$z_1 = g\{[v^1, v^2] - [v^0, v^2]\} + C_1(L)$$

Therefore $Z_1(K, L) = \mathbf{Z}$

$$b_1 \in B_1(K;L) \Rightarrow b_1 \in C_1(K;L) \quad \text{and} \quad b_1 = \bar{\partial} C_2$$

But

$$C_2 \equiv \{0\},$$

therefore

$$b_1 \equiv \{0\}, B_1(K;L) \equiv \{0\}$$

Thus

$$H_1(K;L) = \mathbf{Z}$$

C. $H_k(K;L)$, $k > 1$, since $C_k(K;L) \equiv \{0\}$, $k > 1$ $H_k(K;L) \equiv \{0\}$, $k > 1$

The relative homology groups $H_K(K;L)$ by their construction, are insensitive to L. This suggests the following interesting possibility. If we excise or cut out the interior L_0 of L will $H_k(K - L_0, L - L_0)$ be isomorphic to $H_k(K, L)$? The answer is yes and we have:

Theorem (excision theorem)

Let K be a complex containing a closed subcomplex L. If L_0 is an open subcomplex of L such that $\bar{L}_0$ the closure of L_0 is contained in the interior of L then

$$H_p(K; L) = H_p(K - L_0, L - L_0), \forall p \tag{4.38}$$

This theorem is extremely useful in calculations as we shall see shortly. We next turn to the relationships that exist between the groups $H_p(K)$; $H_p(L)$ and $H_p(K; L)$.

A. Relation between $H_p(L)$ and $H_p(K)$. Since L is a subcomplex of K we can relate L to K by means of the inclusion homomorphism on the corresponding chain groups

$$i: C_p(L) \to C_p(K) \tag{4.39}$$

defined as:

$$i[C_p] = c_p, c_p \in C_p(L) \subset C_p(K)$$

In turn this map induces the group homomorphism:

$$i^*: H_p(L) \to H_p(K)$$

B. Relation between $H_p(K)$ and $H_p(K; L)$. This relationship is established by considering the homomorphism: $j: C_p(K) \to C_p(K; L)$ defined by:

$$j[C_p] = c_p + C_p(L), C_p \in C_p(K)$$

Then j induces a homomorphism: $j^*: H_p(K) \to H_p(K; L)$ as we now demonstrate. Observe that by definition: $j[\partial C_p] = \partial c_p + C_{p-1}(L)$ On the other hand $\bar{\partial}[j(C_p)] = \bar{\partial}[c_p + C_p(L)] = \partial c_p + C_{p-1}(L)$. Where $\bar{\partial}$ represents the relative boundary operator. From these two expressions we see that

$$j\partial = \bar{\partial} j$$

Now suppose $z_p \in H_p(K)$. From the result just established we then have:

$$\bar{\partial} j(z_p) = j(\partial z_p)$$

But $\partial z_p = 0$, since $z_p \in H_p(K)$, thus $\bar{\partial}[j(z_p)] = 0$ which means that $j(z_p) \in Z_p(K; L)$. Thus using j each class $[z_p] \in H_p(K)$ can be mapped into the class $[j(z_p)] \in H_p(K; L)$. It is straightforward to check that such a mapping is a homomorphism. We write:

$$j^*: H_p(K) \to H_p(K; L), \forall p \tag{4.40}$$

C. Relation between $H_p(K; L)$ and $H_{p-1}(L)$. We now come to the most interesting of the inter-relationships between homology groups. This relationship is interesting because it relates two groups which differ in their dimensional index 'p', namely $H_p(K, L)$ and $H_{p-1}(L)$. To see how this comes about consider $z_p \in H_p(K; L)$. By definition this implies that $z_p \in C_p(K; L)$ and $\bar{\partial} z_p = 0$. i.e.

$$\partial z_p + C_{p-1}(L) = 0$$

∂z_p is some element C_{p-1}, in $C_{p-1}(L)$. Furthermore, C_{p-1} is not just a chain but a cycle, since $\partial C_{p-1} = \partial^2 z_p$ and $\partial^2 z_p = 0$, by Theorem (4.12). Thus C_{p-1} determines a unique member of $H_{p-1}(L)$. We write this correspondence (which can be shown to be a group homomorphism) as:

$$\partial^* : H_p(K; L) \to H_{p-1}(L) \tag{4.41}$$

given by: $\bar{\partial}^*([z_p + C_p(L)]) = [\partial z_p]$, where $[z_p + C_p(L)] \in H_p(K; L)$, and $[\partial z_p]$, as we just saw, is an element of $H_{p-1}(L)$.

All of the inter-relationships, which we have described, can be fitted together into a sequence of groups and homomorphisms called the homology sequence of the pair $(K; L)$. This is defined as:

4.5 EXACT SEQUENCES

Definition

The homology sequence of the complex K with subcomplex L is the sequence of groups and homomorphisms:

$$\ldots \xrightarrow{\partial^*} H_p(L) \xrightarrow{i^*} H_p(K) \xrightarrow{j^*} H_p(K; L) \xrightarrow{\partial^*} H_{p-1}(L) \xrightarrow{i^*} \ldots \tag{4.42}$$

The homology sequence has the following important property:

Theorem

The homology sequence of the complex K with subcomplex L is exact, that is, the image of each homomorphism in the sequence is equal to the kernel of the next homomorphism. (4.43)

The proof of this Theorem is straightforward and consists of establishing the following six results:

1. Image $i^* \subset$ Kernel j^*
2. Kernel $j^* \subset$ Image i^*
3. Image $j^* \subset$ Kernel ∂^*

4. Kernel $\partial^* \subset$ Image j^*
5. Image $\partial^* \subset$ Kernel i^*
6. Kernel $i^* \subset$ Image ∂^*

We sketch the proof of (5) and (6) and leave the other results for the reader to establish. Recall that the map ∂^*: $H_p(K, L) \to H_{p-1}(L)$ sent an element $z_p + C_p(L)$ of $H_p(K, L)$ into the element ∂z_p of $C_{p-1}(L)$. By definition this is a boundary element and hence in the kernel of the inclusion homomorphism i^*. This establishes (5). Next suppose b_{p-1} is an element of the kernel of the inclusion homomorphism. Then, by definition, $b_{p-1} = \partial z_p$, $z_p \in H_p(L)$ which, in turn, can be written as an element $z_p + C_p(L)$ of $H_p(K; L)$ as we saw earlier since $\bar{\partial}[z_p + C_p(L)] = 0$. This establishes (6). Establishing Theorem (4.43) is useful because there are many theorems that compare the groups of an exact sequence. We, state, as an example the simplest of these theorems.

Theorem

Suppose that an exact sequence has a section of four groups:

$$\{0\} \to A \xrightarrow{g} B \xrightarrow{h} \{0\}$$

where $\{0\}$ denotes the trivial group. Then g is an isomorphism from A onto B. (4.44)

Proof

The image $f(\{0\}) = \{0\}$ contains only the identity element of A. Exactness of the sequence then means that g has kernel $\{0\}$, so that g is one-to-one. The kernel of h is all of B and this, again by exactness, must be the image $g(A)$. Thus g is an isomorphism.

A few examples illustrating how the exact homology sequence combined with the excision Theorem (4.38) is used to determine homology groups will now be given.

Example 1. As our first example we prove, using the exact homology sequence and the excision theorem, that $H_p(S^1) = \{0\}$ for $p > 1$. Introduce, $A^1 = S^1 - [n]$, where $[n]$ represents the north pole of the unit circle S^1 and $B^1 = S^1 - [s]$, where $[s]$ represents the south pole of S^1. It is clear that: $A^1 \cup B^1 = S^1$, and $X^1 = A^1 \cap B^1 = R_1 \cup R_2$, where $R_1 \cap R_2 = \phi$. Figure 4.4 explains the geometrical ideas involved.

Consider now the following exact homology sequences for $p > 1$

$$\to H_p(A^1) \to H_p(S^1) \to H_p(S^1, A^1) \to H_{p-1}(A^1) \to H_{p-1}(S^1) \to \dots \tag{4.45}$$

Figure 4.4

and

$$\to H_p(X^1) \to H_p(B^1) \to H_p(B^1, X^1) \to H_{p-1}(X^1) \to H_{p-1}(B^1) \to \ldots \tag{4.46}$$

Observe now that A^1 and B^1 are contractible spaces hence:

$$H_p(A^1) = H_p(B^1) = \{0\} \text{ for } p > 0 \tag{4.47}$$

so that (4.45) and (4.46) become for $p > 1$

$$\{0\} \to H_p(S^1) \to H_p(S^1, A) \to \{0\}$$

and

$$\{0\} \to H_p(B^1, X^1) \to H_{p-1}(X^1) \to \{0\}$$

Theorem (4.44) applies and it immediately follows that

$$H_p(S^1) = H_p(S^1, A^1)$$

and

$$H_p(B^1, X^1) = H_{p-1}(X^1) \tag{4.48}$$

Next note by removing a point say, the south pole $[s]$ from S^1 and A^1, S^1 becomes B^1 and A^1 becomes X^1. Thus from the excision theorem (4.38) we have:

$$H_p(S^1, A^1) = H_p(S^1 - [s], A^1 - [s])$$
$$= H_p(B^1, X^1)$$

Hence (4.48) implies:

$$H_p(S^1) = H_{p-1}(X^1), p > 1 \tag{4.49}$$

Finally we note that X^1 can be retracted to two points $[e]$ and $[w]$ say. Hence by Theorem (4.1a)

$$H_{p-1}(X^1) = H_{p-1}([e] \cup [w]) \tag{4.50}$$

The space consisting of the two points $[e]$ and $[w]$ has dimension zero, by definition and thus by definition, $C_p([e] \cup [w]) \equiv \{0\}$ for $p > 0$ so that $H_{p-1}([e] \cup [w]) \equiv \{0\}$ for $p > 1$ which establishes the result we were after.

Example 2. As our second example we prove that if K is a simplicial complex then

$$H_p(K) = H_{p+1}(\Sigma(K)), \text{ for } p > 0$$

where $\Sigma(K)$ is the suspension of K. In general for any compact space X, the suspension of X, written $\Sigma(X)$ is homeomorphic to the topological space $(X \times [-1, 1)])/\sim$, where $\sim$ is the equivalence relation in $X \times [1, -1]$ which identifies all points in $(X \times -1)$ and all points in $(X \times 1)$. Figure 4.5 explains what is involved.

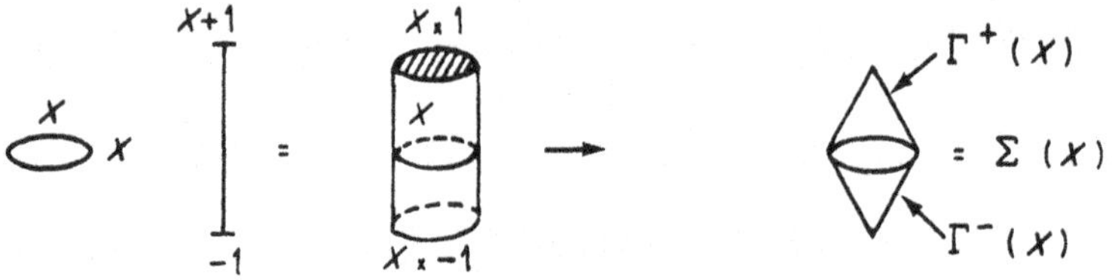

Figure 4.5

In terms of the figure let us call the 'northern hemisphere' of $\Sigma(X)$, $\Gamma^+(X)$ and the 'southern hemisphere' of $\Sigma(X)$, $\Gamma^-(X)$. Now consider the following exact sequences:

$$\to H_p(K) \to H_p(\Gamma^+(K)) \to H_p(\Gamma^+(K), K) \to H_{p-1}(K) \to H_{p-1}(\Gamma^+(K) \to$$

and (4.51)

$$\to H_p(\Gamma^-(K)) \to H_p(\Sigma(K)) \to H_p(\Sigma(K), \Gamma^-(K)) \to H_{p-1}(\Gamma^-(K)) \to H_{p-1}(\Sigma(K)) \to \ldots$$

It is possible to prove that the spaces $\Gamma^+(K)$, $\Gamma^-(K)$ are contractible so that $H_p(\Gamma(K)) = H_p(\Gamma^-(K)) = \{0\}$, for $p > 0$. Equation (4.51) then becomes for $p > 1$

$$\{0\} \to H_p(\Gamma^+(K), K) \to H_{p-1}(K) \to \{0\}$$

and

$$\{0\} \to H_p(\Sigma(K)) \to H_p(\Sigma(K), \Gamma^-(K)) \to \{0\}$$

These imply

$$H_p(\Gamma^+(K), K) = H_{p-1}(K) \tag{4.52}$$

and

$$H_p(\Sigma(K)) = H_p(\Sigma(K), \Gamma^-(K))$$

Finally we note that if the point -1 is removed from $\Sigma(K)$ it is retractible to $\Gamma^+(K)$, while removing the point -1 from $\Gamma^-(K)$ makes it retractible to K. The excision theorem then gives:

$$H_p(\Sigma(K), \Gamma^-(K)) = H_p(\Sigma(K) - (-1), \Gamma^-(K) - [-1])$$
$$= H_p(\Gamma^-(K), K)$$

which combined with (4.52) establishes the result we were after.

Example 3. As our final example we calculate $H_p(S^n)$ using exact homology sequences and the excision Theorem (4.38).

Introduce

$$A^n = S^n - [n], [n] = \text{north polar point}$$

$$B^n = S^n - [s], [s] = \text{the south polar point}$$

Then $A^n \cup B^n = S^n$, and $X^n = A^n \cap B^n$ is retractible to S^{n-1} for $n > 1$. A picture explaining the geometry in the case $n = 2$ is given in the Fig. 4.6.

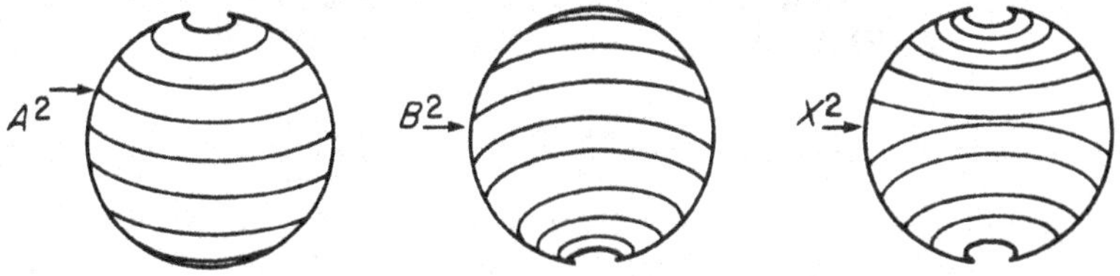

Figure 4.6

Again, we have the exact sequences:

$$\to H_p(A^n) \to H_p(S^n) \to H_p(S^n, A^n) \to H_{p-1}(A^n) \to H_{p-1}(S^n) \to \ldots$$

and

$$\to H_p(X^n) \to H_p(B^n) \to H_p(B^n, X^n) \to H_{p-1}(X^n) \to H_{p-1}(B^n) \to \ldots$$

From these and the observation that $H_p(A^n) = H_p(B^n) = \{0\}$ for $p > 0$ it immediately follows that

$$H_p(S^n) = H_p(S^n, A^n)$$

and

$$H_p(B^n, X^n) = H_{p-1}(X^n), \text{ for } p > 1$$

Now we use the excision theorem (4.38) to prove that

$$H_p(S^n, A^n) = H_p(S^n - [s], A^n - [s])$$
$$= H_p(B^n, X^n)$$

and hence

$$H_p(S^n) = H_{p-1}(X^n)$$

For $n > 1$, X^n is retractible to S^{n-1}, we thus have:

$$H_p(S^n) = H_{p-1}(S^{n-1}) \tag{4.53}$$

valid for $n > 1$, $p > 1$. By repeated use of (4.53) and a direct computation of $H_1(S^1)$ it is possible to show that:

$$H_p(S^n) = \begin{cases} \mathbf{Z} \text{ for } p = 0, p = n \\ \{0\} \text{ otherwise} \end{cases} \tag{4.54}$$

We end our discussion of the homology groups $H_p(K)$ by making a few remarks.

4.6 TORSION, KUNNETH FORMULA, EULER–POINCARÉ FORMULA AND SINGULAR HOMOLOGY

Torsion

The homology groups $H_p(K)$ were obtained from the chain groups $C_p(K)$. The elements of the chain group $C_p(K)$ were formal linear combinations of the oriented p-simplexes of K multiplied by integer co-efficients. It is possible to generalize the chain groups by introducing as co-efficients of the p-simplexes elements of an arbitrary abelian group G rather than the integers i.e. writing $C_p \in C_p(K)$ as:

$$C_p = \sum_{i=1}^{l_p} g_i\sigma_i^p, \, g_i \in G$$

These are the chain groups defined over the abelian group G, written $C_p(K, G)$ and the corresponding homology groups $H_p(K; G)$ can be constructed. It is quite remarkable that the groups $H_p(K, G)$ are not really more general than the groups $H_p(K)$ we have considered. Indeed it is possible to prove that a knowledge of the groups $H_p(K)$ and G is enough to completely determine the groups $H_p(K; G)$. We do not prove this theorem here but point out the following amusing feature of homology theory. If the integer co-efficients are replaced by rational co-efficients Q the corresponding homology groups $H_p(K; Q)$, far from being more general, contain in them less information than the corresponding homology groups $H_p(K)$ with integer co-efficients.

The reason for this is simple. We recall that the integer co-efficient homology groups $H_p(K)$ had the general structure: $H_p(K) = G_p(K) \oplus T_p(K)$, where elements of $T_p(K)$, the torsion subgroup, were finite order cyclic group. This meant that if $t_p \in T_p(K)$ was of order n then $nt_p = 0$, the identity element. For chains with rational rather than integer co-efficients, the equation $nt_p = 0$ would mean that $t_p = 0$. Thus $H_p(K; Q)$ does not have any torsion subgroup, while $H_p(K)$ does.

The Kunneth formula

In the Chapter on the fundamental group we proved that if $K = X \times Y$, then

$$\pi_1(X \times Y, x_0 \times y_0) = \pi_1(X, x_0) \oplus \pi_1(Y, y_0)$$

Is there an analogous formula for the homology groups? There is, but it is more complicated. We write down, the formula for homology groups with rational rather than integer co-efficients so that complications due to the torsion subgroups are not present. The formula, known as the Kunneth formula is:

$$H_p(X \times Y; Q) = \bigoplus_{k+q=p} H_k(X; Q) \otimes H_q(Y; Q) \tag{4.55}$$

As an application of the Kunneth formula we determine $H_p(T^2)$, where $T^2 = S^1 \times S^1$ the 2-torus. We know that

$$H_p(S^1) = \begin{cases} \mathbf{Z}, p = 0, 1 \\ \{0\}, \text{otherwise.} \end{cases}$$

Thus using (4.55) we get

$$H_0(T^2) = H_0(S^1 \otimes S^1) = H_0(S^1) \otimes H_0(S^1) = \mathbf{Z}$$

$$\begin{aligned} H_1(T^2) &= H_0(S^1) \otimes H_1(S^1) \oplus H_1(S^1) \otimes H_0(S^1) \\ &= \mathbf{Z}^2 \end{aligned}$$

$$\begin{aligned} H_2(T^2) &= H_0(S^1) \otimes H_2(S^1) \oplus H_1(S^1) \otimes H_1(S^1) \oplus H_2(S^1) \otimes H_0(S^1) \\ &= \mathbf{Z} \end{aligned}$$

and

$$H_p(T^2) = \{0\}, \text{ for } p > 3.$$

A more general version of (4.55) with the torsion subgroups taken into account is discussed in reference [2].

The Euler-Poincaré formula

We have commented on the fact that the rank of $H_p(K)$, $R_p(K)$ is related to the number of $(p+1)$ dimensional holes present in the space. There is a remarkable formula involving $R_p(K)$ and the number of p-simplexes of K, called the Euler-Poincaré formula, which we will now prove. First we write down the formula:

$$\chi(K) = \sum_{P=0}^{n} (-1)^P lp = \sum_{P=0}^{n} (-1)^P R_p(K) \tag{4.56}$$

where lp, $p = 0, 1, \ldots, n$ denotes the number of p-simplexes present in K. Observe that since the groups $H_p(K)$ are topological invariants the sum $\sum_{P=0}^{n} (-1)^P R_p(K)$ is a topological invariant. Thus $\chi(K)$ which can be calculated simply from a knowledge of the number of different dimensional simplexes present in K is a topological invariant of K! $\chi(K)$ is called the Euler characteristic of K. To prove the formula we note that for each dimension p, $0 \leq p \leq n$ we have:

$$\partial : C_p(K) \to C_{p-1}(K)$$

where C_{-1} is by definition the zero space. From the rank and nullity theorem of linear algebra it follows that:

$$\begin{aligned} lp &= \text{dimension of } C_p(K) \\ &= \text{dimension (kernel } \partial) + \text{dimension (image } \partial) \\ &= \text{dimension of } Z_p(K) + \text{dimension } B_{p-1}(K) \end{aligned}$$

While:

$$\begin{aligned} R_p(K) &= \text{dimension } H_p(K) = \text{dimension } [Z_p(K)/B_p(K)] \\ &= \text{dimension } Z_p(K) - \text{dimension } B_p(K) \end{aligned}$$

Thus:

$$\begin{aligned} \chi(K) &= \sum_{p=0}^{n} (-1)^p R_p(K) \\ &= \sum_{p=0}^{n} (-1)^p [\text{dimension } Z_p(K) \text{-dimension } B_p(K)] \\ &= \sum_{p=0}^{n} (-1)^p \text{ dimension } Z_p(K) + \sum_{p=0}^{n} (-1)^{p+1} \text{ dimension } B_p(K) \end{aligned}$$

Now observe that:

$$\sum_{p=0}^{n} (-1)^{p+1} \text{ dimension } B_p(K) = \sum_{p=0}^{n} (-1)^p \text{ dimension } B_{p-1}(K)$$

Since dimension $B_p = 0$, for $p = n$
and dimension $B_{-1} = 0$. Thus

$$\chi(K) = \sum_{p=0}^{n} (-1)^p \text{ [dimension } Z_p(K) + \text{dimension } B_{p-1}(K)]$$

$$= \sum_{p=0}^{n} (-1)^p lp$$

which establishes the formula.

Singular homology

As our final remark we briefly sketch how the homology groups for an arbitrary topological space X can be constructed. We start by introducing the standard p-simplexes Δ_p. These are the set of points $(x_0, \ldots, x_p)$ in euclidean space with the property: $0 \leq x_i \leq 1$ and $\sum_{i=p}^{p} x_i = 1$. The singular p-simplexes λ_i^p in X can now be defined as continuous mappings: λ_i^p: $\Delta_p \to X$. The term singular is used to indicate that the maps λ_i^p need not be invertible. In terms of the singular p-simplexes λ_i^p the singular chain group $S_p(x)$ can be defined as the set of formal sums:

$S_p = \sum \lambda_i^p g_i$, $g_i \in G$, an abelian groups and the addition of two singular chain elements is defined as:

$$\sum \lambda_i^p g_i + \sum \lambda_i^p h_i \equiv \sum \lambda_i^p (g_i + h_i)$$

when $g_i, h_i \in G$.

To construct the singular homology groups the boundary map ∂ on the singular chain $S_p(x)$ has to be defined. Suppose, as before, that ∂ is linear so that all we really need to know is the action of ∂ on a singular p-simplex λ^p. This is defined as:

$$\partial\lambda^p = \sum_{r=0}^{p} (-1)^r \lambda^p \circ F^r$$

where $\lambda^p \circ F^r$ denotes the rth face of the singular p-simplex λ^p and in turn is defined as follows:

$$\lambda^p \circ F^r : \Delta_{p-1} \to X$$

where

$$F^r : \Delta_{p-1} \to \Delta_p$$

denotes the map:

$$F^r(x_0, \ldots, x_{n-1}) = (x_0, \ldots, \hat{x}_{r_1}, \ldots, x_n)$$

The notation $\hat{x}_r$ means that the vertex x_r has been omitted. It is possible to show that $\partial^2 = 0$ and define the singular homology group for the arbitrary topological space X as:

$$H_p(X; G) = \frac{\text{Ker}\,[\partial : S_p \to S_{p-1}]}{\text{Image}\,[\partial : S_{p+1} \to S_p]}$$

For further information regarding singular homology theory, references 2 and 3 can be consulted.

REFERENCES

1. HILTON, P. J. and WYLIE, S., "Homology Theory". Cambridge University Press, 1962.
2. MASSEY, W. S., "Singular Homology Theory". Springer Verlag, 1980.
3. MAUNDER, C. R. F., "Algebraic Topology". Van Nostrand Reinhold Co., 1972.

CHAPTER 5

The Higher Homotopy Groups

5.1 INTRODUCTION

In this chapter we give an account of the n-dimensional analogues of the fundamental group. These are the higher homotopy groups $\pi_n(X; x_0)$ of the space X based at the point $x_0 \in X$. We recall that the group elements of $\pi_1(X; x_0)$ were homotopy classes of loops in X based at x_0. As a first step in the construction of the groups $\pi_n(X; x_0)$ we generalize the notion of a loop to n-dimensions.

5.2 DEFINITION OF HIGHER HOMOTOPY GROUPS

Let I_n be the closed n-dimensional cube in euclidean space. If α is a continuous map:

$$\alpha : I_n \to X$$

such that α takes the surface ∂I_n of the cube I_n into one point $x_0 \in X$, then α is called an n-loop in X based at the point x_0. (5.1)

Thus we can represent α as $\alpha(t_1, t_2, \ldots, t_n)$ where $t_1, \ldots, t_n$ are the co-ordinates of a point in I_n so that $0 < t_i < 1$, i and $\alpha(t_1, \ldots, t_n) = x_0$ if $t_i = 0$ or $t_i = 1$ for any i since $t_i = 0$ or $t_i = 1$ represent points on ∂I_n.

The geometric content of the definition for $n = 2$ is shown in Fig. 5.1.

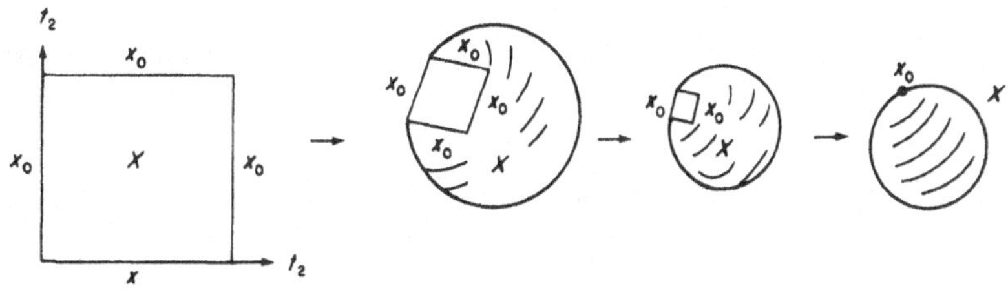

Figure 5.1

The square represents the source of the 2-loop α and the sequence of pictures illustrates how this square, with the points on its boundary identified, is homeomorphic to S^2, the two-dimensional loop.

The product law for two n-loop α and β both based at the point x_0 is next defined.

Definition

Let α and β be n-loops at $x_0 \in X$ then the product $\gamma = \alpha \circ \beta$ is the n-loop at $x_0 \in X$ given by:

$$\gamma(t_1, \ldots, t_n) = \alpha * \beta(t_1, \ldots, t_n) = \begin{cases} \alpha(2t_1, t_2, \ldots, t_n) & 0 \leq t_1 \leq \frac{1}{2} \\ \beta(2t_1 - 1, t_2, \ldots, t_n) & \frac{1}{2} \leq t_1 \leq 1 \end{cases} \tag{5.2}$$

The geometric content of this definition for $n = 2$ is shown in Fig. 5.2

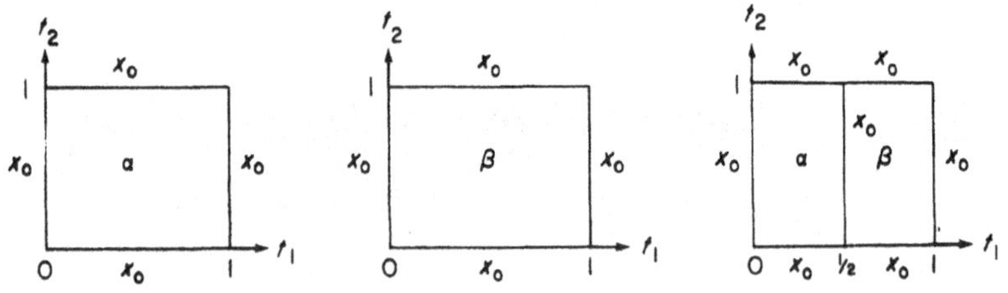

Figure 5.2

Proceeding as in the case of the fundamental group $\pi_1(X; x_0)$ the notion of two n-loops α and β being homotopic, written, as before as $\alpha \simeq \beta$ is next introduced.

Definition

Two n-loops α and β based at x_0 are called homotopic if there exists a continuous map $H(s; t_1, \ldots, t_n)$; $0 \leq s \leq 1$ such that

$$H(0; t_1, \ldots, t_n) = \alpha(t_1, \ldots, t_n)$$

$$H(1; t_1, \ldots, t_n) = \beta(t_1, \ldots, t_n)$$

$$H(s; t_1, \ldots, t_n) = x_0; \text{ if } (t_1, \ldots, t_n) \varepsilon\, \partial I_n \text{ for all } s$$

It is straightforward to check that homotopy is an equivalence relation and hence the space of n-loops can be partitioned into disjoint classes, members in a given class being homotopic to each other. Let us denote, as in Chapter 3, by $[\alpha]$ the equivalence class of n-loops homotopic to the n-loop α. Again, exactly as in the case of the fundamental group, it is possible to prove the following Lemma.

Lemma

If $\alpha, \beta, \gamma \ldots ; \alpha', \beta', \gamma', \ldots, e, \alpha^{-1}, \beta^{-1}, \ldots$ be n-loops based at $x_0 \varepsilon X$ then

i. If $\alpha \simeq \alpha'$ and $\beta \simeq \beta'$ then $\alpha' * \beta' \simeq \alpha * \beta$
ii. $(\alpha * \beta) * \gamma \simeq \alpha * (\beta * \gamma)$
iii. $e * \alpha \simeq \alpha * e$ (5.4)
iv. If $\alpha \simeq \alpha'$ then $\alpha^{-1} \simeq (\alpha')^{-1}$
v. $\alpha * \alpha^{-1} \simeq \alpha^{-1} * \alpha \simeq e$

where the n-loops e and α^{-1} are defined as follows:

Definition

The n-loop which takes I_n into x_0 is called the constant mapping and is denoted by e. Thus:

$$e : I_n \to x_0 \tag{5.5}$$

Definition

The inverse of the *n-loop* α, written α^{-1} is defined as:

$$\alpha^{-1}(t_1, t_2, \ldots, t_n) = \alpha(1 - t_1, t_2, \ldots, t_n) \tag{5.6}$$

Lemma (5.4) shows that the homotopy classes of n-loops based at x_0, $[\alpha]$, $[\beta], \ldots$ form a group with product law: $[\alpha] \circ [\beta] = [\alpha * \beta]$. This is the n-dimensional homotopy group $\pi_n(X; x_0)$ of the topological space X with base point x_0. In contrast to the fundamental group the higher homotopy groups are abelian as we now demonstrate.

5.3 ABELIAN NATURE OF HIGHER HOMOTOPY GROUPS

Theorem

The n-dimensional homotopy group $\pi_n(X; x_0)$ is abelian for $n > 1$. (5.7)

Proof

We start by noting that any n-loop α based at x_0 is homotopic to an n-loop $\tilde{\alpha}$ obtained by taking α and continuously distorting it in the following way: instead of letting just the boundary of I_n be mapped into x_0 by the n-loop α we thicken the boundary and let the thickened portion map into x_0. This defines the n-loop $\tilde{\alpha}$. Figure 5.3 explains what is involved for the case $n = 2$.

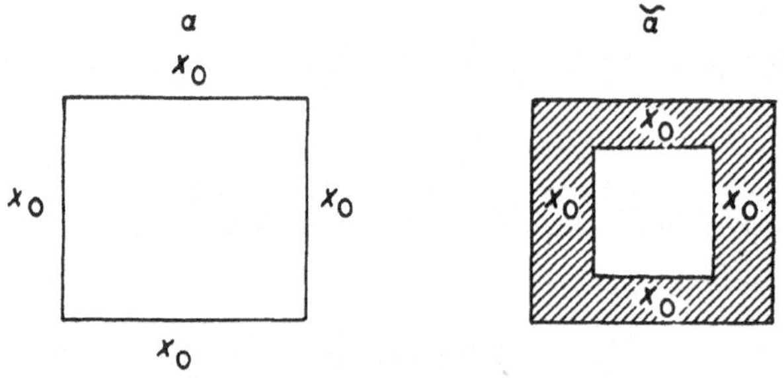

Figure 5.3

The shaded region in the source of $\tilde{\alpha}$ is all mapped into x_0. The proof of Theorem (5.7) is an almost immediate consequence of this observation and is demonstrated pictorially in Fig. 5.4.

Since the product rule for two n-loops involved joining them along the t_1 axis the sequence of pictures in Fig. 5.4 effectively proves Theorem (5.7) for $n > 2$. The essential difference between the case $n = 1$ and $n > 1$ is that the boundary of I_n is disconnected for $n = 1$.

Many of the theorems we proved for the fundamental group in Chapter 3 generalize to the higher homotopy groups. We state a few of them omitting proofs.

Theorem

If X is path connected and x_0 and x_1 are points of X then $\pi_n(X; x_0)$ is isomorphic to $\pi_n(X; x_1)$ for each n. (5.8)

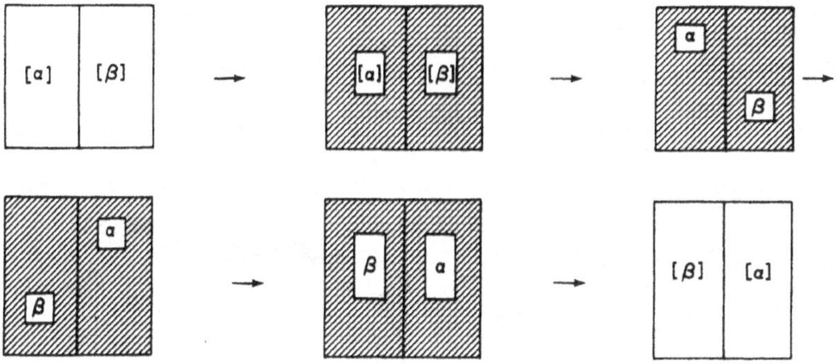

Figure 5.4

As before the isomorphism depends on the path used to connect the points x_0 and x_1.

Theorem

If X is contractible by a homotopy that leaves x_0 fixed then

$$\pi_n(X; x_0) = \{0\}, \text{ for all } n > 1 \tag{5.9}$$

Theorem

Let X and Y be spaces with $x_0 \in X$, $y_0 \in Y$, then

$$\pi_n(X \times Y; (x_0, y_0) \simeq \pi_n(X; x_0) \oplus \pi_n(Y; y_0) \tag{5.10}$$

Unlike the fundamental group $\pi_1(X; x_0)$ or the homology groups $H_p(X)$ there is no algorithm for determining $\pi_n(X; x_0)$ for all n even when X is a triangulable space. Because of this we will only describe one general calculational tool of homotopy theory: the exact homotopy sequence. To do this we have to first define the relative homotopy groups $\pi_n(X, Y; y_0)$ where $y_0 \in Y \subset X$.

5.4 RELATIVE HOMOTOPY GROUPS

Suppose X is a topological space containing a closed subspace Y and suppose $y_0 \in Y$. A relative n-loop with respect to Y in X can be defined as follows:

Definition

A relative n-loop α with respect to a closed subspace Y in X is a continuous map:

$$\alpha : I_n \to X$$

such that α takes all the faces of the n-cube I_n into one point $y_0 \in Y \subset X$ except one of the open faces. The exceptional open face J_{n-1} is mapped into Y while the boundary of J_{n-1}, ∂J_{n-1} is mapped into y_0. Figure 5.5 explains what this means for $n = 2$.

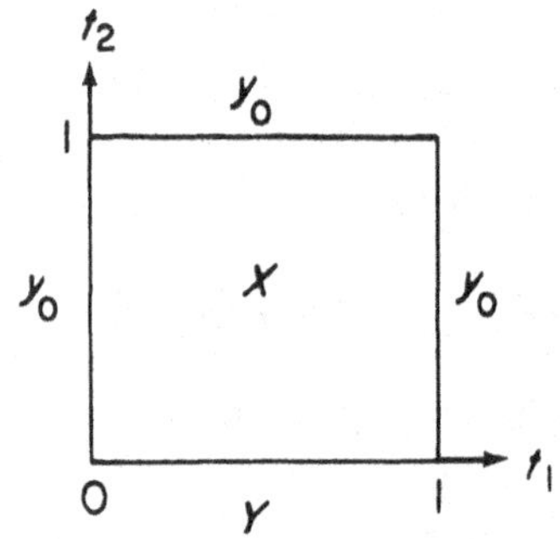

Figure 5.5

The product of 2 relative n-loops can be defined as follows:

Definition

Let α and β be two relative n-loops at $y_0 \in Y$, with respect to a closed subspace Y of the topological space X then the definition: $\gamma = \alpha \circ \beta$ as:

$$\gamma(t_1, \ldots, t_n) = \alpha * \beta(t_1, \ldots, t_n) = \begin{cases} \alpha(2t_1, \ldots, t_n); & 0 \leq t_1 \leq \frac{1}{2} \\ \beta(2t_1 - 1, t_2, \ldots, t_n) & \frac{1}{2} \leq t_1 \leq 1 \end{cases}$$

where we take the exceptional open face of J_n which is mapped into Y to correspond to $0 < t_1 < 1$. Figure 5.6 explains what this means for $n = 2$. (5.12)

The idea of two relative n-loops with respect to Y in X being homotopic can be defined by generalizing Definition (5.3). We have:

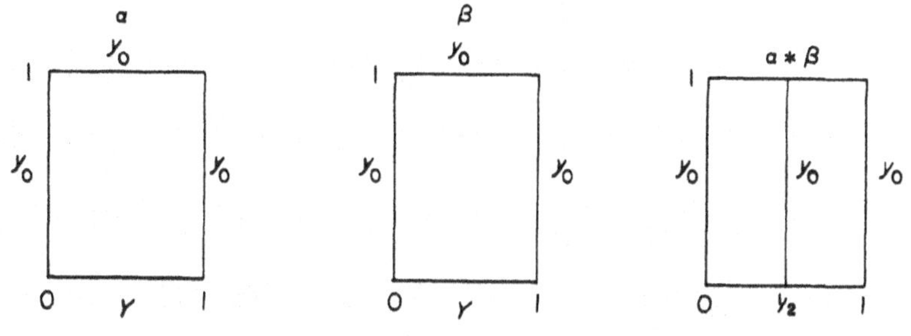

Figure 5.6

Definition

Two relative n-loops α and β with respect to Y in X are called homotopic if there exists a continuous map

$$
\begin{gathered}
H(s; t_1, \ldots, t_n), \quad 0 \leqslant s \leqslant 1 \text{ such that:} \\
H(0; t_1, t_2, \ldots, t_n) = \alpha(t_1, \ldots, t_n) \\
H(1; t_1, \ldots, t_n) = \beta(t_1, \ldots, t_n) \\
H(s, t_1 \ldots t_n) = \begin{cases} y_0 & \text{if } (t_1, \ldots, t_n) \in \partial I_n \text{ and } \partial I_n \neq J_{n-1} \\ Y, & \text{if } (t_1, \ldots, t_n) \in J_{n-1} \\ y_0 & \text{if } (t_1, \ldots, t_n) \in \partial J_{n-1} \end{cases}
\end{gathered}
\tag{5.13}
$$

Thus two relative n-loops α and β are homotopic, written $\alpha \simeq \beta$, if one can be continuously deformed into the other while all the faces of the n-cube I_n except J_{n-1} are open face mapped into y_0. The exceptional open face J_{n-1} is mapped into Y and its boundary ∂J_{n-1} is mapped into y_0. With the help of this homotopy relation it is again possible to show that the homotopy classes $[\alpha]$ of the relative n-loops α with respect to Y in X form a group with product rule: $[\alpha]\circ[\beta]=[\alpha * \beta]$. This is the relative homotopy group: $\pi_n(X; Y; y_0)$ based at $y_0 \in Y$. It should be apparent that the identity element of this group is precisely the homotopy class of all closed n-loops in Y, based at y_0. Thus given a topological space X containing a closed subspace Y with $y_0 \in Y$ we have the homotopy groups $\pi_n(Y, y_0)$; $\pi_n(X; y_0)$ and $\pi_n(X, Y; y_0)$ at our disposal. As in the case of homology theory we try to relate these groups to each other.

A. Relationship between $\pi_n(Y; y_0)$ *and* $\pi_n(X; y_0)$. Since $Y \subset X$ the inclusion map $i: Y \to X$ defined as:

$$i[\alpha]=[\alpha]; \; [\alpha] \in Y \subset X$$

immediately gives the group homomorphism:

$$i*; \pi_n(Y; y_0) \rightarrow \pi_n(X; y_0), \forall n \tag{5.14}$$

B. Relation between $\pi_n(X; y_0)$ and $\pi_n(X, Y; y_0)$. Think of $\pi_n(X; y_0)$ as $\pi_n(X, y_0; y_0)$ then the inclusion map: $j:(X, y_0\, y_0) \rightarrow (X, Y; y_0)$ (remember $y_0 \in Y$) induces a homorphism:

$$j^*: \pi_n(X, y_0; y_0) \rightarrow \pi_n(X, Y; y_0) \tag{5.15}$$

C. Relation between $\pi_n(X, Y; y_0)$ and $\pi_{n-1}(Y, y_0)$. Recall that the elements of $\pi_n(X, Y; y_0)$ were homotopy classes of relative n-loops. These were defined in terms of continuous maps from I_n which had the property that one of the open faces of I_n, J_{n-1} was mapped into Y and the boundary of J_{n-1} was mapped into $y_0 \in Y$, the base point of the relative homotopy group. Thus if any map of I_n is restricted to the face J_{n-1} plus its boundary it is easy to see that a homomorphism between $\pi_n(X, y; y_0)$ and $\pi_{n-1}(Y, y_0)$ would result. This is called the boundary homomorphism and written as:

$$\partial^*: \pi_n(X, Y; y_0) \rightarrow \pi_{n-1}(Y; y_0) \tag{5.16}$$

We can now state the following theorem:

5.5 THE EXACT HOMOTOPY SEQUENCE

Theorem

The sequence:

$$\ldots \rightarrow \pi_n(Y; y_0) \xrightarrow{1*} \pi_n(X; y_0) \xrightarrow{j*} \pi_n(X, Y; y_0) \xrightarrow{\partial*} \pi_{n-1}(Y, y_0) \xrightarrow{i*} \ldots \tag{5.17}$$

is exact.

Proof

The proof consists of verifying the following results:

i. image $l^* = \text{Kernel } j^*$
ii. Image $j^* = \text{Kernel } \partial^*$
iii. Image $\partial^* = \text{Kernel } l^*$

We will sketch the proof of (i) and leave the others for the reader to verify. To prove (i) we first show that Image $l^* \subset$ Kernel j^*. The inclusion map has as its elements n-loops in Y based at y_0. By definition those are homotopic to the identity element of $\pi_n(X, Y; y_0)$. Hence Image $i^* \subset$ Kernel j^*. Next we have to show that Kernel $j^* \subset$ Image i^*. This again follows from the definition of the relative homotopy group $\pi_n(X, Y; y_0)$. We first note that Kernel $j^* \subset$ homotopy class of the identity element of $\pi_n(X, Y; y_0)$ which in turn is equal to the homotopy class of n-loops in Y based at y_0, which is, by definition the Image of i^*.

As an example of how the homotopy sequence may be used we prove the following theorem.

Theorem

$\pi_k(S^n)$ is homomorphic to $\pi_{k+1}(S^{n+1})$ for $k>1, n>1$ (5.18)

We start by introducing, as in Chapter 4, the following spaces:

$A^n = S^n - [s]$, where $[s]$ denotes the south polar point

$B^n = S^n - [n]$, where $[n]$ denotes the north polar point

We observe that the spaces A^n, B^n are contractible, hence by Theorem (5.9), $\pi_k(A^n) = \{0\}$, $\pi_k(B^n) = \{0\}$ for $k>1$.

Now consider the two exact sequences:

$$\ldots \to \pi_{k+1}(A^{n+1}) \xrightarrow{i*} \pi_{k+1}(S^{n+1}) \xrightarrow{j*} \pi_{k+1}(S^{n+1}, A^{n+1}) \xrightarrow{\partial*} \pi_k(A^n) \to \ldots$$

$$\ldots \to \pi_{k+1}(B^n) \xrightarrow{j*} \pi_{k+1}(B^n, S^n) \xrightarrow{\partial*} \pi_k(S^n) \xrightarrow{i} \pi_K(B^n) \to \ldots$$

For $k>0$ we get:

$$\pi_{k+1}(S^{n+1}) = \pi_{k+1}(S^{n+1}, A^{n+1})$$

$$\pi_{k+1}(B^n, S^n) = \pi_k(S^n)$$

The procedure followed up to this point parallels our calculation of $H_k(S^n)$ in Chapter 4. But now we have to proceed differently. There is no excision theorem in homotopy theory which would allow us to directly say that $\pi_{k+1}(S^{n+1}, A^{n+1})$ is isomorphic to $\pi_{k+1}(B^n, S^n)$. However we observe that $B^n \subset S^{n+1}$ and $S^n \subset A^{n+1}$. Thus the inclusion map $i[B^n, S^n] \to [B^n, S^n] \in [S^{n+1}, A^{n+1}]$ leads to the homorphism:

$$i^*: \pi_{k+1}(B^n, S^n) \to \pi_{k+1}(S^{n+1}, A^{n+1})$$

The relationships established can be conveniently summarized as follows:

$$\begin{array}{ccc} \pi_{k+1}(S^{n+1}, A^{n+1}) & \xleftarrow{i*} & \pi_{k+1}(B^n, S^n) \\ \uparrow j_* & & \downarrow \partial_* \\ \pi_{k+1}(S^{n+1}) & & \pi_k(S^n) \end{array}$$

From the diagram, and the fact that ∂^* and j_* in this case are isomorphisms, it follows that there is a homomorphism $E = j_*^{-1} i_* \circ \partial_*^{-1}$ between $\pi_k(S^n)$ and $\pi_{k+1}(S^{n+1})$ which is our theorem.

The theorem proved is a special case of the following general result due to Freudenthal which we state without proof.

Theorem

A homomorphism between $\pi_k(S^n)$ and $\pi_{k+1}(S^{n+1})$ exists for $k < 2n - 1$. When $k < 2n - 1$ the homomorphism is in fact, an isomorphism. When $k = 2n - 1$, the homomorphism is onto. There are some immediate consequences of this theorem which should be noted.

Corollary

If $k < n$, $\pi_k(S^n) = \{0\}$, since $\pi_1(S^p) = \{0\}$ for $p > 1$ (5.20)

Corollary

If $k = n$, we have $\pi_n(S^n) = \pi_2(S^2)$ (5.21)

There is a close formal similarity between the homology groups $H_k(X)$ and the homotopy groups $\pi_k(X)$. Both have relative groups and satisfy exact sequence relations. Can $\pi_k(X)$ be related to $H_k(X)$? This is indeed possible. We have the following theorem, due to Hurewicz:

Theorem

The first non-zero homology group (where $n > 1$) and the first non-zero homotopy group have the same dimensions and are isomorphic. (5.22)

When $n = 1$, the theorem is true provided the fundamental group $\pi_1(X)$ is abelian. A statement of the theorem to cover the non-abelian case is contained in reference [2] listed at the end of this chapter.

As an application of the theorem and we note that $H_1(S^2) = \{0\}$, $H_2(S^2) = \mathbf{Z}$; while $\pi_1(S^2) = \{0\}$. Thus, it follows that $\pi_2(S^2) = \mathbf{Z}$ combining this result with Corollary (5.21) we get:

Corollary

If $k = n$, $\pi_n(S^n) = \mathbf{Z}, \forall_n$ (5.22)

We end this chapter with the following remark. Unlike the homology groups $H_n(S^k)$, the higher homotopy groups $\pi_n(S^k)$ for $n > k$ need not be trivial. For instance it was shown by Hopf that, $\pi_3(S^2) = \mathbf{Z}$. We will prove this result in Chapter 9.

REFERENCES

1. HILTON, P. J., "An Introduction to Homotopy Theory". Cambridge University Press, 1953.
2. HU, S. T., "Homotopy Theory". Academic Press, 1959.
3. STEENROD, N., "Topology of the Fibre Bundles". Princeton University Press, 1970.

CHAPTER 6

Cohomology and De Rham Cohomology

6.1 INTRODUCTION

In this chapter we shall give an account of cohomology groups. The prefix *co* in the word cohomology reminds us that there is a duality between homology and cohomology groups. This duality at first suggests that one may use cohomology or homology with equal ease. In fact cohomology is a more powerful and easier tool to use than homology. We shall show why this is so in this chapter, and shall summarize the superior properties of cohomology at the end of the chapter. In Chapter 4 we finished our discussion of homology groups by introducing $H_n(X; G)$: the singular homology group with coefficients in an arbitrary Abelian group G. The cases dealt with in that chapter covered only the choice $G = \mathbf{Z}$. It will be natural in this chapter to depart from this choice and take $G = \mathbf{R}$.

Consider then, rather than an arbitrary topological space X, a topological space M where M is a compact differentiable manifold. We shall discuss later on how restrictive it is to consider only such spaces M. The key to constructing the cohomology of M is Stokes' theorem for differential forms:

$$\int_M \mathrm{d}\omega = \int_{\partial M} \omega \tag{6.1}$$

where ω is an $(n-1)$-form and $\dim M = n$. Instead of integrating over M itself we can also integrate over some lower dimensional subset of M. Homology theory provides us straightaway with a class of such subsets: the class of all singular p-chains C with $p = 0, \ldots n$. Recall that a p-chain C used in calculations of $H_p(M; \mathbf{Z})$ is a formal finite linear combination:

$$C = a_1\lambda_1 + a_2\lambda_2 + \ldots \tag{6.2}$$

where the a_i are all integers, and the λ_i are singular p-simplexes: i.e. maps $\lambda_i : \Delta_p \to M$, Δ_p being the standard n-simplex in $\mathbf{R}^p$. It is now that we make the transition from integral to real coefficients. Require then only that the a_i be real numbers rather than integers. Require also that the maps $\lambda_i : \Delta_p \to M$ should be C^∞. This means that the (singular) p-chain C can now be called a C^∞ p-chain, and is an appropriate structure to consider on a differentiable manifold.

Now take a p-chain C and a p-form ω with $0 \leq p \leq n$. The integral of ω over C is defined using the maps λ_i^* and is given by (recall that $\lambda_i^*\omega$ is a p-form on Δ_p)

$$\int_C \omega = \sum_{i=1}^{k} a_i \int_{\Delta_p} \lambda_i^* \omega; \qquad C = \sum_{i=1}^{k} a_i \lambda_i \tag{6.3}$$

The integral (6.3) is a real number and thus we can regard ω as producing, from C, a real number. So if we denote by C_p the set of all C^∞ p-chains, then a p-form ω is a map from C_p to $\mathbf{R}$, i.e. an element of the dual of C_p, in other words ω is a co-chain. To summarize we have:

$$\omega : C_p \rightarrow \mathbf{R}$$
$$C \mapsto \langle \omega, C \rangle$$

with

$$\langle \omega, C \rangle = \int_C \omega, C \in C_p, \qquad \omega \in \Omega^p(M) \tag{6.4}$$

The need for real coefficients is now seen to arise from the need to be able to integrate real-valued differential forms. To define the cohomology group† $H^p(M; \mathbf{R})$ we first must have a look at Stokes' theorem. Using the notation introduced in (6.4), Stokes' theorem takes the form:

$$\langle \mathrm{d}\omega, C \rangle = \langle \omega, \partial C \rangle \tag{6.5}$$

where ω is a $(p-1)$-form and C is a p-chain. This notation shows (as it is designed to show) that the boundary operator ∂, and the exterior derivative operator d, are formal adjoints of one another. We use this adjoint relationship of d and ∂ to identify the cohomology class $[\omega]$ of a p-form ω. We know already, from our treatment of homology theory, that the elements of a homology class, which we write as $[C]$, belong to the space

$$Z_p(M)/B_p(M) \tag{6.6}$$

where $Z_p(M)$ are all p-chains C for which $\partial C = \phi$, and $B_p(M)$ are all p-chains C for which $C = \partial C'$ for some $(p+1)$-chain C'. The dual of the space Z_p/B_p we write as

$$Z^p(M)/B^p(M) \tag{6.7}$$

† Notice that an index such as p is always upper for cohomology and lower for homology. This is in analogy with the practise for the covariant and contravariant indices for tensors.

(i.e. we simply move the index p upwards), where $Z^p(M)$ are all cochains or p-forms $\boldsymbol{\omega}$ for which $d\boldsymbol{\omega} = \mathbf{O}$, and $B^p(M)$ are all p-forms $\boldsymbol{\omega}$ for which $\boldsymbol{\omega} = d\boldsymbol{\eta}$ for some $(p-1)$-form $\boldsymbol{\eta}$. To show precisely that the space Z^p/B^p is the dual of the space Z_p/B_p we must define how $[\boldsymbol{\omega}]$ acts on $[C]$ to give a real number. This is quite straightforward since it needs no new technique, $[\boldsymbol{\omega}] \in Z^p/B^p$ acts on $[C] \in Z_p/B_p$ to give the real number $([\boldsymbol{\omega}], [C])$ defined in (6.8)

$$([\boldsymbol{\omega}], [C]) = \int_C \boldsymbol{\omega} \tag{6.8}$$

We should check that if $\boldsymbol{\omega}'$ and C' give rise to the same cohomology class and homology class respectively as $\boldsymbol{\omega}$ and C, i.e. $[\boldsymbol{\omega}] = [\boldsymbol{\omega}']$ and $[C] = [C']$, then the RHS of (6.8) is unchanged. Stokes' theorem ensures that is automatically so. For if $[\boldsymbol{\omega}] = [\boldsymbol{\omega}']$ then we can say that:

$$d\boldsymbol{\omega} = d\boldsymbol{\omega}' = 0$$

and

$$\boldsymbol{\omega} = \boldsymbol{\omega}' + d\boldsymbol{\eta} \tag{6.9}$$

for some $\boldsymbol{\eta}$. Thus

$$\begin{aligned} ([\boldsymbol{\omega}'], [C]) &= \int_C (\boldsymbol{\omega} - d\boldsymbol{\eta}) \\ &= \int_C \boldsymbol{\omega} - \int_{\partial C} \boldsymbol{\eta} \\ &= \int_C \boldsymbol{\omega} = ([\boldsymbol{\omega}], [C]) \end{aligned} \tag{6.10}$$

since $\partial C = \phi$. In a similar fashion if $[C] = [C']$ then we can say that

$$\partial C = \partial C' = \phi$$

and

$$C = C' + \partial C'' \tag{6.11}$$

for some C''. Thus

$$\begin{aligned} ([\boldsymbol{\omega}][C']) &= \int_{C - \partial C''} \boldsymbol{\omega} \\ &= \int_C \boldsymbol{\omega} - \int_{\partial C''} \boldsymbol{\omega} \\ &= \int_C \boldsymbol{\omega} = ([\boldsymbol{\omega}], [C]) \end{aligned} \tag{6.12}$$

since $d\omega = \mathbf{O}$. So Z^p/B^p is the dual of the space Z_p/B_p. Finally the space $Z_p(M)/B_p(M)$ is the pth-homology group of M with real coefficients, and from now on is written $H_p(M; \mathbf{R})$, and $Z^p(M)/B^p(M)$ is the pth-cohomology group of M with real coefficients, and is written $H^p(M; \mathbf{R})$. The content of these constructions of $H^p(M; \mathbf{R})$ and $H_p(M; \mathbf{R})$ can be summarized in one of the usual ubiquitous topological diagrams:

$$\cdots \xleftarrow{\partial_{p-1}} C_{p-1} \xleftarrow{\partial_p} C_p \xleftarrow{\partial_{p+1}} C_{p+1} \xleftarrow{\partial_{p+2}} \cdots$$

$$H_p(M; \mathbf{R}) = Z_p(M)/B_p(M) = \ker \partial_p/\mathrm{im}\, \partial_{p+1}$$

$$\cdots \xrightarrow{d_{p-1}} \Omega^{p-1} \xrightarrow{d_p} \Omega^p \xrightarrow{d_{p+1}} \Omega^{p+1} \xrightarrow{d_{p+2}} \cdots$$

$$H^p(M; \mathbf{R}) = Z^p(M)/B^p(M) = \ker d_{p+1}/\mathrm{im}\, d_p \tag{6.13}$$

Note that we have supplied the operators ∂ and d with indices to denote the spaces they act on. The exactness of the two sequences in (6.13) follows from the properties $\partial^2 = 0$ and $d^2 = 0$ of d and ∂, and using indices it becomes the pair of equations:

$$\begin{aligned} \partial_p \partial_{p+1} &= 0 \\ d_{p+1}\, d_p &= 0 \end{aligned} \tag{6.14}$$

for any p. Two further points to note are (i) the opposite direction of the arrows in the two sequences serves us to remind one of the duality between them, and (ii) each sequence has the same *finite* number of terms, for we know that Ω^p and C^p are non-trivial only for $p = 0, \ldots n$, where $n = \dim M$. Therefore the only cohomology groups which can be non-trivial are: $H^0(M; \mathbf{R})$, $H^1(M; \mathbf{R}) \ldots H^n(M; \mathbf{R})$. This construction of cohomology groups for differentiable manifolds using p-forms ω is due to de Rham [2], the $H^p(M; \mathbf{R})$ are therefore often called de Rham cohomology groups.

6.2 $H^p(M; \mathbf{R})$ AND POINCARÉ'S LEMMA

Now that we have expressed cohomology in terms of forms, we see that to construct $H^p(M; \mathbf{R})$ we are simply looking for those p-forms ω which satisfy $d\omega = \mathbf{O}$, but for which there is no $(p-1)$-form η satisfying $\omega = d\eta$. On a point of nomenclature, if $d\omega = \mathbf{O}$, then ω is called *closed*, and if $\omega = d\eta$, then ω is called *exact*. So $Z^p(M)$ is the space of closed forms and $B^p(M)$ is the space of all exact forms. Clearly all exact forms are closed, i.e. $B^p(M) \subset Z^p(M)$, but the converse is usually not true. If the converse were true, then all cohomology groups $H^p(M; \mathbf{R})$ would be trivial, and we shall see that this is not true. However, it is known that if a space M is contractible to a point, then all closed forms on M are also exact. This result is known as Poincaré's lemma. Before proving it, let us examine a non-exact closed form. A simple example is available in two dimensions. Let M be the manifold $\mathbf{R}^2-\{0\}$ ($\mathbf{R}^2$ minus the origin, or the punctured plane), then $\boldsymbol{\omega}$ is the one-form defined below.

$$\boldsymbol{\omega}=\frac{-y}{(x^2+y^2)}\,\mathrm{d}x+\frac{x}{(x^2+y^2)}\,\mathrm{d}y \tag{6.15}$$

We see from (6.15) that $\boldsymbol{\omega}$ would diverge if x and y were zero, thus the manifold M must not contain the origin. A quick calculation of $\mathrm{d}\boldsymbol{\omega}$ yields

$$\begin{aligned}\mathrm{d}\boldsymbol{\omega}&=\frac{\partial}{\partial y}\left\{\frac{-y}{(x^2+y^2)}\right\}\mathrm{d}y\wedge\mathrm{d}x+\frac{\partial}{\partial x}\left\{\frac{x}{x^2+y^2}\right\}\mathrm{d}x\wedge\mathrm{d}y\\&=\frac{(y^2-x^2)}{(x^2+y^2)^2}\,\mathrm{d}y\wedge\mathrm{d}x+\frac{(y^2-x^2)}{(x^2+y^2)^2}\,\mathrm{d}x\wedge\mathrm{d}y\\&=\mathbf{0}\end{aligned} \tag{6.16}$$

so that $\boldsymbol{\omega}$ is closed. Trial and error leads one to consider the function $\eta(x, y)$ given by:

$$\eta(x,y)=\arctan(y/x) \tag{6.17}$$

This is because

$$\frac{\partial\eta}{\partial x}=\frac{-y}{x^2+y^2};\qquad \frac{\partial\eta}{\partial y}=\frac{x}{x^2+y^2} \tag{6.18}$$

thus it might appear that $\boldsymbol{\omega}=\mathrm{d}\eta$. This argument fails because η is actually the angle θ of polar coordinates and single valuedness requires that the domain of definition of η be restricted to $\mathbf{R}^2-\mathbf{R}_+$, where $\mathbf{R}_+$ is the non-negative x-axis, $\mathbf{R}_+=\{x: x\geqslant 0\}$. So on this domain $\mathbf{R}^2-\mathbf{R}_+$ we *do* have $\boldsymbol{\omega}=\mathrm{d}\eta$. Unfortunately there is no other function f which is defined on all of $\mathbf{R}^2-\{0\}$, and for which $\boldsymbol{\omega}=\mathrm{d}f$. To see this note that if this were so, on $\mathbf{R}^2-\mathbf{R}_+$ we would have

$$\begin{gathered}\boldsymbol{\omega}=\mathrm{d}f=\mathrm{d}\eta\\ \Rightarrow\mathrm{d}(f-\eta)=\mathbf{0}\\ \Rightarrow\frac{\partial f}{\partial x}=\frac{\partial\eta}{\partial x};\\ \frac{\partial f}{\partial y}=\frac{\partial\eta}{\partial y}\Rightarrow\\ f=\eta+C,\ C\ \text{const}\end{gathered} \tag{6.19}$$

But if η can only be defined on $\mathbf{R}^2-\mathbf{R}_+$, then $\eta+C=f$ can also only be defined on $\mathbf{R}^2-\mathbf{R}_+$. Thus $\boldsymbol{\omega}$ is closed but not exact. However, we can still say that $\boldsymbol{\omega}$ is exact locally, if we take a neighbourhood of $\mathbf{R}^2$ not containing zero, then $\boldsymbol{\omega}=\mathrm{d}\eta$ in this neighbourhood. Note that the punctured plane $\mathbf{R}^2-\{0\}=M$ cannot be contracted to a point, and so M is not covered by Poincaré's lemma.

6.3 POINCARÉ'S LEMMA

To prove Poincaré's lemma we first define the term contractible. M is *smoothly* contractible if there is a C^∞ map $\alpha: M\times[0,1]\rightarrow M$ such that:

$$\begin{aligned} &(x,t)\mapsto x \text{ if } t=0; \qquad x\in M,\ t\in[0,1] \\ &(x,t)\mapsto x_0 \textit{ if } t=1 \end{aligned} \tag{6.20}$$

so as t varies smoothly from 0 to 1, the map α shrinks M down to the single point x_0. The next stage in the proof is to derive a certain formula concerning forms on $M\times[0,1]$. Having done this, we find that the information that this formula implies for forms on M is just Poincaré's lemma. To this end let $\boldsymbol{\omega}$ be an r-form on $M\times[0,1]$, i.e. $\boldsymbol{\omega}\in\Omega^r(M\times[0,1])$; then if t is the local coordinate for $[0,1]$, and $x_1,\ldots,x_n$ are the local coordinates for M, it is possible to decompose $\boldsymbol{\omega}$ in the way shown below:

$$\begin{aligned}\boldsymbol{\omega}(t,x)=&f_{i_1\ldots i_{r-1}}(t,x)\,\mathrm{d}t\wedge\mathrm{d}x^{i_1}\wedge\ldots\wedge\mathrm{d}x^{i_{r-1}}\\ &+g_{i_1\ldots i_r}(t,x)\,\mathrm{d}x^{i_1}\wedge\ldots\wedge\mathrm{d}x^{i_r}\end{aligned} \tag{6.21}$$

We have simply singled out the term in $\boldsymbol{\omega}$ involving $\mathrm{d}t$. Next we introduce an operator denoted by P which does two things: it reduces an r-form to an $(r-1)$-form and it transfers the form from $M\times[0,1]$ to M, i.e. $\mathrm{P}\boldsymbol{\omega}\in\Omega^{r-1}(M)$. P is defined by:

$$\mathrm{P}\boldsymbol{\omega}=\left\{\int_0^1\mathrm{d}t'\,f_{i_1\ldots i_{r-1}}(t',x)\right\}\mathrm{d}x^{i_1}\wedge\ldots\wedge\mathrm{d}x^{i_{r-1}} \tag{6.22}$$

Points to check for oneself in (6.22) are that $\mathrm{P}\boldsymbol{\omega}$ is now an $(r-1)$-form, the differential $\mathrm{d}t$ having been deleted, and that the variable t has been integrated out so that $\mathrm{P}\boldsymbol{\omega}$ is a form on M. A further technical preliminary concerns an alternative way of obtaining the second term in (6.21). This requires us to define the map β_t given below:

$$\begin{aligned}&\beta_t: M\mapsto M\times[0,1]\\ &x\mapsto(x,t)\end{aligned} \tag{6.23}$$

β_t may not appear to be of much interest. However, recall that drawing on Chapter 2, we know that β_t induces the map $\beta_t^*: \Omega^r(M\times[0,1])\mapsto\Omega^r(M)$. In fact it is the case that

$$\beta_t^*\omega(t,x)=g_{i_1\ldots i_r}(t,x)\,\mathrm{d}x^{i_1}\wedge\ldots\wedge\mathrm{d}x^{i_r} \tag{6.24}$$

To prove this use carefully the definition of Chapter 2 to construct β_t^* and the result is then almost automatic because of the simple form of β_t. Finally, the formula we discussed above, and which we shall now derive is:

$$\mathrm{dP}\omega+\mathrm{Pd}\omega=\beta_1^*\omega-\beta_0^*\omega \tag{6.25}$$

The proof is a routine calculation. Calculating $\mathrm{dP}\omega$ first we obtain:

$$\begin{aligned}\mathrm{dP}\omega&=\mathrm{d}\left\{\int_0^1\mathrm{d}t'\,f_{i_1\ldots i_{r-1}}(t',x)\right\}\mathrm{d}x^{i_1}\wedge\ldots\wedge\mathrm{d}x^{i_{r-1}}\\&=\left\{\int_0^1\mathrm{d}t'\,\frac{\partial f_{i_1\ldots i_{r-1}}(t',x)}{\partial x^{i_r}}\right\}\mathrm{d}x^{i_r}\wedge\mathrm{d}x^{i_1}\wedge\ldots\wedge\mathrm{d}x^{i_{r-1}}\end{aligned} \tag{6.26}$$

Turning to $\mathrm{Pd}\omega$ we find that:

$$\begin{aligned}\mathrm{d}\omega=&\frac{\partial f_{i_1\ldots i_{r-1}}(t,x)}{\partial x^{i_r}}\mathrm{d}x^{i_r}\wedge\mathrm{d}t\wedge\mathrm{d}x^{i_1}\wedge\ldots\wedge\mathrm{d}x^{i_{r-1}}\\&+\frac{\partial g_{i_1\ldots i_r}(t,x)}{\partial t}\mathrm{d}t\wedge\mathrm{d}x^{i_1}\wedge\ldots\wedge\mathrm{d}x^{i_r}\\&+\frac{\partial g_{i_1\ldots i_r}(t,x)}{\partial x^{i_{r+1}}}\mathrm{d}x^{i_{r+1}}\wedge\mathrm{d}x^{i_1}\wedge\ldots\wedge\mathrm{d}x^{i_r}\end{aligned} \tag{6.27}$$

So that

$$\begin{aligned}\mathrm{Pd}\omega=&-\left\{\int_0^1\mathrm{d}t'\,\frac{\partial f_{i_1\ldots i_{r-1}}(t,x)}{\partial x^{i_r}}\right\}\mathrm{d}x^{i_r}\wedge\mathrm{d}x^{i_1}\wedge\ldots\wedge\mathrm{d}x^{i_{r-1}}\\&+\left\{\int_0^1\mathrm{d}t'\,\frac{\partial g_{i_1\ldots i_r}(t',x)}{\partial t'}\right\}\mathrm{d}x^{i_1}\wedge\ldots\wedge\mathrm{d}x^{i_r}\end{aligned} \tag{6.28}$$

Adding (6.26) to (6.28) we finally obtain the promised result.

$$\begin{aligned}\mathrm{dP}\omega+\mathrm{Pd}\omega&=\left\{\int_0^1\mathrm{d}t'\,\frac{\partial g_{i_1\ldots i_r}(t',x)}{\partial t'}\right\}\mathrm{d}x^{i_1}\wedge\ldots\wedge\mathrm{d}x^{i_r}\\&=(g_{i_1\ldots i_r}(1,x)-g_{i_1\ldots i_r}(0,x))\,\mathrm{d}x^{i_1}\wedge\ldots\wedge\mathrm{d}x^{i_r}\\&=\beta_1^*\omega-\beta_0^*\omega\end{aligned} \tag{6.29}$$

Now to prove Poincaré's lemma we simply show that if $\boldsymbol{\omega}$ is an r-form on M with $\mathrm{d}\boldsymbol{\omega} = 0$ and M is contractible, then

$$\boldsymbol{\omega} = \mathrm{d}\boldsymbol{\eta}$$

where

$$\boldsymbol{\eta} = -\mathrm{P}\alpha^*\boldsymbol{\omega} \tag{6.30}$$

The result follows by noting that if $\boldsymbol{\omega} \in \Omega^r(M)$ then $\alpha^*\boldsymbol{\omega} \in \Omega^r(M \times [0, 1])$. Now we apply (6.29) to the form $\alpha^*\boldsymbol{\omega}$. This says that

$$\mathrm{dP}\alpha^*\boldsymbol{\omega} + \mathrm{Pd}\alpha^*\boldsymbol{\omega} = \beta_1^*\alpha^*\boldsymbol{\omega} - \beta_0^*\alpha^*\boldsymbol{\omega} \tag{6.31}$$

The RHS of (6.31) is equal to

$$(\alpha\beta_1)^*\boldsymbol{\omega} - (\alpha\beta_0)^*\boldsymbol{\omega}$$

Using the definitions of α and β_t it is immediate that $\alpha\beta_1$ is the constant map from M to M under which $x \mapsto x_0$, thus $(\alpha\beta_1)^* = 0$; and $\alpha\beta_0$ is the identity map from M to M. The RHS of (6.31) is therefore $-\boldsymbol{\omega}$. The term $\mathrm{Pd}\alpha^*\boldsymbol{\omega}$ in the LHS of (6.31) is zero since (for any α),

$$\mathrm{d}\alpha^*\boldsymbol{\omega} = \alpha^*\mathrm{d}\boldsymbol{\omega} \tag{6.32}$$

this is just an application of the chain rule for partial derivatives. Further since $\mathrm{d}\boldsymbol{\omega} = \mathbf{O}$ by assumption, then $\mathrm{d}\alpha^*\boldsymbol{\omega} = \mathbf{O}$. Putting this information all into the same equation gives

$$\boldsymbol{\omega} = -\mathrm{dP}\alpha^*\boldsymbol{\omega} \tag{6.33}$$

as claimed above. This proves Poincaré's lemma.

6.4 CALCULATION OF $H^p(M; \mathbf{R})$

Our first calculation of a cohomology group is encouragingly easy. We simply observe that since $\mathbf{R}^n$ is contractible†, all closed p-forms on $\mathbf{R}^n$ are exact, and we have (we remind the reader that $H^p(M; \mathbf{R})$ is always trivial for $p > \dim M$)

$$H^p(\mathbf{R}^n; \mathbf{R}) = 0, \qquad p = 1, \ldots n \tag{6.34}$$

However, we also have

$$H^0(\mathbf{R}^n; \mathbf{R}) = \mathbf{R} \tag{6.35}$$

† The map $\alpha : \mathbf{R}^n \times [0, 1] \to \mathbf{R}^n$ under which $(x, t) \mapsto (1-t)x$ contracts $\mathbf{R}^n$ to the point 0.

To understand (6.35) remember that 0-forms on M are simply functions on M, and that a closed 0-form is simply a function f satisfying $\mathrm{d}f = 0$. There are no -1-forms, so f can never be exact. Finally the solution to

$$\mathrm{d}f = 0$$

is

$$f = c, \qquad c \in \mathbf{R} \tag{6.36}$$

So to each real number c there corresponds an element of the zeroth cohomology class, i.e. an element of $H^0(\mathbf{R}^n; \mathbf{R})$. We would like to calculate the groups $H^p(S^n; \mathbf{R})$, $p = 0, \ldots, n$. Before we do this we establish two simple but important properties of maps between manifolds. For the first property we have a smooth map α between two manifolds

$$\alpha : M \to N$$

which gives rise to

$$\alpha^* : \Omega^p(N) \to \Omega^p(M) \tag{6.37}$$

the property of α^* that we have in mind is that α^* induces a map, for which we shall use the *same* symbol α^*, between the cohomology groups of M and N.

$$\begin{aligned} \alpha^* : H^p(N; \mathbf{R}) &\to H^p(M; \mathbf{R}) \\ [\omega] &\mapsto [\alpha^* \omega] \end{aligned} \tag{6.38}$$

This new map α^* is defined in (6.38). To check that (6.38) is a well defined map between cohomology groups, is to check that α^* takes closed forms to closed forms, and exact forms to exact forms. This is immediate since

$$\begin{aligned} \mathrm{d}\alpha^* \omega &= \alpha^* \mathrm{d}\omega \\ &= 0 \quad \text{if } \mathrm{d}\omega = 0 \end{aligned}$$

and

$$\alpha^* \mathrm{d}\eta = \mathrm{d}\alpha^* \eta \tag{6.39}$$

The second property is that if we have two maps α and β, $\alpha, \beta : M \to N$, and we have also that α and β are homotopic, then the two maps α^* and β^* defined in (6.38) are *equal*. Homotopy of the maps α and β, as we defined in Chapter 1, equations (1.41, 42), means that we have a map F such that

$$F : M \times [0, 1] \to N \tag{6.40}$$

satisfying

$$F(x,0)=\alpha(x)$$
$$F(x,1)=\beta(x) \tag{6.41}$$

For differentiable manifolds we further demand that F be C^{∞}, or that the homotopy be smooth. Using the map β_t of (6.23) we can express α and β in terms of β_t and F:

$$\alpha=F\circ\beta_0$$
$$\beta=F\circ\beta_1 \tag{6.42}$$

To prove the desired result we must show that if $d\boldsymbol{\omega}=\mathbf{O}$ then the difference $\alpha^*\boldsymbol{\omega}-\beta^*\boldsymbol{\omega}$ is exact i.e.

$$\alpha^*\boldsymbol{\omega}-\beta^*\boldsymbol{\omega}=d\boldsymbol{\eta} \tag{6.43}$$

For if we then take cohomology equivalence classes on both sides of (6.43) we obtain

$$[\alpha^*\boldsymbol{\omega}-\beta^*\boldsymbol{\omega}]=[d\boldsymbol{\eta}]$$
$$=0$$

i.e.

$$[\alpha^*\boldsymbol{\omega}]=[\beta^*\boldsymbol{\omega}] \tag{6.44}$$

Using (6.42) the LHS of (6.43) is equal to

$$(F\circ\beta_0)^*\boldsymbol{\omega}-(F\circ\beta_1)^*\boldsymbol{\omega}$$
$$=\beta_0^*(F^*\boldsymbol{\omega})-\beta_1^*(F^*\boldsymbol{\omega})$$
$$=-dPF^*\boldsymbol{\omega}-PdF^*\boldsymbol{\omega} \quad \text{(using (6.25))}$$
$$=-dPF^*\boldsymbol{\omega}-PF^*d\boldsymbol{\omega} \quad \text{(using (6.23))}$$
$$=-dPF^*\boldsymbol{\omega} \quad (\text{since } d\boldsymbol{\omega}=\mathbf{O})$$

Thus we have verified (6.43) with $\boldsymbol{\eta}=-PF^*\boldsymbol{\omega}$. We can now make use of these two properties to compute $H^1(M;\mathbf{R})$ when M is simply connected. The result will be

$$H^1(M;\mathbf{R})=0 \tag{6.45}$$

for M simply connected.

If a connected manifold M is simply connected then every closed curve on M can be contracted to a point. When M is a simply connected differentiable manifold the differentiable structure allows one to say that every closed curve on M can be smoothly contracted to a point. An alternative way of stating this contractibility of curves is to say that every closed curve on M is smoothly homotopic to the constant curve. We now write this down in terms of maps. Let Γ be any closed curve on M and Γ_0 be the constant curve. We have therefore two C^∞ maps:

$$\begin{aligned} \Gamma &: S^1 \to M \\ \Gamma_0 &: S^1 \to M \end{aligned} \tag{6.46}$$

under Γ_0 all elements $\theta \in S^1$ are mapped onto the same point x_0 of M. Next choose any closed 1-form $\boldsymbol{\omega}$ on M then we write the integral of $\boldsymbol{\omega}$ round a curve Γ as $\int_\Gamma \boldsymbol{\omega}$, and define it in (6.47) below by

$$\int_\Gamma \boldsymbol{\omega} = \int_{S^1} \Gamma^* \boldsymbol{\omega}. \tag{6.47}$$

But since Γ is homotopic to Γ_0, and since $\boldsymbol{\omega}$ is closed, then we have just proved that

$$\Gamma^* \boldsymbol{\omega} = \Gamma_0^* \boldsymbol{\omega} + d\boldsymbol{\eta} \tag{6.48}$$

for some $\boldsymbol{\eta}$. Further since Γ_0 is the constant curve it follows immediately that $\Gamma_0^* = 0$. Thus

$$\int_\Gamma \boldsymbol{\omega} = \int_{S^1} \Gamma^* \boldsymbol{\omega} = \int_{S^1} d\boldsymbol{\eta} = 0 \tag{6.49}$$

since S^1 has no boundary. In summary, we have proved that if M is simply connected, then $\int_\Gamma \boldsymbol{\omega} = 0$ for all closed 1-forms $\boldsymbol{\omega}$. It remains to show that $\boldsymbol{\omega}$ is also exact. This is so because $\boldsymbol{\omega} = df$, where f is the 0-form or function given by

$$f(x) = \int_{x_0}^{x} \boldsymbol{\omega} \tag{6.50}$$

(the RHS of (6.50) is an integral along a curve, of necessity non-closed, beginning at x_0 and ending at x. The integral is independent of the shape of the curve joining x to x_0 because $\int_\Gamma \boldsymbol{\omega} = 0$ for any *closed* curve Γ.) Thus

$$H^1(M; \mathbf{R}) = 0$$

as claimed. In particular if $M = S^n$, then S^n is simply connected for $n > 1$. So we have our first result for spheres namely

$$H^1(S^n; \mathbf{R}) = 0, \qquad n > 1 \tag{6.51}$$

Speaking intuitively, and with even less rigour than is the average in this book, one can see that homotopy divides closed curves on an arbitrary manifold into equivalence classes—one for each hole in the manifold. Each closed curve round a particular hole is homotopic to another round the same hole. This provides us with the useful result that

$$\dim H^1(M; \mathbf{R}) = \text{the number of holes in } M \qquad (6.52)$$

Our second result for spheres is the calculation of $H^1(S^1; \mathbf{R})$. We shall find that

$$H^1(S^1; \mathbf{R}) = \mathbf{R} \qquad (6.53)$$

To begin with note that because S^1 has dimension one, all 1-forms ω on S^1 are closed. The key to proving (6.53) is to show that

$$\int_{S^1} \omega = 0 \Rightarrow \omega \text{ exact} \qquad (6.54)$$

for 1-forms ω. The converse is also true being an immediate consequence of Stokes' theorem. To show that (6.54) is true consider S^1 with a single point p deleted, i.e. consider $S^1 - \{p\}$. Since $S^1 - \{p\}$ is contractible, then ω restricted to $S^1 - \{p\}$ is exact i.e.

$$\omega = \mathrm{df} \text{ on } S^1 - \{p\}$$

But since $\{p\}$ is a set of measure zero then

$$\int_{S^1 - \{p\}} \omega = \int_{S^1} \omega = 0 \qquad (6.55)$$

Let p be the point corresponding to $\theta = 0$ in polar coordinates, then the LHS of (6.55) is

$$\int_{S^1 - \{p\}} \mathrm{df} = \lim_{\varepsilon \to 0} \{f(2\pi - \varepsilon) - f(\varepsilon)\} = 0 \qquad (6.56)$$

Thus because f is C^∞, f must be periodic over the whole of S^1; and therefore $\omega = \mathrm{df}$ extends to the whole of S^1 making ω exact. Now in constructing $H^1(S^1; \mathbf{R})$ we are looking for all closed, non-exact, 1-forms. Because of (6.54) all such forms have non-zero integral over S^1. Suppose then ω and ω' are two closed non-exact 1-forms on S^1. The statement that $H^1(S^1, \mathbf{R}) = \mathbf{R}$ means that there is some $c \in \mathbf{R}$ for which $\omega - c\omega'$ is exact. We can prove this immediately for consider the form $\omega - c\omega'$ where

$$c = \left\{\int_{S^1} \omega\right\} \Big/ \left\{\int_{S^1} \omega'\right\} \qquad (6.57)$$

(c is always non-zero since we have shown that $\int_{S^1} \omega$, $\int_{S^1} \omega'$ are non-zero). Integrating $\omega - c\omega'$ we obtain

$$\begin{aligned}\int_{S^1} (\omega - c\omega') &= \int_{S^1} \omega - c \int_{S^1} \omega' \\ &= 0\end{aligned} \tag{6.58}$$

by (6.57). Thus (6.54) shows that $\omega - c\omega'$ is exact and we have proved (6.53). In case the reader is wondering if there are any 1-forms ω on S^1 with $\int_{S^1} \omega \neq 0$, we remind him that there is always the volume form, or in the case of S^1, the form whose integral gives the length of the circumference of S^1.

The calculation of $H^p(S^n; \mathbf{R})$ for the remaining values of p and n takes little extra work. When $p = 0$ we have simply $\omega = \mathrm{f}$, where f is a constant function on S^n. So we obtain

$$H^0(S^n; \mathbf{R}) = \mathbf{R} \tag{6.59}$$

a result which we saw also holds for $\mathbf{R}^n$. For $p > 1$ we need only to prove the inductive formula

$$H^p(S^n; \mathbf{R}) = H^{p-1}(S^{n-1}; \mathbf{R}), p > 1 \tag{6.60}$$

The argument used to verify (6.60) is quite straightforward, and is worth spelling out in detail: let A_1 be S^n with the north pole deleted, and let A_2 be S^n with the south pole deleted. Now take a closed, non-exact, p-form ω on S^n. A_1 and A_2 are both contractible so we have

$$\begin{aligned}\omega &= \mathrm{d}\eta_1 \text{ on } A_1 \\ \omega &= \mathrm{d}\eta_2 \text{ on } A_2\end{aligned} \tag{6.61}$$

for some $(p-1)$-forms, η_1 on A_1, and η_2 on A_2. Also on $A_1 \cap A_2$

$$\mathrm{d}\eta_1 = \mathrm{d}\eta_2 \Rightarrow \mathrm{d}(\eta_1 - \eta_2) = 0$$

Therefore the $(p-1)$-form $\rho = \eta_1 - \eta_2$ is a closed $(p-1)$-form on $A_1 \cap A_2$. Note also that if ω' is cohomologous to ω, i.e. $[\omega] = [\omega']$, then because $\omega' = \omega + \mathrm{d}\eta$, for some η, the construction above yields for ω' the *same* $(p-1)$-form ρ. So we associate with the pth cohomology class $[\omega]$, the $(p-1)$-form ρ on $A_1 \cap A_2$.

We now reverse the procedure and start with closed $(p-1)$-forms on $A_1 \cap A_2$ and construct closed p-forms on $A_1 \cup A_2 = S^n$. Let ρ be a closed $(p-1)$-form on $A_1 \cap A_2$. Now choose a partition of unity on S^n with just two terms, i.e. we have two open sets U_1 and U_2, and two differentiable functions $e_1(x)$ and $e_2(x)$, such that:

$$\begin{aligned} &U_1 \cup U_2 = S^n \\ &e_1(x) + e_2(x) = 1 \\ &e_i(x) = 0 \text{ if } x \notin U_i; \qquad i = 1, 2 \end{aligned} \tag{6.62}$$

Also require that U_1 contains only the north pole of S^n, and U_2 contains only the south pole of S^n. Now given our closed $(p-1)$-form $\boldsymbol{\rho}$ on $A_1 \cap A_2$, define $\boldsymbol{\eta}_1$ on A_1, and $\boldsymbol{\eta}_2$ on A_2, to be

$$\begin{aligned} \boldsymbol{\eta}_1 &= e_1 \boldsymbol{\rho} \\ \boldsymbol{\eta}_2 &= -e_2 \boldsymbol{\rho} \end{aligned} \tag{6.63}$$

The definition (6.63) gives rise to a p-form $\boldsymbol{\omega}$ on $A_1 \cup A_2 = S^n$ if we choose

$$\begin{aligned} \boldsymbol{\omega} &= \mathrm{d}\boldsymbol{\eta}_1 \quad \text{on } A_1 \\ \boldsymbol{\omega} &= \mathrm{d}\boldsymbol{\eta}_2 \quad \text{on } A_2 \end{aligned} \tag{6.64}$$

Note that $\boldsymbol{\omega}$ is closed, and that $\boldsymbol{\eta}_1 - \boldsymbol{\eta}_2 = \boldsymbol{\rho}$ on $A_1 \cap A_2$. Further if $[\boldsymbol{\rho}'] = [\boldsymbol{\rho}]$ for some other closed $(p-1)$-form $\boldsymbol{\rho}'$, so that $\rho' = \rho + \mathrm{d}\boldsymbol{\nu}$ with $\boldsymbol{\nu}$ a $(p-2)$-form, then (6.64) gives the same p-form $\boldsymbol{\omega}$ as that given using $\boldsymbol{\rho}$. To summarize what we have achieved, we say that we have constructed a one-to-one correspondence, α say, between closed p-forms on $A_1 \cup A_2 = S^n$ and closed $(p-1)$-forms on $A_1 \cap A_2$; more important still, α induces a one-to-one correspondence between the cohomology groups $H^p(S^n; \mathbf{R})$ and $H^{p-1}(A_1 \cap A_2; \mathbf{R})$. These groups are therefore isomorphic:

$$H^p(S^n; \mathbf{R}) \simeq H^{p-1}(A_1 \cap A_2; \mathbf{R}) \tag{6.65}$$

The final stage in our calculation is to show that

$$H^{p-1}(A_1 \cap A_2; \mathbf{R}) \simeq H^{p-1}(S^{n-1}; \mathbf{R}) \tag{6.66}$$

To prove (6.66) we first of all point out that $A_1 \cap A_2$ is S^n with both north and south poles deleted. This means that stereographic projection can be used to map homeomorphically $A_1 \cap A_2$ onto $\mathbf{R}^n - \{0\}$. We give explicitly the projection in (6.67)

$$\begin{aligned} &p_i = \frac{x_i}{1 + x^2}; \qquad p_{n+1} = \frac{1 - x^2}{1 + x^2} \\ &i = 1, \ldots, n; \qquad x^2 = x_1^2 + \ldots + x_n^2 \end{aligned}$$

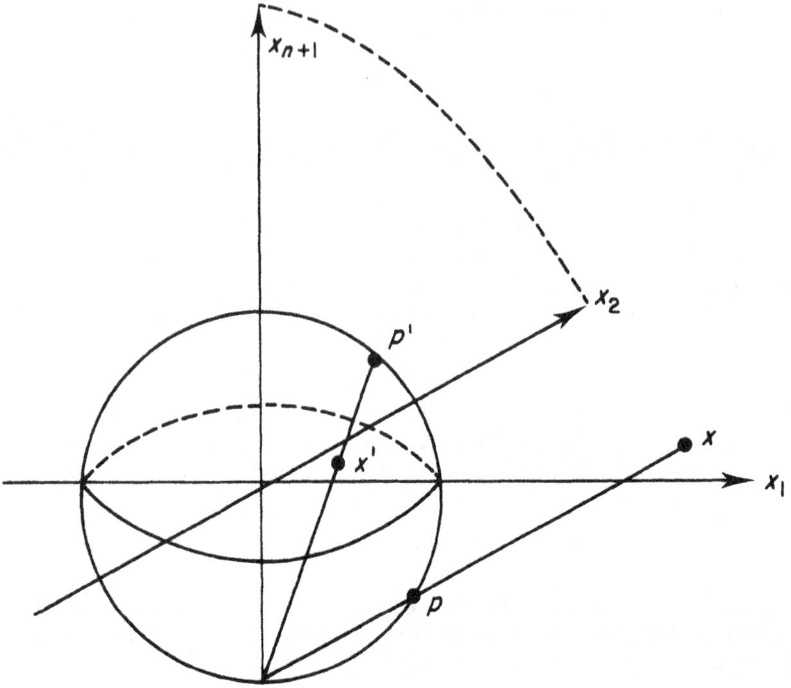

Figure 6.1

or equivalently

$$x_i = \frac{2p_i}{(1+p_{n+1})}, \qquad p_1^2+p_2^2+\ldots+p_{n+1}^2$$
$$i=1,\ldots n. \tag{6.67}$$

In (6.67) p denotes a point on S^n and x denotes a point in $\mathbf{R}^n$. The projection takes place from the south pole $p_{n+1}=-1$. The points $x=\infty$ and $x=0$ of $\mathbf{R}^n$ correspond to the south and north pole of S^n respectively. Since $A_1 \cap A_2$ contains neither of these poles then (6.67) defines a map f say from $A_1 \cap A_2$ to $\mathbf{R}^n - \{0\}$. One can see from (6.67) that f is a homeomorphism, in fact since f is C^∞ on $A_1 \cap A_2$, it is also a diffeomorphism. In any case this means that

$$H^{p-1}(A_1 \cap A_2; \mathbf{R}) \simeq H^{p-1}(\mathbf{R}^n - \{0\}; \mathbf{R}) \tag{6.68}$$

The last calculation is to show that $H^{p-1}(\mathbf{R}^n-\{0\};\mathbf{R})$ and $H^{p-1}(S^{n-1};\mathbf{R})$ are isomorphic, and this is done by explicitly constructing maps:

$$\alpha^*:H^{p-1}(\mathbf{R}^n-\{0\};\mathbf{R})\rightarrow H^{p-1}(S^{n-1};\mathbf{R})$$

and

$$\beta^*:H^{p-1}(S^{n-1};\mathbf{R})\rightarrow H^{p-1}(\mathbf{R}^n-\{0\};\mathbf{R}) \tag{6.69}$$

such that α^* and β^* are inverses of one another. The maps α^* and β^* are defined by giving the maps α and β defined below:

$$\alpha:S^{n-1}\rightarrow\mathbf{R}^n-\{0\}$$

$$p\mapsto p$$

$$\beta:\mathbf{R}^n-\{0\}\rightarrow S^{n-1}$$

$$x_i\mapsto\frac{x_i}{(x^2)^{1/2}} \tag{6.70}$$

α is just the inclusion map which sends $p\in S^{n-1}$ to the same point $p\in\mathbf{R}^{n-1}-\{0\}$. β is an example of a deformation retraction—it deforms $\mathbf{R}^n-\{0\}$ into S^{n-1}. The map $\beta\circ\alpha$ is the identity on S^{n-1}, while the map $\alpha\circ\beta$, though not the identity on $\mathbf{R}^n-\{0\}$, is *homotopic* to the identity on $\mathbf{R}^n-\{0\}$. The homotopy $F(x,t)$ is defined by

$$F(x,t)=xt+(1-t)\alpha(\beta(x)) \tag{6.71}$$

This means, using the property of homotopic maps proved in (6.43), that the maps $(\alpha\circ\beta)^*$ and $(\beta\circ\alpha)^*$ that are induced on the homology groups are both the identity:

$$\begin{aligned}(\alpha\circ\beta)^*&:H^p(\mathbf{R}^n-\{0\};\mathbf{R})\rightarrow H^p(\mathbf{R}^n-\{0\};\mathbf{R})\\(\beta\circ\alpha)^*&:\bar{H}^p(S^{n-1};\mathbf{R})\rightarrow H^p(S^{n-1};\mathbf{R})\end{aligned} \tag{6.72}$$

That is to say that α^* and β^* are inverses of one another as we claimed above. So we have proved that

$$H^p(S^n;\mathbf{R})\simeq H^{p-1}(\mathbf{R}^n-\{0\};\mathbf{R})\simeq H^{p-1}(S^{n-1};\mathbf{R})$$

or simply

$$H^p(S^n,\mathbf{R})\simeq H^{p-1}(S^{n-1};\mathbf{R});\, p>1 \tag{6.73}$$

If we combine the inductive result (6.73) with our earlier results for spheres namely

$$\begin{aligned}&H^0(S^n;\mathbf{R})=\mathbf{R}\\&H^1(S^1;\mathbf{R})=\mathbf{R}\\&H^1(S^n;\mathbf{R})=0,\qquad n>1\end{aligned} \tag{6.74}$$

then we see at once that we know $H^p(S^n; \mathbf{R})$ completely, the result is that

$$\begin{aligned} &H^p(S^n; \mathbf{R}) = 0, && p > n \\ &H^p(S^n; \mathbf{R}) = \mathbf{R}, && p = n \\ &H^p(S^n; \mathbf{R}) = 0, && 1 \leq p < n \\ &H^0(S^n; \mathbf{R}) = \mathbf{R} \end{aligned} \tag{6.75}$$

Having followed and understood the calculation of $H^p(S^n; \mathbf{R})$, the reader should be able to calculate $H^p(T^n; \mathbf{R})$ where T^n is the torus $S^1 \times S^1 \ldots S^1$ (n factors). One needs essentially $H^p(S^1; \mathbf{R})$ and a hint that the method of proof is by use of the wedge product. In contrast to the result for spheres, for which only the $p = 0$ and $p = n$ cohomology groups are non-trivial, $H^p(T^n; \mathbf{R})$ are all non-trivial. As a further incentive to calculation we quote the result

$$H^p(T^n; \mathbf{R}) = \mathbf{R}^\alpha, \qquad \alpha = \binom{n}{p} \tag{6.76}$$

6.5 GENERAL REMARKS

To end our calculations of cohomology groups we quote some general results. We shall not prove them but just comment on the sort of method used to prove them. In each case the proof is not difficult, but for readers who have a limited amount of time available the proofs may be found in Spivak [4]. There are three of these results and they are:

A. If M is a compact, connected, orientable manifold, and $\dim M = n$, then

$$H^n(M; \mathbf{R}) = \mathbf{R} \tag{6.77}$$

B. If M is a compact, connected, non-orientable manifold, and $\dim M = n$, then

$$H^n(M; \mathbf{R}) = 0 \tag{6.78}$$

C. If M is a non-compact, connected, manifold, orientable *or* non-orientable, and $\dim M = n$, then

$$H^n(M; \mathbf{R}) = 0 \tag{6.79}$$

The proof of (A) relies on the fact that the result holds for M in $\mathbf{R}^n$. Then by considering a covering of M with open sets and an associated partition of unity, one maps these open sets into $\mathbf{R}^n$ where the result holds, and then one pieces together the result over the whole of M.

The proof of (B) imitates the proof of (A) until lack of orientability causes a breakdown of the procedure. This happens because, in the covering of M with open sets, non-orientability means that there is a sign change of a Jacobian somewhere in passing from the coordinates of one open set to those of another. The consequence of this sign change is only felt globally when one pieces together a closed n-form ω from its values on each open set. The closed n-form ω is decomposed locally into a non-exact and an exact part, but the change of sign causes a global cancellation of all the non-exact parts, so that any closed n-form is always exact, hence $H^n(M; \mathbf{R}) = 0$.

The proof of (C) again uses a covering of M with open sets, but this time, because M is non-compact, an infinite covering is used. This results in a partition of unity with an infinite number of terms. Comparison with the proof of (A) then shows straightaway that any closed n-form on M is always exact, thus $H^n(M; \mathbf{R}) = 0$ as claimed.

In commenting on the result (A), we see that we now know the nth-cohomology group of an arbitrary compact, connected, orientable manifold regardless of how many holes and unusual features it has. The result is always the same, and can be remembered by simply memorizing the result for spheres, $H^n(S^n; \mathbf{R}) = \mathbf{R}$.

As regards result (B) we can immediately say that

$$H^2\,(\text{Möbius strip}; \mathbf{R}) = 0 \tag{6.80}$$

However, H^1 (Möbius strip; $\mathbf{R}$) is not zero as can be checked using the arguments which were used in establishing (6.44). Another example of non-orientable type is all real projective spaces P^n for n even, for which we can write

$$H^{2n}(P^{2n}; \mathbf{R}) = 0 \tag{6.81}$$

(P^n is defined in the usual way to be the quotient space obtained from S^n by identifying antipodal points. It is easy to check that P^n is, orientable for n odd, and non-orientable for n even.)

Result (C) is fairly general and one can readily produce examples of non-compact connected manifolds.

6.6 THE CUP PRODUCT

There is a very useful algebraic structure which is present in cohomology theory. This is called the cup product. It works as follows. Given $[\omega] \in H^p(M; \mathbf{R})$ and $[\nu] \in H^q(M; \mathbf{R})$, then we define the cup product of $[\omega]$ and

$[\nu]$, written $[\omega]\cup[\nu]$, by

$$[\omega]\cup[\nu]=[\omega\wedge\nu] \tag{6.82}$$

The RHS of (6.82) is a $(p+q)$-form so that $[\omega\wedge\nu]\in H^{p+q}(M;\mathbf{R})$. One can verify immediately, using the properties of closed and exact forms that (6.82) is a well defined product of cohomology classes. The cup product $\bigcup$ is therefore a map of the form:

$$\bigcup : H^p(M;\mathbf{R})\times H^q(M;\mathbf{R})\to H^{p+q}(M;\mathbf{R}) \tag{6.83}$$

If we define the sum of all the cohomology groups $H^*(M;\mathbf{R})$, by

$$H^*(M;\mathbf{R})=\bigoplus_{p>0} H^p(M;\mathbf{R}) \tag{6.84}$$

Then the cup product has the neater looking form

$$\bigcup : H^*(M;\mathbf{R})\times H^*(M;\mathbf{R})\to H^*(M;\mathbf{R}) \tag{6.85}$$

The product on $H^*(M;\mathbf{R})$ defined in (6.85) makes $H^*(M;\mathbf{R})$ into a ring. This is an asset worth having in cohomology theory. For example, it can happen that two spaces M and N have the same cohomology groups and yet are not topologically the same. This can be demonstrated by computing the cup product $\bigcup$ for $H^*(M;\mathbf{R})$ and $H^*(N;\mathbf{R})$, and showing that the resulting rings are different. We choose, for reasons of space, not to do this here. However, an example of what we are referring to is provided by choosing $M=S^2\times S^4$, and $N=\mathbf{C}P^3$, complex projective 3-space. For the details see Maunder [3].

6.7 SUPERIORITY OF COHOMOLOGY OVER HOMOLOGY

We would like to close this chapter with some remarks favouring the use of cohomology rather than homology.

The first point is that homology is constructed using the boundary operator ∂, while cohomology is constructed using the exterior derivative operator d. The operator ∂ which assigns a manifold to its boundary is a global operator, and is therefore difficult to use without certain global information about the manifold. In contrast the exterior derivative d is a first order partial differential operator and so is a local operator requiring no global information in general. Thus the passage from homology, to its dual cohomology, pays a handsome dividend in rephrasing a problem written in global language into one written in local language. The second point is that in cohomology there is the extra structure given by the cup product which, as we have seen above, can be used to distinguish spaces from one another. It is worth noting here that since $\cup$ is a map of the form

$$\cup : H^*(M; \mathbf{R}) \times H^*(M; \mathbf{R}) \to H^*(M; \mathbf{R})$$

then there is a map $\cup^*$ say, dual, in a formal sense, to $\cup$, and given in terms of homology. However, because duality reverses the direction of maps it is of the form:

$$\cup^* : H_*(M; \mathbf{R}) \to H_*(M; \mathbf{R}) \times H_*(M; \mathbf{R}) \tag{6.86}$$

(where $H_*(M; \mathbf{R}) = \bigoplus_{p \geq 0} H_p(M; \mathbf{R})$).

Thus $\cup^*$ does not provide homology theory with a ring structure. Actually there is also what is called a cap product [3], but this involves homology *and* cohomology—it is a map formed from a product of homology and cohomology classes. We do not discuss it in this book.

Finally there are important generalized cohomology theories such as sheaf cohomology [5] and K-theory [1]. Their dual homology theories are not usually discussed since they do not, in general, seem to be needed.

REFERENCES

1. ATIYAH, M. F., "K-Theory". W. A. Benjamin Inc., 1967.
2. DE RHAM, G., "Variétés Differentiable". Hermann, 1960.
3. MAUNDER, C. R. F., "Algebraic Topology". Van Nostrand Reinhold, 1972.
4. SPIVAK, M., "Differential Geometry". Publish or Perish Inc., 1970.
5. WELLS, R. O., "Differential Analysis on Complex Manifolds". Springer Verlag, 1980.

CHAPTER 7

Fibre Bundles and Further Differential Geometry

7.1 INTRODUCTION

In this chapter we first introduce fibre bundles and then further develop our discussion of differential geometry. Most of the topics covered in previous chapters come to fruition in this chapter. In particular we shall find topology and geometry coming together to establish some important results.

The topology discussed will range over fibre bundles, vector bundles and the topological invariants which are needed to characterize them. These invariants are cohomology classes: the Chern class, the Pontrjagin class and the Stiefel–Whitney class to mention the most important ones. The cohomology class used will depend on the details of the bundle. They are also, as we shall see, given in terms of the local geometry so that they are an example of a coming together of topology and geometry.

The geometry discussed will take up from where we left off at the end of Chapter 2. We shall deal with connections, covariant derivatives, curvature and torsion. Riemannian geometry, and the pseudo-Riemannian geometry needed for general relativity, will be introduced; as will the geometry of Lie groups which is relevant for the non-Abelian gauge theories or Yang–Mills theories of physics. To mention another example of the expression of a topological invariant in terms of local geometry, there will be the generalized Gauss–Bonnet theorem relating an integral over the curvature to the Euler–Poincaré characteristic.

Having given the flavour of the chapter by some selective naming of topics we now turn to something specific—bundles.

The word bundle occurs in topology with various words before it—there are vector bundles, tangent bundles, and fibre bundles, to mention the more commonly occurring usages. However, as we shall see, the term fibre bundle is the most general; i.e. vector bundles and tangent bundles are just certain sorts of fibre bundles. We shall explain the nomenclature for the various sorts of fibre bundle as we meet them. To begin with though we must define a fibre bundle.

7.2 FIBRE BUNDLE

A fibre bundle is, in intuitive terms, a topological space T which is locally, *but not necessarily globally*, a product of two spaces. For example let us take the rectangular strip of Fig. 7.1a and join the ends as shown to form a cylinder T.

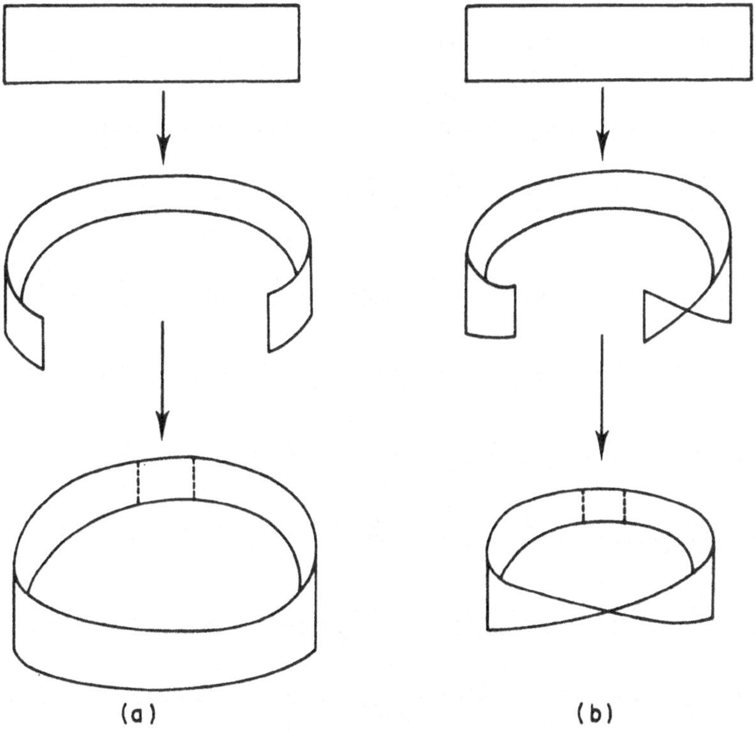

Figure 7.1

The cylinder T is both locally and globally a product. Globally $T = L \times S^1$, L being a line segment and S^1 a circle. On the other hand, Fig. 7.1b shows the Möbius strip; locally the Möbius strip and the cylinder are both products of line segments, c.f. the rectangular regions defined by the dotted lines in Fig. 7.1a and 7.1b, however, the Möbius strip is *not* globally a product, while the cylinder is. Now the idea in fibre bundle theory is to study spaces which are not globally products but are only locally products.

All spaces which are global products are fibre bundles, but they are called *trivial bundles*. Various definitions seeking to axiomatise these simple intuitive concepts were experimented with in the mathematical literature between 1935 and 1940 [c.f. ref. 16]. The definition of a fibre bundle can now be given, a fibre bundle (E, Π, F, G, X) is the following collection of requirements:

i. A topological space E (sometimes called the *total space*).
ii. A topological space X called the *base space*, and a *projection* $\Pi : E \to X$ of E onto X.
iii. A topological space F called the *fibre*.
iv. A *group G* of *homeomorphisms* of the fibre F.
v. A set of open coordinate neighbourhoods U_α covering X which reflect the local triviality of the bundle E. More specifically with each U_α there is given a homeomorphism

$$\phi_\alpha : \Pi^{-1}(U_\alpha) \to U_\alpha \times F \tag{7.1}$$

where ϕ_α^{-1} satisfies

$$\Pi\phi_\alpha^{-1}(x, f) = x \text{ with } x \in U_\alpha, f \in F$$

The definition (7.1) of a fibre bundle E over X needs some further discussion to show how it reflects the intuitive notions introduced above. First let us have in mind a specific bundle: the Möbius strip shown in Fig. 7.2 with some of its bundle structure displayed.

In Fig. 7.2 we see that E is the whole Möbius strip, that the base space X is S^1, and that the fibre F is a line segment. Two sample fibres $\Pi^{-1}(x)$ and $\Pi^{-1}(y)$ over the points x and y of S^1 are shown, the dotted lines indicating the action of the projection Π. A typical open set U_α of X is depicted together with its inverse image $\Pi^{-1}(U_\alpha)$; then one follows the homeomorphism ϕ_α, and allows the diagram to show how ϕ_α untwists $\Pi^{-1}(U_\alpha)$ into the product $U_\alpha \times F$. We have not yet described the group G. G arises in considering the transition from one set of local coordinates given by ϕ_α, U_α say to another set given by ϕ_β, U_β. Suppose then that U_α and U_β overlap, $U_\alpha \cap U_\beta \neq \phi$. Then $\phi_\alpha \circ \phi_\beta^{-1}$ is a continuous invertible map of the form:

$$\phi_\alpha \circ \phi_\beta^{-1} : (U_\alpha \cap U_\beta) \times F \to (U_\alpha \cap U_\beta) \times F \tag{7.2}$$

Now let $x \in U_\alpha \cap U_\beta$ and $f \in F$, then fix x and only allow f to vary, the map $\phi_\alpha \circ \phi_\beta^{-1}$ for fixed x is now just a map from F to F. Let us denote this map by $g_{\alpha\beta}(x)$, $g_{\alpha\beta}(x)$ is called a transition function and is a homeomorphism of the fibre F. The set of all these homeomorphisms for all choices of U_α, ϕ_α form a group and this is the group G. G is called the structure group of the fibre bundle E. The group G for the Möbius strip is the group

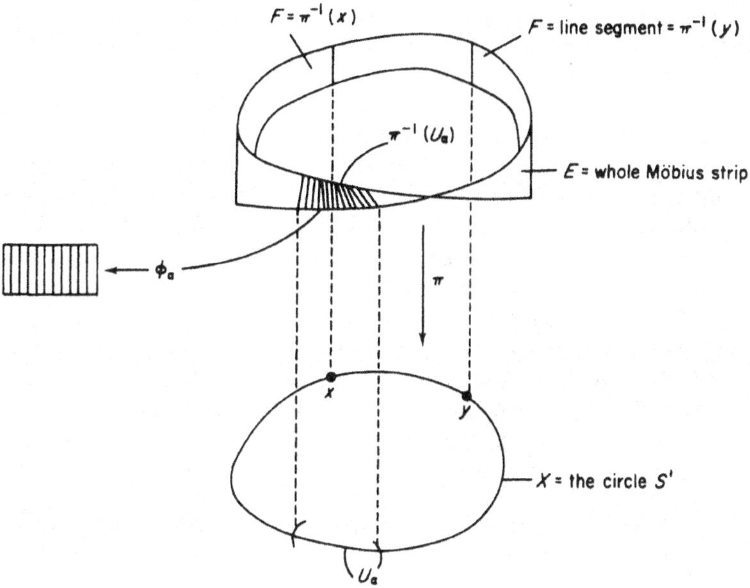

Figure 7.2

$\{e, g\}$ on two elements. To see this one must be given the open covering $\{U_\alpha\}$ of coordinate neighbourhoods specified in part (v) of (7.1). The covering contains only two elements U_1 and U_2 where U_1 and U_2 are both open arcs of S^1 chosen so that they overlap. Thus on drawing a picture for oneself one sees that $U_1 \cap U_2$ is made up of two *disjoint* open arcs A and B say. We will find that $g_{\alpha\beta}(x)$ is given by:

$$g_{12}(x) = \begin{cases} e & \text{if } x \in A \\ g & \text{if } x \in B \end{cases}$$

$$g_{21}(x) = g_{21}^{-1}(x)$$

$$g_{11} = g_{22} = e \tag{7.3}$$

In the terms of Fig. 7.2 we can detect the presence of the group $G = \{e, g\}$ by moving a fibre F once round the Möbius strip. If we do this we find that F has been reflected in its midpoint, this reflection, R say, is a manifestation of the group element g; clearly R^2 = identity, so that we end up with the group $\{e, g\}$ of just two elements. The point is that if one moves once round the Möbius strip once, then one must have passed through A and B once. Then in changing from coordinates in A to coordinates in B

(or vice versa), one must use the transition function g_{12} or g_{21}, and hence the fibre F is acted on by the group element g; or in geometrical language F is reflected about its midpoint.

If one compares the definition of a fibre bundle E with the definition of a manifold M, one sees that there are similarities between the transition functions $g_{\alpha\beta}(x)$ of a bundle E, and the maps $\phi_\beta \circ \phi_\alpha^{-1}$ used to change coordinates on a manifold M. A manifold is locally like $\mathbf{R}^n$, a fibre bundle is locally like a product. In fact much of the structure of a manifold is determined by giving the covering $\{U_\alpha\}$ and the maps $\phi_\beta \circ \phi_\alpha^{-1}$. For fibre bundles a similar situation obtains: if one is given X, the transition functions $g_{\alpha\beta}(x)$, the fibre F, and the group G, then one can reconstruct the bundle E to which these objects belong. Let us see how this happens. A glance at the Definition (7.1) shows that if we are given X, $g_{\alpha\beta}(x)$, F, and G, then we need to find ϕ_α, Π, and E. First of all E will be constructed by applying an equivalence relation to the set $\tilde{E}$ where $\tilde{E}$ consists of all products of the form $U_\alpha \times F$ or more formally

$$\tilde{E} = \bigcup_\alpha U_\alpha \times F \tag{7.4}$$

An element of $\tilde{E}$ is written (x, f), $x \in U_\alpha$. The equivalence relation is denoted by $\sim$ and is defined as follows: let $(x, f) \in U_\alpha \times F$ and $(x', f') \in U_\beta \times F$, then

$$(x, f) \sim (x', f')$$

if

$$x = x' \quad \text{and} \quad g_{\alpha\beta}(x) f = f' \tag{7.5}$$

E is simply the set of all equivalence classes under this equivalence relation, symbolically

$$E = \tilde{E}/\sim \tag{7.6}$$

If we denote the equivalence class containing $(x, f) \in U_\alpha \times F$, by $[(x, f)]$, then we can now define the projection Π by

$$\begin{aligned} \Pi &: E \to X \\ [(x, f)] &\mapsto x \end{aligned} \tag{7.7}$$

Finally, the function ϕ_α is defined by defining its inverse ϕ_α^{-1}, c.f. (7.8)

$$\begin{aligned} \phi_\alpha^{-1} &: U_\alpha \times F \to \Pi^{-1}(U_\alpha) \\ (x, f) &\mapsto ([x, f)] \end{aligned} \tag{7.8}$$

This definition automatically satisfies $\Pi\phi^{-1}(x, f) = x$ as it should. We have hence constructed the bundle whose transition functions are $g_{\alpha\beta}(x)$. In practise it is usual when thinking about a bundle, and when doing calculations, to use the transitions functions $g_{\alpha\beta}(x)$—thus from now on we shall nearly always specify a bundle by giving only X, F, G and the transition functions $g_{\alpha\beta}(x)$.

In order to understand clearly the reconstruction process of a bundle from its transition functions, we shall work through the appropriate steps for the Möbius strip. Recall the material that we are given is: $X = S^1$, F = a line segment, $G = \{e, g\}$ where g acts on F by reflecting the line segment about its midpoint, and $\{U_\alpha\}$ together with $g_{\alpha\beta}(x)$. The covering $\{U_\alpha\}$ consists of two overlapping open arcs of S^1 so that $U_1 \cap U_2 = A \cup B$, where A and B are two *disjoint* open arcs. The transition functions are, as before, given by

$$g_{11}(x) = g_{22}(x) = e$$

$$g_{12}(x) = g_{21}^{-1}(x) = \begin{cases} e & \text{if } x \in A \\ g & \text{if } x \in B \end{cases} \tag{7.9}$$

The object is to construct the bundle E. Well E is constructed by taking equivalence classes. In particular

$$(x, f) \sim (x', f')$$

if we have

$$(x, f) \in U_\alpha \times F,\ (x', f') \in U_\beta \times F$$

satisfying

$$x = x' \quad \text{and} \quad g_{\alpha\beta}(x) f = f' \tag{7.10}$$

We only need to consider $g_{\alpha\beta} = g_{12} = g$. In that case $g_{\alpha\beta}(x) f = f'$ becomes

$$f = f' \quad \text{if } x \in A$$

and

$$gf = f' \quad \text{if } x \in B \tag{7.11}$$

Thus if $x \in A$ then the equivalence class $[(x, f)]$ containing (x, f) contains only the single element (x, f). However, if $x \in B$ then $([x, f)]$ has just two elements: (x, f) and (x, gf). With the aid of the pictures in Fig. 7.3, we can complete our account of the construction of E. Figure 7.3a shows $X = S^1$ split into two, so as to display U_1, U_2 and the intersection regions A and

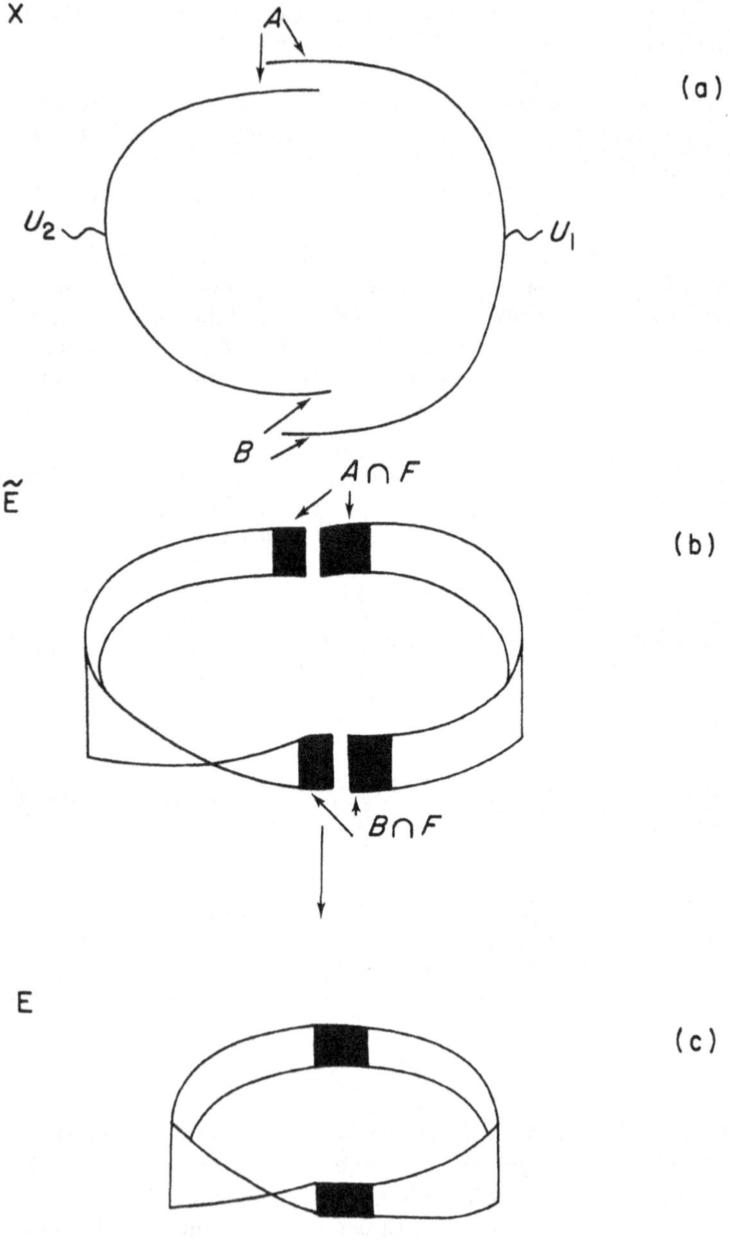

Figure 7.3

B. Figure 7.3b shows $\tilde{E}$ or the Möbius strip E split into two; the shaded regions will be glued together as Fig. 7.3c indicates. Note that the regions labelled $A \cap F$ are glued together *without* a twist; this corresponds to the equivalence class $[(x, f)]$ having only the single element (x, f) if $x \in A$, i.e. $g_{12}(x) = e$ if $x \in A$. In contrast though note that the regions labelled $B \cap F$ will be glued together *with* twist, i.e. $g_{12}(x) = g$ if $x \in B$. This corresponds to the equivalence class $[(x, f)]$ having two elements if $x \in B$. Now these elements are (x, f) and (x, gf), where g is reflection in the midpoint of the fibre F. Since the equivalence relation *identifies* these two elements, that is why we must glue them together; the way to do this for all $x \in B$ and all $f \in F$ is to glue the shaded regions together with the twist as shown. Finally Fig. 7.3c shows the bundle E completely assembled, the shaded regions are shown glued together so that they are now only isomorphic to $A \cap F$ and $B \cap F$.

Having gone through all this detailed discussion of the content of the definition of a bundle, there are two properties of the transition functions $g_{\alpha\beta}(x)$ which we have yet to mention. The first property follows immediately from the definition of $g_{\alpha\beta}(x)$ and is the set of relations listed in (7.12) (no summation over repeated indices).

$$
\begin{aligned}
g_{\alpha\alpha}(x) &= e, \; x \in U_\alpha \\
g_{\alpha\beta}(x) &= (g_{\beta\alpha}(x))^{-1}, \; x \in U_\alpha \cap U_\beta \\
g_{\alpha\beta}(x) g_{\beta\gamma}(x) &= g_{\alpha\gamma}(x), \; x \in U_\alpha \cap U_\beta \cap U_\gamma
\end{aligned}
\tag{7.12}
$$

These are often called compatibility relations. The second property concerns a certain lack of uniqueness concerning the $g_{\alpha\beta}(x)$. In other words there will be more than one set of transition function $g_{\alpha\beta}(x)$ giving rise to the same bundle E. To see how this is so consider the following set of circumstances: take two bundles E and E' with the same base space, fibre, and group. Let their coordinates and coverings be given by $\{\phi_\alpha, U_\alpha\}$ and $(\psi_\alpha, U_\alpha\}$ respectively. Further *require* that the map $\phi_\alpha \circ \psi_\alpha^{-1} : U_\alpha \times F \to U_\alpha \times F$ (which when restricted to a fixed $x \in U_\alpha$ is a homeomorphism $\lambda_\alpha(x)$ say of the fibre F), gives rise to a homeomorphism $\lambda_\alpha(x)$ of F *which belongs to the structure group* G. Given this set of circumstances we find immediately that the corresponding transition functions $g_{\alpha\beta}(x)$ and $g'_{\alpha\beta}(x)$ are related by:

$$
g'_{\alpha\beta}(x) = \lambda_\alpha^{-1}(x) g_{\alpha\beta}(x) \lambda_\beta(x) \qquad x \in U_\alpha \cap U_\beta \tag{7.13}
$$

This is immediate since by definition

$$
\lambda_\alpha(x) = \phi_\alpha \circ \psi_\alpha^{-1}(x), \qquad g_{\alpha\beta}(x) = \phi_\alpha \circ \phi_\beta^{-1}(x) \qquad g'_{\alpha\beta}(x) = \psi_\alpha \circ \psi_\beta^{-1}(x) \tag{7.14}
$$

Thus the RHS of (7.13)

$$
\begin{aligned}
&= \{\phi_\alpha \circ \psi_\alpha^{-1}(x)\}^{-1} \circ \{\phi_\alpha \circ \phi_\beta^{-1}(x)\} \circ \{\phi_\beta \circ \psi_\beta^{-1}(x)\} \\
&= \psi_\alpha \circ \psi_\beta^{-1}(x) \qquad (7.15) \\
&= g'_{\alpha\beta}(x)
\end{aligned}
$$

The insistance that $\lambda_\alpha(x)$ belong to the structure group G means that as $g_{\alpha\beta}(x)$ varies $\lambda_\alpha^{-1}(x)g_{\alpha\beta}(x)\lambda_\beta(x)$ generates all elements of G and thus that $g'_{\alpha\beta}(x)$ does so too. Such bundles as E and E' differ only in their assignments ϕ_α and ψ_α of coordinates, and are topologically the same. We therefore say that bundles E and E' with the same base space, fibre, and group are equivalent if their transition functions $g_{\alpha\beta}(x)$ and $g'_{\alpha\beta}(x)$ are related by an equation of the form of (7.13). Thus when in the rest of this book we use the term bundle we actually mean an equivalence class of bundles, however, in conformity with standard practise we shall not continually use the term equivalence class. It is important to remember though that given a transition matrix $g_{\alpha\beta}(x)$ for a bundle E, that $g_{\alpha\beta}(x)$ may be multipled on the left and on the right by two maps according to (7.13), without changing the bundle. When we discuss Yang-Mills gauge theories we shall see that this freedom to change $g_{\alpha\beta}(x)$ is the freedom to make a gauge transformation.

In the equivalence of the bundles E and E' discussed above the coordinates are given by $\{\phi_\alpha, U_\alpha\}$ and $\{\psi_\alpha, U_\alpha\}$ respectively. Notice that the coverings U_α of the base space are the same in both cases, this of course is not necessary. If the coordinates were given by $\{\phi_\alpha, U_\alpha\}$ and $\{\psi_\alpha, V_\alpha\}$, then we shall say that E and E' are equivalent if the homeomorphism $\phi_\alpha \circ \psi_\beta^{-1}(x)$ coincides with an element of the structure group G for $x \in U_\alpha \cap V_\beta$. (An analogue of the formula (7.13) is a little untidy since we would have to consider the four sets U_α, U_β, V_α and V_β simultaneously; this is why we only state the equivalence in this slightly more general case.)

7.3 MORE EXAMPLES OF BUNDLES

A whole class of examples of bundles is discovered when one works with the tangent spaces $T_p(M)$ of a differentiable manifold. For each manifold M there is a bundle $T(M)$ called the tangent bundle where

$$T(M) = \bigcup_{p \in M} T_p(M) \qquad (7.16)$$

The base space of $T(M)$ is the manifold M, the fibre at any point $p \in M$ is the tangent space $T_p(M)$. The projection Π is defined by:

$$\begin{aligned} \Pi: T(M) &\to M \\ \mathbf{V} \in T_p(M) &\mapsto p \end{aligned} \tag{7.17}$$

The fibre $T_p(M)$ at the point p is a vector space of dimension n ($n = \dim M$), and clearly all the fibres are copies of the vector space $\mathbf{R}^n$. Now to exhibit the local product structure of $T(M)$ we have to produce a homeomorphism $\phi_\alpha : \Pi^{-1}(U_\alpha) \to U_\alpha \times \mathbf{R}^n$. We can do this by using the local coordinates, x^i say, of a point $p \in M$. In particular, let $p \in U_\alpha$, $U_\alpha \subset M$; let the coordinates in $\mathbf{R}^n$ for U_α be x^i, and let $\mathbf{V} \in \Pi^{-1}(U_\alpha)$ be a tangent vector at p. Then if we express $\mathbf{V}$ in terms of these coordinates we get

$$\mathbf{V} = a^i(p) \frac{\partial}{\partial x^i}\bigg|_p \tag{7.18}$$

This means that ϕ_α can be defined as given in (7.19) below:

$$\begin{aligned} \phi_\alpha : \Pi^{-1}(U_\alpha) &\to U_\alpha \times \mathbf{R}^n \\ \mathbf{V} &\mapsto (p, a^i(p)) \end{aligned} \tag{7.19}$$

The coordinates $(p, a^i(p))$ demonstrate the local product structure required. The structure group of $T(M)$ is easy to identify, it is G, where $g = Gl(n, \mathbf{R})$ the group of real $n \times n$ invertible matrices. Because the fibres of $T(M)$ are vector spaces, the action of an element of $Gl(n\ \mathbf{R})$ on an element of the fibre is just the usual action of a matrix on a vector. T(M) is an example (the classic example) of a vector bundle. In fact a fibre bundle E is said to be a vector bundle if its fibre F is a vector space. We have only dealt so far with real manifolds so that the vector space F is $\mathbf{R}^n$. In general, the fibre of a vector bundle could be a complex vector space $\mathbf{C}^n$. An example of this is $T(M)$ where M is a complex analytic manifold such as the Riemann sphere. A vector bundle E is said to be real of rank n if $F \simeq \mathbf{R}^n$ and complex of rank n if $F \simeq \mathbf{C}^n$. The transition matrix $g_{\alpha\beta}(p)$ of $T(M)$ is easily worked out: let U_α and U_β be overlapping open sets in M with local coordinates x^i_α and x^i_β respectively. Then one can express $\mathbf{V}$ in terms of either of these coordinates†

$$\mathbf{V} = \begin{cases} a^i_\alpha(p) \dfrac{\partial}{\partial x^i_\alpha}\bigg|_p \\ a^i_\beta(p) \dfrac{\partial}{\partial x^i_\beta}\bigg|_p \end{cases}, \quad p \in U_\alpha \cap U_\beta \tag{7.20}$$

† α and β are of course just labels, not tensorial indices.

One can then verify from the Definition (7.19) of ϕ_α that $g_{\alpha\beta}(p)$ is the map given in (7.21) below.

$$\begin{aligned} g_{\alpha\beta}(p) &: \mathbf{R}^n \to \mathbf{R}^n \\ a^i_\alpha(p) &\mapsto a^i_\beta(p) \end{aligned} \tag{7.21}$$

But from the chain rule for partial derivatives it follows that

$$a^i_\alpha = \frac{\partial x^i_\alpha}{\partial x^j_\beta} a^j_\beta \tag{7.22}$$

i.e. that $g_{\alpha\beta}(p)$ is none other than the Jacobian matrix for the change of variables

$$g_{\alpha\beta}(p) = \left[\frac{\partial x^i_\alpha}{\partial x^j_\beta}\right] \tag{7.23}$$

where $\partial x^i_\alpha/\partial x^j_\beta$ is the i, j^{th} entry in the $n \times n$ matrix $g_{\alpha\beta}(p)$. The compatibility relations are immediately satisfied with this form for $g_{\alpha\beta}(p)$.

Another important bundle arising from a manifold M is the cotangent bundle $T^*(M)$. This bundle is constructed analogously to the tangent bundle T(M) except that the fibre at each point $p \in M$ is the cotangent space $T^*_p(M)$. $T^*(M)$ is again a vector bundle, however, its fibres $T^*_p(M)$ are vector spaces of covariant vectors (1-forms), whereas the fibres $T_p(M)$ of $T(M)$ are vector spaces of contravariant vectors.

The transition matrix $g_{\alpha\beta}(p)$ of $T^*(M)$ is found by a similar method to that used for $T(M)$. One expresses a 1-form ω in terms of two sets of local coordinates in the usual way:

$$\omega = \begin{cases} a^\alpha_i(p)\,\mathrm{d}x^i_\alpha \\ a^\beta_i(p)\,\mathrm{d}x^i_\beta, \end{cases} \qquad p \in U_\alpha \cap U_\beta \tag{7.24}$$

Then using the chain rule we have

$$\begin{aligned} a^\alpha_i\,\mathrm{d}x^i_\alpha &= a^\beta_i\,\mathrm{d}x^i_\beta \\ &= a^\beta_i \frac{\partial x^i_\beta}{\partial x^j_\alpha}\mathrm{d}x^j_\alpha \end{aligned} \tag{7.25}$$

Thus

$$g_{\alpha\beta}(p) = \left[\frac{\partial x^j_\beta}{\partial x^i_\alpha}\right]_{n\times n} \tag{7.26}$$

where $\partial x^j_\beta/\partial x^i_\alpha$ is the i, j^{th} entry in the $n \times n$ matrix $g_{\alpha\beta}(p)$. Notice that the transition matrix for $T^*(M)$ is the transpose of the inverse of the transition matrix for $T^*(M)$. We shall see later that this leads to $T(M)$ and $T^*(M)$ being equivalent bundles. Continuing along these lines there is the tensor bundle $T^a_b(M)$ whose fibre at p is the space $\mathbf{T}^a_b$ of tensors of type (a, b) at p. The space $\mathbf{T}^a_b$ being given by

$$\mathbf{T}^a_b = T_p(M)\otimes \ldots \otimes T_p(M) \otimes T^*_p(M) \otimes \ldots \otimes T^*_p(M) \tag{7.27}$$

where there are a factors of $T(M)$ and b factors of $T^*(M)$. The transition matrix $g_{\alpha\beta}(p)$ for $T^a_b(M)$ is the tensor product of a factors of $J_{\alpha\beta}(p)$ and b factors of $(J^{-1}_{\alpha\beta})^T$, where $J_{\alpha\beta}(p)$ is the transition matrix for $T(M)$. Note that the fibre F of $T^a_b(M)$ is $\mathbf{R}^{n^{a+b}}$. This means that the transition matrix $g_{\alpha\beta}(p) \in Gl(n^{a+b}, \mathbf{R})$. However, because of the special form of $g_{\alpha\beta}(p)$ it is clear that the structure group G of $T^a_b(M)$ is only a subgroup H of $Gl(n^{a+b}, \mathbf{R})$ rather than the whole of $Gl(n^{a+b}, \mathbf{R})$. This subgroup H, is of course isomorphic to $Gl\ (n, \mathbf{R})$.

Bundle	Base space	Fibre	Structure group	Transition functions
E Mobius/strip 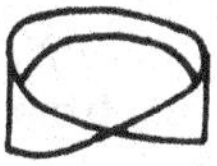	X Circle 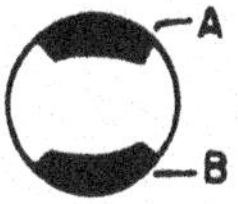	F line segment	G the group $\{e, g\}$ on two elements	$g_{\alpha\beta}(x)$ $g_{12}(x) = g^{-1}_{21}(x)$ $= \begin{cases} e, & x \in A \\ g, & x \in B \end{cases}$ otherwise $g_{\alpha\beta}(x) = e$
Tangent bundle $T(M)$	Differential manifold M	$T_p(M) \simeq \mathbf{R}^n$	$Gl(n, \mathbf{R})$	$\left[\dfrac{\partial x^i_\alpha}{\partial x^j_\beta}\right]_{n\times n} = J_{\alpha\beta}(p)$
Cotangent bundle $T^*(M)$	M	$T^*_p(M) \simeq \mathbf{R}^n$	$Gl(n, \mathbf{R})$	$\{J^{-1}_{\alpha\beta}(p)\}^T$
Tensor bundle $T^a_b(M)$	M	$T^b_a(M) \simeq \mathbf{R}^{n^{a+b}}$	A certain subgroup H of $Gl(n^{a+b}, \mathbf{R})$ $H \simeq Gl(n, \mathbf{R})$	$\otimes^a J_{\alpha\beta}(p)$ $\otimes^b \{J^{-1}_{\alpha\beta}(p)\}^T$
Frame bundle $F(M)$	M	all ordered bases of $T_p(M)$ $\Big\} \simeq Gl(n, \mathbf{R})$	$Gl(n, \mathbf{R})$	$J_{\alpha\beta}(p)$

(7.28)

Our last example of a bundle in this section is called the frame bundle $F(M)$. The base space of this bundle is, as the notation suggests, the manifold M. The fibre at the point $p \in M$ is the collection of all ordered bases or frames of the tangent space $T_p(M)$. Because all frames can be reached by acting with an element of $Gl(n, \mathbf{R})$ on one fixed frame, then the fibre F is $Gl(n, \mathbf{R})$ itself. A bundle such as this one where the fibre F is the structure group is called a principal fibre bundle. $F(M)$ is thus our first example of a principal bundle. We shall see that principal bundles play a very important part in deciding on the triviality of any bundle, principal or otherwise. Before going on in our discussion let us make a list of the bundles treated so far (see (7.28)).

7.4 WHEN IS A BUNDLE TRIVIAL?

A primary objective in working with bundles must be to acquire the ability to recognize when a bundle E is trivial. It turns out that the triviality of a bundle (E, Π, F, G, X) can always be answered by examining a certain principal bundle associated with E. To this end then, we introduce this principal bundle, and give the criteria on which depends the triviality of E.

Given any bundle E with fibre F, structure group G, and base space X, together with its transition functions $g_{\alpha\beta}(x)$; one can always construct from E, a principal bundle. One takes the same base space X, and the same transition functions $g_{\alpha\beta}(x)$, but one replaces the fibre F by the group G itself. Then one constructs the principal bundle, which we denote by $P(E)$, using the construction procedure given in (7.4 to 7.8). For example, if $E = T(M)$ then $P(E) = F(M)$; or if $E =$ Möbius strip, then $P(E)$ is the bundle depicted in Fig. 7.4.

Figure 7.4 shows P (Möbius strip) with the projection Π and the base space S^1. In fact P (Möbius strip) can be seen to give a double covering of the circle S^1; the fibres above an arbitrary pair of points in S^1 are shown in Fig. 7.4 and they consist of two points each, this corresponds to the fact that the fibre is the group $G = \{e, g\}$.

$P(E)$ is called the principal bundle associated with E. Then the triviality of both E and $P(E)$ can be answered by the following theorem:

$$\begin{gathered} P(E) \textit{ and } E \textit{ are trivial if and only if} \\ P(E) \textit{ has a section} \end{gathered} \tag{7.29}$$

We have not used the term section before, so we now define it. A cross-section, or simply, a section of a bundle E is a continuous map

$$s : X \to E$$

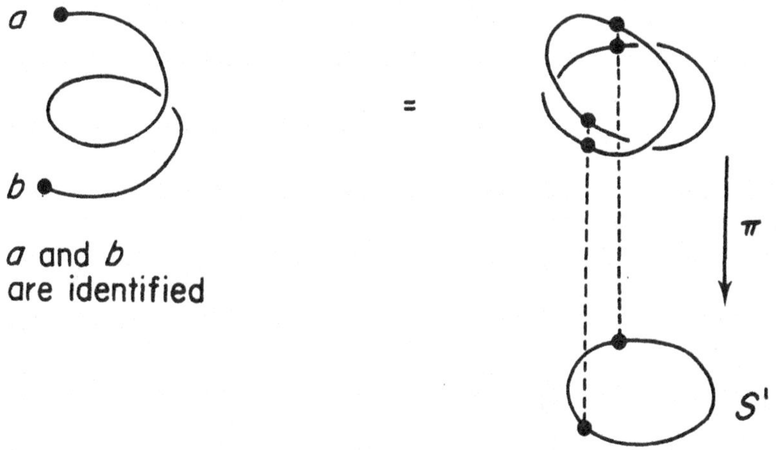

Figure 7.4

satisfying

$$\Pi s(x) = x, \qquad x \in X \tag{7.30}$$

In other words a section $s: X \to E$ is a continuous undoing of the work of the projection $\Pi: E \to X$. *Now the existence of a section s for a principal bundle P(E) implies the triviality of P(E) and E.* (The proof of the converse is so short and easy that we leave it to the reader.) The way one proves this is to use s to construct a homeomorphism between $P(E)$ and $X \times G$—i.e. we show that $P(E)$ is globally a product of its fibre G and its base space X. The construction is as follows: because $P(E)$ is a principal bundle $s(x) \in G$, also if g is any element of G then $gs(x)$ belongs to the fibre over the point $x \in X$. Further, all elements of the fibre over x are of the form $gs(x)$, for some $g \in G$. Now if we let x vary, since a bundle is the union of all its fibres, then we see that all elements $e \in P(E)$ can be written as $gs(x)$ for some $g \in G$ and some $x \in X$. Thus the desired homeomorphism from $P(E)$ to $X \times G$ is given by:

$$\begin{aligned} \phi &: P(E) \to X \times G \\ e &= gs(x) \mapsto (x, g) \end{aligned} \tag{7.31}$$

It is a straightforward matter to check the continuity and invertibility of ϕ, thus ϕ is a homeomorphism, and $P(E)$ is trivial. It now follows that E is also trivial. This is because, by the definition of $P(E)$, E and $P(E)$ have the *same* transition functions, and by looking at the transition functions one can always tell whether or not a bundle is trivial. We now explain how this is done. We claim that if a bundle E is trivial then its transition functions $g_{\alpha\beta}(x)$ can always be written as, using the notation of (7.13)

$$g_{\alpha\beta}(x)=\lambda_\alpha^{-1}(x)\lambda_\beta(x) \tag{7.32}$$

where, as in (7.13), $\lambda_\alpha(x)$ is a homeomorphism of the fibre F belonging to the structure group G. Then by reference to (7.13) we see that this means that the alternative transition function $g'_{\alpha\beta}(x)=\mathrm{id}$ (id denotes the identity map on F), describes the same bundle. But if the transition functions of a bundle E are just the identity on F, then it is immediate that E is a product and therefore trivial—in simpler language, if $g'_{\alpha\beta}(x)=\mathrm{id}$, then the fibres F are glued together with no twist.

It remains to prove (7.32) for a trivial bundle. Well, if a bundle E is trivial then there is a (global) homeomorphism

$$\phi: E \to X \times F \tag{7.33}$$

Then if $\{\phi_\alpha, U_\alpha\}$ are the usual coordinates for E, we can choose $\lambda_\alpha(x)=\phi\circ\phi_\alpha^{-1}(x)$, for a fixed x $\lambda_\alpha(x)$ is a homeomorphism of F belonging to the structure group G. Now we simply note that

$$\lambda_\alpha(x)g_{\alpha\beta}(x)\lambda_\beta^{-1}(x)=\mathrm{id} \tag{7.34}$$

for the LHS of (7.34) is

$$\{\phi\circ\phi_\alpha^{-1}(x)\}\circ\{\phi_\alpha\circ\phi_\beta^{-1}(x)\}\circ\{\phi\circ\phi_\beta^{-1}(x)\}^{-1}=\mathrm{id} \tag{7.35}$$

Hence

$$g_{\alpha\beta}(x)=\lambda_\alpha^{-1}(x)\lambda_\beta(x) \tag{7.36}$$

and we have proved (7.32). The form (7.32) is an important one for the transition function, we should remember then that if $g_{\alpha\beta}(x)$, for some bundle E, factorizes according to (7.32), then E is trivial. Finally, we repeat that $P(E)$ trivial $\Rightarrow E$ trivial, since they have the same transition functions; and also that $P(E)$ is trivial $\Leftrightarrow$ it has a section s.

We can easily see, as our intuition has already led us to believe, that a Möbius strip is a non-trivial bundle. This is done by simply examining Fig. 7.5 below.

The point is that if θ is a local coordinate on S^1, then we must have $s(\theta)=s(\theta+2\pi)$ (actually we should not really allow local coordinates on S^1 to exceed the value 2π, c.f. (2.8, 9), but we ignore this detail here). But if we study $P(E)$ in Fig. 7.5, we see that because $P(E)$ is a double covering of S^1, we cannot have $s(\theta)=s(\theta+2\pi)$ unless s jumps (i.e. is discontinuous). Thus no continuous section s of $P(E)$ exists and therefore $P(E)$ and E are

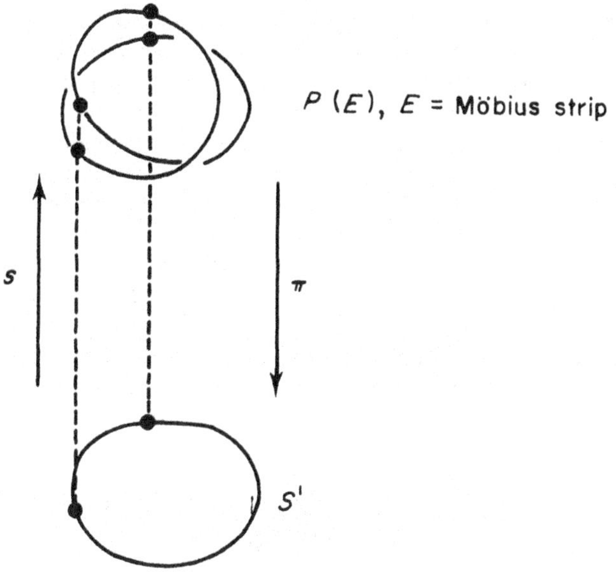

Figure 7.5

non-trivial†. It is also important to realize that even though $P(E)$ does not have a section, there is nothing to prevent E itself having a section. We display a section of the Möbius strip in Fig. 7.6.

In Fig. 7.6, A is identified with A' and B with B', also the dotted line aa' is the centre of the strip. By examining Fig. 7.6 we can see that provided the distance ab is the same as the distance $a'b'$, then the curve drawn will have matching end-points b and b' when the Möbius strip is assembled. Any such curve for which $ab = a'b'$ is a section of the Möbius strip. Thus

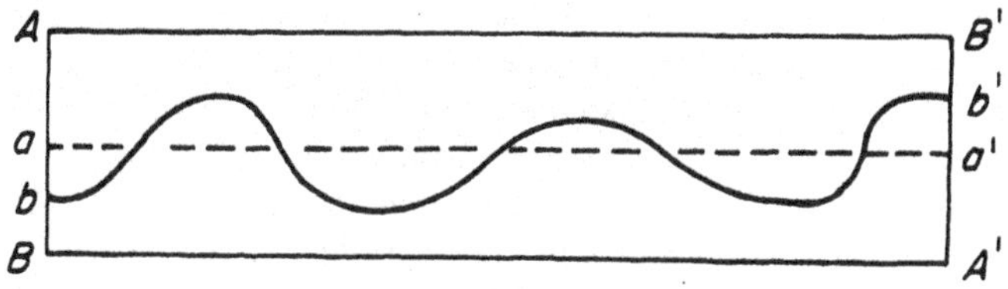

Figure 7.6

† The non-triviality of E and $p(E)$ can also be seen in this case by demonstrating that $g_{12}(x)$ does not factorize into the form $\lambda_1^{-1}(x)\lambda_2(x)$ given by (7.36).

one must remember that it is only the existence of a section in a principal bundle $P(E)$ that determines the triviality or otherwise of $P(E)$ and E. E itself may have a section†, as does the Möbius strip and yet still be non-trivial.

We learn from this discussion that the principal bundle $P(E)$ has an importance which raises it above the bundle E. The extra structure that the bundle E may carry when compared with $P(E)$ can be seen to be a kind of mathematical disguise. The triviality of E rests entirely on a result expressed in terms of $P(E)$ only—namely the existence of a section of $P(E)$. Finally, we point out that there may be several bundles E, E', E'' say, all with the same principal bundle i.e. $P(E)=P(E')=P(E'')$; their principal bundles will coincide if they simply have in common their bases spaces, their transition functions, and their structural groups, but not their fibres.

7.5 SECTIONS OF BUNDLES AND SINGULARITIES OF VECTOR FIELDS

Let us have a look at the intuitive content of a section of some of the bundles that we have introduced so far.

A section of $T(M)$ is a continuous (or indeed C_∞) assignment of an element of the fibre $T_x(M)$ to each point $x \in M$. Since an element of $T_x(M)$ is simply a contravariant vector, then a section s of M is none other than a continuous contravariant vector field on M. In the same way a section of $T^*(M)$ is a covariant vector field on M, or it is a 1-form $\boldsymbol{\omega}$ where $\boldsymbol{\omega} \in \Omega^1(M)$. A section of $T^a_b(M)$ is a tensor field on M.

We know from our general mathematical knowledge that some of these vector and tensor fields have zeros on M, and that their zeros play an important rôle in geometry and topology. One can see a little of this rôle emerging when one considers what it means for the frame bundle $F(M)$ to have a section: a section s of $F(M)$ is a continuous assignment of a basis to $T(M)$. To be more explicit $s(x)$ is a basis for $T_x(M)$ the tangent space at x, and as x varies $s(x)$ varies continuously thus providing a bases for $T_x(M)$ for any $x \in M$. Thus the passage from a basis of $T_x(M)$ to a basis of $T_{x'}(M)$ can be made in a continuous manner for any $x, x' \in M$. The connection that the existence of $s(x)$ has with singularities of vector fields is developed in what follows.

† We shall not include the adjective continuous before the word section but we shall intend that all sections be continuous. Further when the base spaces are, as they usually will be, differentiable manifolds, the bulk of our mathematics will be true if we replace continuous maps by C^∞ maps.

Suppose one succeeds in finding one continuous vector field $\mathbf{V}(x)$ with no zeros on M, then let us ask for more: let us now ask for two vector fields, $\mathbf{V}(x)$ and $\mathbf{W}(x)$ say, such that $\mathbf{V}(x)$ and $\mathbf{W}(x)$ are linearly independent on M. Note that the generalization, for two vector fields, of a single vector field being non-vanishing on M, is that the two vector fields be linearly independent on M—a zero, also called a singular point, of the set $\{\mathbf{V}(x), \mathbf{W}(x)\}$ is a point of linear dependence. This process can be continued until one asks what is the largest number p of linearly independent vector fields $\mathbf{V}(x) \dots \mathbf{V}_p(x)$ that one can have on M. The answer is that the largest value that p can have is n where $n = \dim M$. However, usually $p < n$. If p does attain its maximum value n then $\mathbf{V}_1(x), \dots, \mathbf{V}_n(x)$ form a basis of $T_x(M)$, and $F(M)$ has a section s given by

$$\begin{aligned} s: M &\to F(M) \\ x &\mapsto \{\mathbf{V}_1(x), \dots, \mathbf{V}_n(x)\} \end{aligned} \tag{7.37}$$

When $p = n$ the manifold M is called parallelizable. Thus the statements: M is parallelizable†, and $F(M)$ has a section, mean the same thing. Now since a vector field on M is a section of $T(M)$, then we see that the existence of a section of the principal bundle $F(M)$, is equivalent to the existence of *n sections* of the bundle $T(M)$ *linearly independent at each point* $x \in M$.

Thus one can rephrase the parallelizability of a manifold M or the existence of a section of $F(M)$ in terms of the singularities of vector fields on M. Parallelizable manifolds are the exception rather than the rule. Nevertheless, we can readily exhibit examples of parallelizable manifolds: a Lie group itself may be regarded as a differentiable manifold, when this is done one can see that a Lie group G is always parallelizable. This is because if $\mathfrak{g}$ denotes the Lie algebra of G, then $\mathfrak{g}$, (regarded only in its rôle of being a vector space rather than an algebra), is isomorphic to the tangent space $T_e(G)$ where e is the identity of G. Further if $\{X_1, \dots, X_n\}$ is a basis for $\mathfrak{g}$, i.e. a frame at e, then this frame at e can be continuously mapped onto a frame at any other point $s \in G$ by the action of the group element s. More specifically the group multiplication allows one, given the element s say, to translate elements $t \in G$ to the left or the right, i.e. to obtain st and ts respectively. Under left translation by s say, the basis $\{X_1, \dots, X_n\}$ of $T_e(G)$ is mapped to a basis of $T_s(G)$ which we denote by $\{s \cdot X_1, \dots, s \cdot X_n\}$. In this way we obtain, in a continuous manner, a frame at any point $s \in G$ and hence a section of $F(G)$; thus G is parallelizable.

† The section s provides an isomorphism between all the tangent spaces $T_x(M)$ of M. This isomorphism can be used to define a notion of parallelism between vectors belonging to different tangent spaces. This is the origin of the word parallelizability.

The Lie group $U(1)$ of complex numbers has as its manifold S^1 thus S^1 is parallelizable. The product $U(1)\times U(1)\ldots\times U(1)$ (n factors) is a Lie group with manifold $S^1\times S^1\ldots\times S^1$ (n factors). This manifold is the torus T^n, thus T^n is parallelizable.

However, if one takes the manifold S^n for arbitrary n and asks when it is parallelizable, the answer is hardly ever. In fact using the generalized cohomology theory called K-theory [2], Adams [1] has shown that the only parallelizable spheres are S^1, S^3, and S^7: one first shows using K-theory that the maximum number p of linearly independent vector fields on S^n is given by

$$p=2^a+8b-1, \qquad 0<a\leqslant 3, \qquad b\geqslant 0 \quad a, b, \text{integers} \tag{7.38}$$

Then we use elementary reasoning to write the integer n as

$$2^{a+4b}(2c+1)-1 \tag{7.39}$$

with a and b as before and $c\geqslant 0$, c an integer. If S^n is parallelizaable then $p=n$, i.e.

$$2^a+8b=2^{a+4b}(2c+1) \tag{7.40}$$

It is then immediate to check that the only solutions to this equation are given by $c=b=0$ and $a=1, 2, 3$ so that $n=1, 3$ or 7 as claimed above. The parallelizability of S^1, S^3 and S^7 has to do with the existence of the complex numbers, the quaternions and the Cayley numbers respectively: a complex number may be written as an ordered pair of real numbers, then the unit complex numbers are topologically a circle S^1; a quaternion may be written as an ordered pair of complex numbers, the unit quaternions (isomorphic incidentally to the group $SU(2)$) are topologically an S^3; finally the Cayley numbers may be written as an ordered pair of quaternions; the unit Cayley numbers are topologically an S^7—however, here this construction procedure ends. For more details c.f. reference [16].

To conclude this section, we expand a little on our remark in the middle of the section on the rarity of parallelizable manifolds. To the extent that Lie groups are plentiful then parallelizable manifolds are also plentiful; however given an arbitrary manifold M it is most unlikely that it is the manifold of some Lie group and hence parallelizable. Indeed we have seen that given a simple class of manifolds such as $M=S^n$, $n=1, 2, \ldots$ then the only parallelizable manifolds are S^1, S^3 and S^7. It is in this latter sense that parallelizable manifolds are the exception rather than the rule.

7.6 CUTTING A BUNDLE DOWN TO SIZE: REDUCTION OF THE GROUP AND CONTRACTION OF THE BASE SPACE

We shall see in this section that many bundles E are too big in the sense that they may be replaced by smaller bundles without losing the topological information contained in them. There are two main ways in which one can make a bundle smaller: one can make the fibre smaller, or one can make the base space smaller, or perhaps one can do both. The key notion in both cases is that of homotopy.

We deal first with the base space, and we begin by quoting the result and then outline its proof. The result is that if the base space X of a bundle E is contractible to a point then E is always trivial—clearly a bundle over a point is trivial so the result means that the topological nature of a bundle is unchanged by contracting the base space.

To use the simple homotopy arguments that provide the proof of this result it is necessary to introduce a bundle called the induced bundle or the pullback bundle. We shall use the term pullback bundle. If one has a bundle F over a base space Y, and one also has a continuous map f from another space X to Y, then the pullback argument enables one to construct a certain bundle E over X. The construction of E is completely straightforward: let the coordinates neighbourhoods for Y be V_i and the transition functions for the bundle F be $g_{ij}(Y)$, $y \in V_i \cap V_j$; then the coordinate neighbourhoods of the new bundle E are defined to be $f^{-1}V_i$ which we denote by U_i, and the transition functions of E are defined to be $g_{ij}(f(x))$ which we denote by $g'_{ij}(x)$. Now since we have defined both the coordinates neighbourhoods and the transition functions of the bundle E the definition of E is complete, and E is constructed as described in (7.4 to 7.8). In summary

$$\begin{aligned} U_i &= f^{-1}V_i \\ g'_{ij}(x) &= g_{ij}(f(x)) \end{aligned} \tag{7.41}$$

and we can also draw the commutative diagram displayed in Fig. 7.7.

In Fig. 7.7 the pullback bundle E is denoted by f^*F and its projection by Π_f. The map $f_*F \to F$ is induced by the map f and the construction procedure for f^*F given above†.

† The full title of f^*F is the pullback of F by f. The reader will notice that we used the notation $f^*\omega$ in a similar context when discussing forms, $f^*\omega$ can also be called a pullback form.

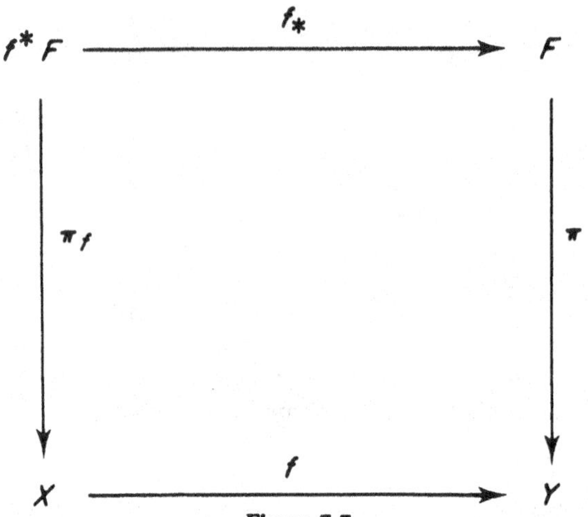

Figure 7.7

In fact it is a simple exercise for the reader to check that the expressions giving f^*F, f_* and Π_f explicitly are:

$$
\begin{aligned}
f^*F &= \{\text{those } (x,f)\times X\times F \text{ satisfying } \Pi(f)=f(x)\}\\
f_* &: f^*F \to F\\
&\quad (x,f)\to f\\
\Pi_f &: f^*F \to X\\
&\quad (x,f)\to x
\end{aligned}
\tag{7.42}
$$

The expressions in (7.42) also demonstrate the commutativity of the diagram in Fig. 7.7. Homotopy now enters with the following plausible result which we state but do not prove, a straightforward proof may be found in reference [16].

Let F be a bundle over Y and let f^*F and g^*F be two pullbacks of F both over the same space X. Then if the maps $f: X\to Y$ and $g: X\to Y$ are homotopic the bundles f^*F and g^*F are equivalent, c.f. Fig. 7.8.

If $f\simeq g$ then f^*F and g^*F are equivalent bundles. Now we specialize to the case where $X=Y$ and where E is a bundle over a contractible base space X. Contractibility of X to the point x_0 means that the maps α_0 and α_1 defined in (7.43) are homotopic

$$
\begin{aligned}
\alpha_0 &: X\to X\\
&\quad x\to x\\
\alpha_1 &: X\to X\\
&\quad x\to x_0
\end{aligned}
\tag{7.43}
$$

X is contractible to x_0 means $\alpha_1\simeq\alpha_0$.

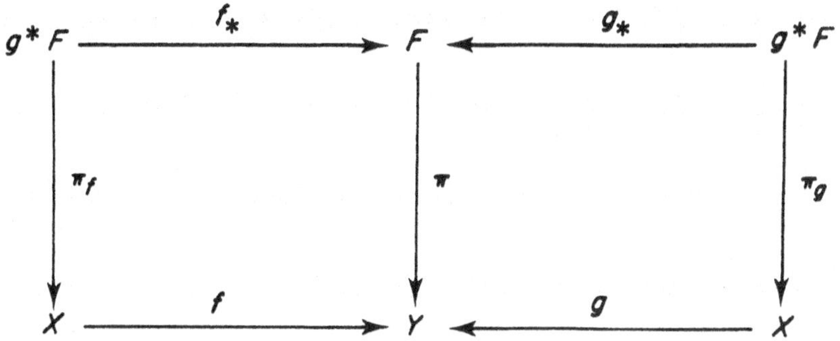

Figure 7.8

Evidently α_1 is the identity map and α_0 is the constant map. Our result says, for this pair of maps, that the bundles $\alpha_0^* E$ and $\alpha_1^* E$ are equivalent. However, $\alpha_1^* E$ is a bundle over the single point x_0 and is therefore trivial, on the other hand $\alpha_0^* E = E$ since α_0 is the identity, thus E itself is trivial. Thus any bundle over a contractible base space X is trivial. Since spaces like $\mathbf{R}^n$ are contractible then one concludes that the simplest non-contractible spaces to choose for the base spaces of bundles are the spheres S^n. Even for these spaces a bundle may still be trivial, for we have seen in our discussion of parallelizability that the frame bundles $F(S^1)$, $F(S^3)$ and $F(S^7)$ are trivial. This result may also be of use in reducing the size of the base space X. For example if X is a cylinder, i.e. $X = S^1 \times L$ where L is a line segment, then although X itself is not contractible part of it is, namely the line segment L. Thus the base space X may be reduced to S^1 without altering the bundle E, more precisely one can produce a map $f: S^1 \to S^1 \times L$ such that the pullback bundle f^*F over S^1 is equivalent to the original bundle E over $S^1 \times L$.

We now turn to the structure group G of a bundle E. An analogous result [16] helps us here: it is that if the fibre F of a bundle E is contractable then E always has a section S. The utility of this result is that if E is a principal bundle so that the fibre F coincides with the structure group G, then the existence of the sections implies the triviality of the bundle. *Thus if either X or G are contractible then the bundle E is trivial.* Let us look at some of the consequences of G being contractible. We take a specific bundle namely the principal bundle $F(M)$, the group is therefore $Gl(n, \mathbf{R})$. This group is actually not contractible but it does have a contractible piece: any $A \in Gl(n, \mathbf{R})$ may be written [6]

$$A = OP \tag{7.44}$$

where O is an $n\times n$ orthogonal matrix, i.e. $O\in O(n)$, and P is a positive definite symmetric matrix, in terms of sets we have:

$$Gl(n, \mathbf{R})=O(n)\times C \tag{7.45}$$

where C is the set of all positive definite symmetric matrices (note C is neither a vector space nor a group). The set C is contractible so that the group $Gl(n, \mathbf{R})$ may be contracted down to the group $O(n)$. Let us first show that C is contractible and then discuss the consequences of its contractibility. If C were a vector space it would be automatically contractible since a vector space of dimension n is homeomorphic to $\mathbf{R}^n$ which is contractible. In fact C, although not a vector space, is homeomorphic to the vector space S of all symmetric matrices†.

To see this let $s\in S$ then it is immediate to check that the matrix $\exp[s]\in C$, i.e. $\exp[s]$ is positive definite as well as symmetric; conversely a standard diagonalization argument of linear algebra shows that all $c\in C$ are of the form $\exp[s]$ for some $s\in S$. Thus the map α defined in (7.46) below can be seen to be a homeomorphism from S to C, since S is a vector space it is contractible, thus C is contractible.

$$\begin{aligned}\alpha: S&\rightarrow C\\ s&\mapsto\exp[s]\end{aligned} \tag{7.46}$$

Now we discuss the consequences of this result. First of all the result means the (non-compact) structure group $Gl(n, \mathbf{R})$ of $F(M)$ may be replaced by, or reduced to, the (compact) subgroup $O(n)$. This is a considerable simplification. Secondly, from the point of view of geometry we have shown that every manifold M always admits a Riemannian metric. This second point needs some further elaboration: A Riemannian metric on M is a continuous assignment of a positive definite symmetric matrix $g(x)$, $x\in M$ to M. (In bundle language we could also say that it is a positive definite symmetric section of the tensor bundle $T^2_0(M)$.) The existence of $g(x)$ can be thought of as being equivalent to the provision of a inner product‡ for vectors in the tangent space $T_x(M)$; this inner product moreover will be invariant under $O(n)$. Given this statement the reduction of the group $Gl(n, \mathbf{R})$ to $O(n)$ amounts to a continuous assignment of an orthogonal frame at each point $x\in M$ i.e. to the provision of an inner product in $T_x(M)$ at each $x\in M$. This reduction of the group $Gl(n, \mathbf{R})$ of $F(M)$ to the compact subgroup $O(n)$ generalizes and this generalization is of some interest to

† We remind the reader that the matrices in all these examples are all real.

‡ The inner product being defined by $\langle\mathbf{A}, \mathbf{B}\rangle=g_{ij}A^iB^j$ where $\mathbf{A}=A^i(\partial/\partial x^i)$, $\mathbf{B}=B^j(\partial/\partial x^j)$ belong to $T_x(M)$.

us: the reduction of $Gl(n, \mathbf{R})$ to $O(n)$ is simply the provision of a Riemannian metric for M, however it may be that $Gl(n, \mathbf{R})$ may be reducible to some other subgroup, G say, of $Gl(n, \mathbf{R})$. When this occurs M is said to have a G-structure. Further commonly encountered examples of G-structures are provided when the dimension n of the manifold M is even. These examples, of which both have non-compact G, are given by:

$$\text{(i)} \quad G = Sp(n, \mathbf{R}), \quad n \text{ even} \tag{7.47}$$

in this case M is said to have an almost Hamiltonian or symplectic structure ($Sp(n, \mathbf{R})$ is the group which leaves invariant a certain antisymmetric quadratic form and is also sometimes written as $Sp(n/2, \mathbf{R})$; we shall define $Sp(n, \mathbf{R})$ shortly).

$$\text{(ii)} \quad G = Gl(n/2, \mathbf{C}), \quad n \text{ even} \tag{7.48}$$

here $Gl(n/2, \mathbf{C})$ mean the group of complex $n/2 \times n/2$ invertible matrices which can be regarded as a subgroup of the group $Gl(n, \mathbf{R})$ of real $n \times n$ invertible matrices. In this case M is said to have an almost complex structure. We shall provide a short discussion of symplectic and almost complex structures later in this chapter.

To return to the case where $G = O(n)$ we see that the reduction of $Gl(n, \mathbf{R})$ to $O(n)$ has consequences for the bundles $T(M)$ and $T^*(M)$. The transition matrices $g_{\alpha\beta}(x)$ for these bundles now take their values in $O(n)$ rather than $Gl(n, \mathbf{R})$. Since the transition matrices for $T^*(M)$ are the transpose inverse of those for $T(M)$, and since these matrices now lie in $O(n)$, we see that their transition matrices are identical, thus the bundles $T(M)$ and $T^*(M)$ are equivalent bundles. Appropriate simplifications also apply to the tensor bundles $T^a_b(M)$ for the various values of a and b.

The reduction of $Gl(n, \mathbf{R})$ to a subgroup G is a special case of a more general result. More generally if one has any principal bundle $P(E)$ with fibre G, rather than the present case $P(E) = Gl(n, \mathbf{R})$ and fibre $Gl(n, \mathbf{R})$, then if G is a connected Lie group then [8]

$$G = H \times D \tag{7.49}$$

where H is a maximal compact subgroup of G and D is topologically a Euclidean (and hence contractible) space. Thus G may be reduced to the *compact* subgroup H resulting in a much simpler principal bundle $P'(E)$ say, which is of course equivalent to the bundle $P(E)$ with the larger fibre G. An example of this is given by taking an n-dimensional complex manifold so that the frame bundle $F(M)$ has fibre $Gl(n, \mathbf{C})$. Then in analogy with (7.45) there is the result

$$Gl(n, \mathbf{C}) = U(n) \times C \tag{7.49}$$

where $U(n)$ is the usual unitary group and C is the set of all positive definite Hermitian matrices. As before C is contractible and the group $Gl(n, \mathbf{C})$ can be reduced to the maximal compact subgroup $U(n)$.

We have learned then that contractibility of X or G for a bundle E implies its triviality. This leads to the simplest choices of X being S^n; while for G more commonly one contracts down to a compact subgroup, which for $F(M)$, leads to a demonstration that a manifold always admits a Riemannian metric, it leads also to the more general idea of G-structures explained above. The reduction of $Gl(n, \mathbf{R})$ to $O(n)$ is not a typical example of a G-structure in the following sense: $Gl(n, \mathbf{R})$ is always reducible to $O(n)$ because of (8.45), however, $Gl(n, \mathbf{R})$ is not always reducible to $Sp(n, \mathbf{R})$ or $Gl(n/2, \mathbf{C})$—the reduction in these two latter cases is only possible when some global topological conditions consisting of relations satisfied by characteristic classes are fulfilled. It is this requirement of satisfying certain global topological conditions that is typical of G-structures, the case when $G = O(n)$ is simply fortunate. For example when we change from Riemannian geometry to pseudo-Riemannian geometry so that the metric is no longer positive definite, then the corresponding reduction is from $Gl(n, \mathbf{R})$ to $O(n-p, p)$ where $O<p<n$ and $(n-p, p)$ is the signature of the metric†. For some manifolds this reduction is not possible, and hence such manifolds may not admit a non-positive definite metric at all, or they may admit metrics of signature $(n-p, p)$ for only certain positive p. For example take $M = S^2$, then the only value of p is $p = 1$. We can then make use of a result given in reference [10]: if M is compact and 2-dimensional and admits a metric of signature $(1, 1)$ then either $M = T^2$ (torus) or $M =$ the Klein bottle. Thus S^2 does not admit a pseudo-Riemannian metric even though it admits an infinite number of Riemannian metrics. The topological restriction for the case of two dimensional or indeed n dimensional compact manifolds with $p = 1$ is the requirement that the Euler number χ be zero [16].

Another global topological condition which outlaws some G-structures is simply that M be non-orientable. This condition prevents $Gl(n, \mathbf{R})$ being reduced to $G = SL(n, \mathbf{R})$ the group of $n \times n$ matrices of unit determinant—a G-structure with $G = SL(n, \mathbf{R})$ would provide an oriented frame at each point of M, and thus endow M with an orientation. Finally, we should recall that even the requirement that the dimension n of M be even, which

† The group $O(n-p, p)$ is the group of $n \times n$ matrices which the non-positive definite quadratic form $x_1^2 + x_2^2 + \ldots + x_{n-p}^2 - x_{n-p+1}^2 - \ldots - x_n^2$ invariant. In pseudo-Riemannian geometry the signature of the metric is usually defined as the difference of the number of positive and negative eigenvalues of the metric. Thus the signature is $n-2p$, also if p is 1, then the metric is called a Lorentz metric.

is needed for the cases $G = Sp(n, \mathbf{R})$ and $G = Gl(n/2, \mathbf{C})$, is a topological requirement since the dimension n of M is a topological invariant.

7.7 REMARKS ON ALMOST HAMILTONIAN AND ALMOST COMPLEX STRUCTURES

We deal first with almost Hamiltonian structures. Therefore we must define the group $Sp(n, \mathbf{R})$ introduced above. Let n be even, $n = 2m$, then take the space $\mathbf{R}^{2m}$ and write the coordinates of a vector $x \in \mathbf{R}^{2m}$ in the following way: $\mathbf{x} = (x_1, \ldots, x_m, x'_1, \ldots, x'_m)$. The group $Sp(n, \mathbf{R})$ can then be defined as the (non-compact) subgroup of $Gl(2m, \mathbf{R})$ which leaves invariant the antisymmetric bilinear† form $A(\mathbf{x}, \mathbf{y})$ given in (7.50) below

$$\begin{aligned} A(\mathbf{x}, \mathbf{y}) &= (x_1 x'_1 - x'_1 y_1) + (x_2 y'_2 - x'_2 y_2) + \ldots + (x_m y'_m - x'_m y_m) \\ &= \mathbf{x}^T J \mathbf{y} \end{aligned} \tag{7.50}$$

where J is the matrix given in (7.51)

$$J = \begin{bmatrix} 0 & I_m \\ -I_m & 0 \end{bmatrix}_{2m \times 2m}$$

with

$$I_m = \begin{bmatrix} 1 & & & 0 \\ & 1 & & \\ & & \ddots & \\ 0 & & & 1 \end{bmatrix}_{m \times m} \tag{7.51}$$

To obtain a symplectic or almost Hamiltonian structure one first needs a $2m$-dimensional manifold M. Then, having noted that the matrix J above is antisymmetric, one looks for a 2-form $\boldsymbol{\omega}$ on M. One does this because the coefficients ω_{ij} say of the 2-form provide an anti-symmetric matrix $[\omega_{ij}]_{2m \times 2m}$ on $T^*_x(M)$. Now if one insists that Det $[\omega_{ij}]_{2m \times 2m}$ be non-zero, and also one normalizes the value of this determinant to be unity; then, with respect to a suitable basis for $T^*_x(M)$, $[\omega_{ij}]_{2m \times 2m}$ may always [15] be written in the standard form J of (7.51). A 2-form $\boldsymbol{\omega}$ with non-zero determinant is called non-degenerate. If the 2-form $\boldsymbol{\omega}(x)$ is non-degenerate for all $x \in M$ then M has an almost Hamiltonian structure. This existence of the non-degenerate 2-form $\boldsymbol{\omega}(x)$ provides us with a section of the

---

† The adjective bilinear is used rather than the adjective quadratic, because the term quadratic is reserved for the case when $\mathbf{x} = \mathbf{y}$, which here would be zero and hence trivial.

covariant tensor bundle $T_2^0(M)$. So if we choose bases for $T_x^*(M)$ in the way indicated above, then one reduces the group $Gl(2m, \mathbf{R})$ of $F(M)$ to $Sp(2m, \mathbf{R})$ in a way exactly analogous to the reduction of $Gl(n, \mathbf{R})$ to $O(n)$ already described. An almost Hamiltonian structure is called a Hamiltonian structure (the manifold M is then also called a symplectic manifold or a Hamiltonian manifold) if $\boldsymbol{\omega}$ satisfies:

$$\mathrm{d}\boldsymbol{\omega} = O \tag{7.52}$$

This integrability condition $\mathrm{d}\boldsymbol{\omega} = \mathbf{O}$ guarantees the *local* existence of bases which reduce $[\omega_{ij}]_{2m\times 2m}$ to the standard form J—without the condition these bases can only be introduced separately for each point $x \in M$. A simple example of a Hamiltonian structure is provided by taking $M = S^2$, and by taking $\boldsymbol{\omega}$ to be the volume element of S^2. Because dim $M = 2$ then $\mathrm{d}\boldsymbol{\omega} = 0$, and because $\boldsymbol{\omega}$ is the volume element then $\boldsymbol{\omega}$ is a 2-form and is non-degenerate. However, the standard way that Hamiltonian structures arise is the following: take an n-dimensional manifold M with n odd or even, then construct the cotangent bundle $T^*(M)$. $T^*(M)$ itself, although it is a vector bundle, can also be considered to be a differentiable manifold. When this is done one automatically obtains a manifold of dimension $2n$—the base space has dimension n and so does the fibre. Let $q_1, \ldots, q_n$ be local coordinate for M and let $p_1, \ldots, p_n$ be coordinates in the fibre, then local coordinates for $T^*(M)$ are simply $q_1, \ldots, q_n, p_1, \ldots, p_n$. The choice of p_i and q_i for these coordinates is suggested by classical mechanics. If M is the configuration space for some classical dynamical system with n degrees of freedom, then $T^*(M)$ is the classical phase space for this dynamical system, hence its coordinates are $q_1, \ldots, q_n, p_1, \ldots, p_n$. Further the desired 2-form $\boldsymbol{\omega}$ is immediately available in a canonical way—simply set

$$\boldsymbol{\omega} = \mathrm{d}\boldsymbol{\kappa} \tag{7.53}$$

where κ is the 1-form on $T^*(M)$ given locally by

$$\boldsymbol{\kappa} = p_i dq^i \tag{7.54}$$

Since $\boldsymbol{\omega}$ is exact it automatically satisfies $\mathrm{d}\boldsymbol{\omega} = \mathbf{0}$. Also computing $\boldsymbol{\omega}$ from (7.53) gives

$$\boldsymbol{\omega} = \mathrm{d}p^1 \wedge \mathrm{d}q^1 + \ldots + \mathrm{d}p^n \wedge \mathrm{d}q^n \tag{7.55}$$

Thus on checking the bookkeeping of indices one sees that the matrix $[\omega_{ij}]_{2n\times 2n}$ is already in the form J of equation (7.51). So given any classical dynamical system one automatically has a Hamiltonian structure. One simply takes the classical phase space $T^*(M)$ and considers it as a $2n$-dimensional manifold with local coordinates $q_1, \ldots, q_n, p_1, \ldots, p_n$ rather than a bundle, $\boldsymbol{\omega}$ is then chosen to be $\mathrm{d}\boldsymbol{\kappa}$ where $\boldsymbol{\kappa} = p_i \mathrm{d}q^i$.

To find examples of global topological obstructions to having Hamiltonian and almost Hamiltonian structures is relatively easy. Firstly, a non-orientable $2m$-dimensional manifold M can never have an almost Hamiltonian structure. This is because a change in the orientation of the local coordinates $x_1, \ldots, x_m, x'_1, \ldots, x'_m$ of M would change the sign of the bilinear form $A(\mathbf{x}, \mathbf{y})$ of (7.50). Thus the Möbius strip and the real projective spaces P^{2m} can never have almost Hamiltonian structures.

Secondly, examples of such manifolds are the spheres $S^4, S^6, S^8, \ldots$. To see this assume that $\boldsymbol{\omega}$ is a non-degenerate 2-form on S^{2m}, $m \geqslant 2$, and

$$\mathrm{d}\boldsymbol{\omega} = 0 \tag{7.56}$$

Then (7.56) guarantees that locally

$$\boldsymbol{\omega} = \mathrm{d}p^1 \wedge \mathrm{d}q^1 + \ldots \mathrm{d}p^m \wedge \mathrm{d}q^m \tag{7.57}$$

where local coordinates on S^{2m} are $(q_1, \ldots, q_m, p_1, \ldots, p_m)$. Now note that $\boldsymbol{\theta} = \boldsymbol{\omega} \wedge \boldsymbol{\omega} \ldots \wedge \boldsymbol{\omega}$ (m times) is a $2m$-form on S^{2m} and is in fact a volume element on S^{2m}. To be a volume element on S^{2m} it merely has to never have a zero, i.e. be of fixed sign. But, in local coordinates in some neighbourhood on S^{2m}

$$\begin{aligned}\boldsymbol{\theta} &= (\mathrm{d}p^1 \wedge \mathrm{d}q^1 + \ldots \mathrm{d}p^m \wedge \mathrm{d}q^m) \wedge \ldots \wedge (\mathrm{d}p^1 \wedge \mathrm{d}q^1 + \ldots \mathrm{d}p^m \wedge \mathrm{d}q^m) \\ &= (m!)\, \mathrm{d}p^1 \wedge \mathrm{d}q^1 \wedge \mathrm{d}p^2 \wedge \mathrm{d}q^2 \ldots \wedge \mathrm{d}p^m \wedge \mathrm{d}q^m\end{aligned} \tag{7.58}$$

In any other overlapping neighbourhood on S^{2m} with local coordinates $\tilde{q}_1, \ldots, \tilde{q}_m, \tilde{p}_1, \ldots, \tilde{p}_m$ we have

$$\boldsymbol{\theta} = (\mathrm{Det\ f})m!\, \mathrm{d}\tilde{p}^1 \wedge \mathrm{d}\tilde{q}^1 \wedge \mathrm{d}\tilde{p}^2 \wedge \mathrm{d}\tilde{q}^2 \ldots \wedge \mathrm{d}\tilde{p}^m \wedge \mathrm{d}\tilde{q}^m \tag{7.59}$$

Where Det f is the Jacobian of the transformation for the change of variables. But since S^{2m} is orientable Det f is always positive thus $\boldsymbol{\theta}$ never changes sign it merely gets multiplied by a positive number. Thus we assert that

$$\int \boldsymbol{\theta} \neq 0 \tag{7.60}$$

However, (7.60) must be false. This is due to the fact that since, as we showed in chapter 6, $H^2(S^{2m}, \mathbf{R}) = 0$, $m \geqslant 2$, then all closed 2-forms on S^{2m}, $m \geqslant 2$ are exact hence

$$\boldsymbol{\omega} = \mathrm{d}\boldsymbol{\kappa} \tag{7.61}$$

for some $\boldsymbol{\kappa}$. But this means that $\boldsymbol{\theta}$ is exact, for (7.61) implies that

$$\begin{aligned}\boldsymbol{\theta} &= \boldsymbol{\omega} \wedge \boldsymbol{\omega} \wedge \ldots \wedge \boldsymbol{\omega} \\ &= \mathrm{d}\boldsymbol{\rho}\end{aligned}$$

where

$$\rho = \kappa \wedge \omega \wedge \ldots \wedge \omega \tag{7.62}$$

Thus

$$\int_{S^{2m}} \theta = \int_{S^{2m}} \mathrm{d}\rho = 0 \tag{7.63}$$

using Stokes' theorem and the fact that S^{2m} has no boundary. Thus the assumption that such an ω existed is false. Further we see that to prevent this argument applying in other cases one should require any compact manifold M which is a candidate for having a Hamiltonian structure to have a non-zero second cohomology group $H^2(M; \mathbf{R})$. Note that the classical phase space example above of $T^*(M)$ is a non-compact manifold—non-compact because the p_i coordinates being the coordinates of a vector space are unbounded. The non-compactness prevents us making the above argument since there is nothing wrong with the volume element of a non-compact manifold being exact. Indeed, it always is, for we saw in Chapter 6 that $H^n(M; \mathbf{R}) = 0$ for an n-dimensional non-compact manifold.

Finally, we show the connection of Hamiltonian structures with the Hamiltonian equations of classical mechanics [17]. Let H be the Hamiltonian of a classical dynamical system with n degrees of freedom. Then consider the vector field $\mathbf{X}(t)$ where, in terms of local coordinates,

$$\mathbf{X}(t) = a^i(t)\frac{\partial}{\partial q^i} + b^i(t)\frac{\partial}{\partial p^i} \tag{7.64}$$

and we choose

$$\begin{aligned} a^i &= \dot{q}^i \\ b^i &= \dot{p}^i \end{aligned} \tag{7.65}$$

Also let ω be the 2-form $\mathrm{d}\kappa$ of equation (7.61). Now the contraction of ω with $\mathbf{X}$, which we shall denote by $\mathbf{X} \lrcorner\, \omega$, is a 1-form. We have therefore

$$\mathbf{X} \lrcorner\, \omega = c_i \mathrm{d}q^i + \mathrm{d}_i \mathrm{d}p^i$$

where

$$\begin{aligned} c_i &= \omega_{ij} a^j \\ d_i &= \omega_{ij} b^j \end{aligned} \tag{7.66}$$

But since ω defines a Hamiltonian structure then

$$[\omega_{ij}]_{2n \times 2n} = \begin{bmatrix} 0 & I_n \\ -I_n & 0 \end{bmatrix}_{2n \times 2n} \tag{7.67}$$

We immediately obtain

$$\begin{aligned} c_i &= b^i \\ d_i &= -a^i \end{aligned} \tag{7.68}$$

Thus

$$\mathbf{X} \lrcorner \boldsymbol{\omega} = \dot{p}^i \mathrm{d}q^i - \dot{q}^i \mathrm{d}p^i \tag{7.69}$$

On the other hand, Hamiltonian's equations are [17]

$$\begin{aligned} \dot{q}^i &= \frac{\partial H}{\partial p^i} \\ \dot{p}^i &= -\frac{\partial H}{\partial q^i} \end{aligned} \tag{7.70}$$

So

$$\mathbf{X} \lrcorner \boldsymbol{\omega} = -\frac{\partial H}{\partial q^i} \mathrm{d}q^i - \frac{\partial H}{\partial p^i} \mathrm{d}p^i \tag{7.71}$$

But the RHS of (7.71) is simply $\mathrm{d}H$ and we can write

$$\mathbf{X} \lrcorner \boldsymbol{\omega} = -\mathrm{d}H \tag{7.72}$$

(this is in fact an expression which is valid everywhere despite our use of local coordinates).

Now we compute the $\mathbf{X}H$ the action of the vector $\mathbf{X}$ on the function H;

$$\begin{aligned} \mathbf{X}H &= \dot{q}^i \frac{\partial H}{\partial q^i} + \dot{p}^i \frac{\partial H}{\partial p^i} \\ &= 0 \end{aligned} \tag{7.73}$$

using Hamilton's equations (7.70). Thus H is constant along the flow of $\mathbf{X}$. (A curve $\phi(t)$ on a manifold M induces a vector field $\mathbf{X}(t)$ on M by defining $\mathbf{X}(t)$ to be always tangent to this curve. When a curve bears this relationship to a vector field it is said to be an integral curve of $\mathbf{X}$. The collection of these curves form a one parameter family of mappings from M to M and is called a flow.) The constancy of H along the flow is simply the conservation of energy, other quantities such as linear momentum and angular momentum will also be constant along the flow. Finally, the volume element $\boldsymbol{\theta} = \boldsymbol{\omega} \wedge \ldots \wedge \boldsymbol{\omega}$ (n-times) is simply the volume element of classical phase space and is also an invariant of the flow. In fact the canonical transformations of classical dynamics [17] which also preserve $\boldsymbol{\theta}$ are simply the symplectic transformations $Sp(2m, \mathbf{R})$ which preserve $\boldsymbol{\omega}$ and hence $\boldsymbol{\theta}$.

Thus the symplectic properties of the 2-form ω are an encoding of many of the properties of classical dynamics. Next we make some remarks on complex structures. First we need to describe how $Gl(m, \mathbf{C})$ may be regarded as a subgroup of $Gl(2m, \mathbf{R})$. A similarity with the remarks on almost Hamiltonian structures is that an antisymmetric matrix plays a fundamental rôle. Let J be an antisymmetric matrix as given in (7.51) then note that

$$J^2 = I \tag{7.74}$$

where I is the $2m \times 2m$ identity matrix. We can think of J loosely as a matrix representation of the square root of -1. Now consider the real vector space $\mathbf{R}^{2m}$. $\mathbf{R}^{2m}$ can be endowed with the structure of a complex vector space if we define scalar multiplication by complex numbers. This is done by taking a complex number $\lambda = \alpha + i\beta$ and a vector $\mathbf{x} \in \mathbf{R}^{2m}$ and defining $\lambda \mathbf{x}$ by the equation below

$$\lambda \mathbf{x} = \alpha \mathbf{x} + \beta J \mathbf{x} \tag{7.75}$$

(the RHS is the definition of the LHS).

Thus the matrix J plays the rôle of multiplication by i. Further, if we denote the resulting complex vector space by $\mathbf{C}^m$, and its basis by $\{\mathbf{e}_1, \ldots, \mathbf{e}_m\}$, then a basis for $\mathbf{R}^{2m}$ can immediately be seen to be $\{\mathbf{e}_1, \ldots, \mathbf{e}_m, J\mathbf{e}_1, \ldots, J\mathbf{e}_m\}$. Finally the group $Gl(m, \mathbf{C})$, regarded as a subgroup of $Gl(2m, \mathbf{R})$, is simply all matrices $M \in Gl(2m, \mathbf{R})$ which commute with J (and so do not spoil the linearity of complex scalar multiplication).

Now if M is a $2m$-dimensional manifold an almost complex structure is an equiping of the tangent spaces $T_x(M) \simeq \mathbf{R}^{2m}$ with the structure of $\mathbf{C}^m$ in the manner just described. Topological requirements for the existence of an almost complex structure on M are at least that M be even-dimensional and orientable. Thus S^{2n+1} and P^{2n} are never endowed with complex structures. If M has dimension $2m$ it may also be a complex analytic manifold of dimension m, e.g. if $m = 1$ and $M = S^2$, then M may also be regarded as the Riemann sphere which is a one-dimensional complex analytic manifold. Such complex analytic manifolds automatically endow their tangent spaces $T_x(M)$ with an almost complex structures. However, just as an almost Hamiltonian structure may not also be Hamiltonian, an almost complex structure may not also be a complex structure i.e. it may not be induced by the fact that M is a complex analytic manifold. The condition for an almost complex structure to come from a complex structure is an integrability condition the details of which we cannot pursue here, the reader is referred to references [4, 10, 18].

For information we point out the even dimensionality and orientability of M are not enough to ensure the existence of an almost complex structure on M: if we take $M = S^{2n}$, $n \geq 1$ then only S^2 and S^6 admit almost complex structures, this has to do with the fact that S^3 and S^7 are parallelisable, the reader is referred to reference [3]. Also the almost complex structure on S^6 is not integrable [3]. We summarise some of the properties of the G structures discussed in the following Table in Section 7.8.

7.8 G-STRUCTURES ON A COMPACT CLOSED MANIFOLD

Name	Subgroup G of $Gl(n, R)$	Necessary conditions for existence
Riemannian	$O(n)$	none
pseudo-Riemannian	$O(n-p, p)$	c.f. ref. 16
Lorentzian	$O(n-1, 1)$	$\chi(M)=0$
orientable	$SL(n, \mathbf{R})$	$H^n(M; \mathbf{R}) \neq 0$
almost Hamiltonian	$Sp(n, \mathbf{R})$	$n=2m, H^n(M; \mathbf{R}) \neq 0$
or symplectic (Hamiltonian)		$H^2(M; \mathbf{R}) \neq 0$
almost complex	$Gl(n/2, \mathbf{C})$	$n=2m, H^n(M; \mathbf{R}) \neq 0$
(Complex)		(c.f. refs 4, 10, 18)

(7.76)

In (7.76) the topological existence conditions are necessary but not in all cases sufficient. Also if the manifold M is *non-compact* then one can add to the list the following remarks: that

i. there are always Lorentz structures on M.
ii. classical phase space $T^*(M)$ is an important example of a *non-compact* manifold which always carries a Hamiltonian structure c.f. (7.53 to 7.55).

7.9 LIE DERIVATIVE

This section marks a return to the topic of geometry rather than topology. We have met, in Chapter 2, two kinds of differential operations: partial differentiation and exterior differentiation. We now introduce a third kind called Lie differentiation. Loosely speaking Lie differentiation is differentiation along a curve in a manifold.

More precisely let $\mathbf{X}$ be a vector field on a manifold M. Then there is a flow on M defined by the integral curves of $\mathbf{X}$—those curves to which $\mathbf{X}$ is always tangent. In local coordinates these curves are the solutions to the system of equations:

$$\frac{dx^i(t)}{dt} = X^i(x^1(t), \ldots, x^n(t)) \qquad i=1, \ldots, n \tag{7.77}$$

where $\mathbf{X}=X^i(\partial/\partial x^i)$ and t is the parameter along one of the curves. We wish to introduce the notion of differentiation along these integral curves. Consider an overlapping family of neighbourhoods indexed by t and denoted by U_t. Then many integral curves pass through the U_t, c.f. Fig. 7.9. (Fig. 7.9 shows three integral curves of the flow generated by **X** and

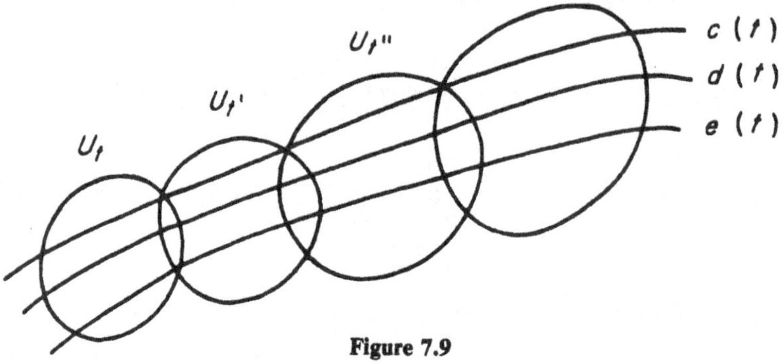

Figure 7.9

the associated one-parameter family of neighbourhoods). The integral curves and the neighbourhoods U_t together determine a one parameter family of diffeomorphism $\phi^t: U_t \to M$, where ϕ_t simply pushes a point $p \in U_t$ a parameter distance t along the flow. Given ϕ_t, the Lie derivative with respect to **X** at p of any multilinear object **T** (tensor vector etc.) is defined by

$$\lim_{t\to 0} \frac{\mathbf{T}(p)-\phi^t_*\mathbf{T}(p)}{t} \tag{7.78}$$

and is written $L_{\mathbf{X}}\mathbf{T}$. ($\phi^t_*\mathbf{T}(p)$ denotes, in accordance with the notation in Chapter 2, the tensor at $\phi^t(p)$ given rise to by the map $\phi^t: U \to M$, or simply the tensor **T** pushed a distance t along the flow.) If **T** was simply an ordinary function f then one can see immediately that

$$\phi^t_* f(p) = f(x^i(t)) \tag{7.79}$$

so that

$$\begin{aligned} L_{\mathbf{X}}f &= \frac{\mathrm{d}f}{\mathrm{d}t}(x^i(t)) \\ &= \frac{\mathrm{d}x^i}{\mathrm{d}t}\frac{\partial f}{\partial x^i} \\ &= \mathbf{X}^i \frac{\partial f}{\partial x^i} \\ &= \mathbf{X}f \end{aligned} \tag{7.80}$$

so the Lie derivative of a function f with respect to **X** is simply the vector field **X** evaluated on f. From the definitions, in particular from the properties of ϕ^t_*, it follows that $L_\mathbf{X}$ is (multi)linear and obeys an exact analogue of Leibnitz's product rule

$$L_\mathbf{X}(\mathbf{T}\otimes\mathbf{W}) = (L_\mathbf{X}\mathbf{T})\otimes W + \mathbf{T}\otimes(L_\mathbf{X}\mathbf{W}) \tag{7.81}$$

To compute $L_\mathbf{X}$ of an aribtrary tensor one thus only needs to $L_\mathbf{X}$ for a vector and a 1-form. For example if **Y** is an arbitrary vector and f is any function then from the definition of $L_\mathbf{X}$ it follows immediately that

$$L_\mathbf{X}(\mathbf{Y}\mathrm{f}) = (L_\mathbf{X}\mathbf{Y})\mathrm{f} + \mathbf{Y}L_\mathbf{X}\mathrm{f} \tag{7.82}$$

But from (7.80) we have that

$$\begin{aligned} L_\mathbf{X} &= \mathbf{X}\mathrm{f} \\ L_\mathbf{X}(\mathbf{Y}\mathrm{f}) &= \mathbf{X}\mathbf{Y}\mathrm{f} \end{aligned} \tag{7.83}$$

Thus (7.82) becomes

$$\mathbf{X}\mathbf{Y}\mathrm{f} = (L_\mathbf{X}\mathbf{Y})\mathrm{f} + \mathbf{Y}\mathbf{X}\mathrm{f}$$

so

$$\begin{aligned} (L_\mathbf{X}\mathbf{Y}) &= (\mathbf{X}\mathbf{Y} - \mathbf{Y}\mathbf{X})\mathrm{f} \\ &= [\mathbf{X}, \mathbf{Y}]\mathrm{f} \end{aligned} \tag{7.84}$$

or, in view of the arbitrariness of **Y** and *f*,

$$L_\mathbf{X} Y = [\mathbf{X}, \mathbf{Y}] \tag{7.85}$$

To find the Lie derivative of a 1-form $\boldsymbol{\omega}$ one takes another vector **Y** and considers the function $\langle\boldsymbol{\omega}, \mathbf{Y}\rangle$ and uses the equation:

$$L_\mathbf{X}\langle\boldsymbol{\omega}, \mathbf{Y}\rangle = \langle L_\mathbf{X}\boldsymbol{\omega}, \mathbf{Y}\rangle + \langle\boldsymbol{\omega}, L_\mathbf{X}\mathbf{Y}\rangle \tag{7.86}$$

From this we find at once that

$$L_\mathbf{X}\boldsymbol{\omega} = \left\{\frac{\partial\omega_i}{\partial x^j}x^j + \omega_j\frac{\partial x^j}{\partial x^i}\right\}\mathrm{d}x^i \tag{7.87}$$

where $\boldsymbol{\omega} = \omega_i\,\mathrm{d}x^i$.

In fluid mechanics in a steady flow the streamlines are the integral curves of the flow generated by the velocity vector **V**. It is common then to introduce the notion of differentiation moving with the fluid or the stream derivative of a physical quantity such as temperature, density, etc. For steady flow this is given by

$$\frac{\mathrm{df}}{\mathrm{d}t} = v^i\frac{\partial\mathrm{f}}{\delta x^i} \tag{7.88}$$

where f is the function representing the physical quantity and v^i are the three components of velocity. Evidently

$$\frac{\mathrm{df}}{\mathrm{d}t}=L_{\mathbf{v}}\mathrm{f} \tag{7.89}$$

7.10 CONNECTION AND CURVATURE

The mathematical notions of connection and curvature are fundamental to modern differential geometry. They are also becoming fundamental to modern physics. For physicists the theory of general relativity is the most familiar place to find the notions of connection and curvature used—however they are also fundamental to electromagnetic theory and Yang-Mills theories. The type of connection used in relativity differs from that used in electromagnetic and Yang–Mills theories. We shall give a general treatment of connections which allows one, at will, to specialize to the particular geometry or geometries appropriate to the particular physical problem at hand. There are a variety of different approaches to describing connections in the mathematical literature, considerations of space limit us to the use of one such approach.

We set the scene with some preliminary intuitive remarks about Riemannian connections. The subject begins when one considers some curved manifold, M say, and wishes to characterize, in some way its curvature. The fact that M is curved means that the tangent spaces $T_p(M)$, $T_{p'}(M)$ at two neighbouring points p and p' change as one moves from p to p'. A connection is essentially a structure which endows one with the ability to compare two such tangent spaces at a pair of infinitesimally separated points. The connection is given by defining what is called parallel transport or parallel translation in M. Consider $T_p(M)$ and $T_{p'}(M)$ and any curve, γ say, joining p to p'. Let $\mathbf{X}\in T_p(M)$ and let $\mathbf{X}$ be a tangent to the curve γ at p, then $\mathbf{X}$ is said to be parallely transported along the curve γ if $\mathbf{X}$ is pushed from p to p' in such a way as to always remain parallel to itself. If t is the parameter for the curve γ then the covariant derivative of $\mathbf{X}$ is the rate of change of $\mathbf{X}$ with respect to t. This covariant derivative will differ from the ordinary partial derivative, the quantity that measures this difference is the connection. Further, this covariant differentiation, being unlike partial differentiation, will not in general be commutative, the curvature is the measure of the non-commutativity of covariant differentiation. Lastly, it may be for some special curves γ, that as $\mathbf{X}$ is parallely transported along the curve, the parallel translate of $\mathbf{X}$ exactly coincides with the tangent to the curve. When this happens the curve γ is called a geodesic. The general mathematical setting for all this is a principal bundle P with fibre G over a manifold M. Now since both the Lie group G and the manifold M are differentiable structures, then the bundle P is also a differentiable manifold. Since this is so we can talk about the tangent and cotangent bundles TP and T^*P of P. Consider then TP and at any point $u \in P$. At u there is a tangent space $T_u(P)$. A sub-space of $T_u(P)$ is those vectors which are only tangent to the fibre passing through u. This subspace is called the vertical subspace at u and we denote it by $V_u(P)$. Together with this vertical subspace $V_u(P)$ there goes a horizontal subspace $H_u(P)$ defined so that

$$T_u(P) = V_u(P) \oplus H_u(P) \tag{7.90}$$

The equation (7.90) above does not define uniquely $H_u(P)$. $H_u(P)$ is made unique by considering what happens as u varies in P. We require that, as u varies, say

$$u \mapsto u' = ug, \qquad g \in G \tag{7.91}$$

(G being the group which is the fibre of P), then $H_{u'} = H_{ug}$ is related to H_u by

$$H_{ug} = R_g H_u \tag{7.92}$$

where $R_g H_u$ denotes the action on H_u of the linear map which is induced on the tangent space $T_u(P)$ by the mapping $u' = ug$ of fibre elements. In other words one can get from one horizontal space to another by the action (on the right) of the group G.

Now given these horizontal and vertical subspaces of $T_u(P)$ we define a general notion of parallel transport and the connection associated with it. One takes any curve γ in M passing through the points p and p'. The parallel transport that we are defining transports or maps the fibre $\pi^{-1}(p)$ above p onto the fibre $\pi^{-1}(p')$ above p'. The rule for parallel transport of fibres along γ from p to p' is as follows. Let $\bar{\gamma}$ be a curve in P which lies above (i.e. $\Pi(\bar{\gamma}) = \gamma$), and which is such that its tangent vectors at any point u on $\bar{\gamma}$ are horizontal, i.e. they belong to $H_u(P)$.

Given γ, and given a point $u_0 \in P$ there is only one such $\bar{\gamma}$ with initial point u_0. The curves $\bar{\gamma}$ are called horizontal lifts of γ.

Now to parallel transport a fibre from p to p' along γ, one takes each $u \in \Pi^{-1}(p)$, constructs the unique lift $\bar{\gamma}$ of γ with initial point u, and maps u to $u' \in \Pi^{-1}(p')$, where u' is the point on $\bar{\gamma}$ which lies above p'. Having defined parallel transport one can go on to construct the connection giving rise to this parallel transport. However now we stop and rest for a moment. Any normal reader can expect to be a little at sea at this point in our discussion of connections. A reader for whom this treatment of connections is new is advised to read the preceding section a couple of times before reading on, alternatively he might prefer to move on to the specific examples of connections discussed between equations 7.142 and 7.197.

It turns out that every connection can be shown to arise from a certain 1-form ω belonging to T^*P, where ω is required to have values in the Lie Algebra $\mathfrak{g}$ of G. That is to say the provision of horizontal and vertical subspaces of TP satisfying the requirements introduced above can be accomplished given this 1-form ω. We first give ω and then explain how it accomplishes the tasks required of it. Let the local coordinates in the bundle P be given by (x, g), $x \in M$, $g \in G$. Then

$$\omega = g^{-1}\mathrm{d}g + g^{-1}\mathbf{A}g \tag{7.93}$$

where $\mathbf{A} = A^a_\mu(x)(\lambda_a/2i)\,\mathrm{d}x^\mu$ and the basis $\lambda_a/2i$ satisfy in the usual way

$$\left[\frac{\lambda_a}{2i}, \frac{\lambda_b}{2i}\right] = f_{abc}\frac{\lambda_c}{2i} \tag{7.94}$$

with f_{abc} the structure constants of $\mathfrak{g}$. Now given the decomposition of $T(P)$ into vertical and horizontal subspaces†

$$T(P) = V(P) \oplus H(P) \tag{7.95}$$

since $T(P)$ has dimension $n+d$, (d is the dimension of G), and $V(P)$ has dimension d then $H(P)$ has dimension n. Now let its basis be

$$\frac{\partial}{\partial x^\mu} + C_{\mu ij}\frac{\partial}{\partial g_{ij}}, \qquad \mu = 1, \ldots, n, \quad i, j = 1, \ldots, d. \tag{7.96}$$

then $H(P)$ is specified once the $C_{\mu ij}$ are known. Well the $C_{\mu ij}$ can be specified by demanding that for any $\mathbf{X} \in TP$

$$\langle \omega, \mathbf{X} \rangle = 0$$

if $\mathbf{X} \in H(P)$. (Recall from Chapter 2 that, if $\omega = a_i dy^i$ and $\mathbf{X} = b_i\ (\partial/\partial y^i)$ then $\langle \omega, \mathbf{X} \rangle = a_i b_i$.) to see how this comes about take any $\mathbf{X} \in T(P)$ then

$$\mathbf{X} = \alpha_{ij}\frac{\partial}{\partial g_{ij}} + \beta^\mu\left(\frac{\partial}{\partial x^\mu} + C_{\mu ij}\frac{\partial}{\partial g_{ij}}\right) \tag{7.97}$$

$\dfrac{\partial}{\partial g_{ij}}$ and $\dfrac{\partial}{\partial x^\mu} + C_{\mu ij}\dfrac{\partial}{\partial g_{ij}}$ being the basis

† We now drop the suffix u on $T(P)$, $V(P)$ and $H(P)$, no confusion should result.

vectors for $V(P)$ and $H(P)$ respectively. Now when written out fully

$$\omega = g_{ij}^{-1}(dg)_{ij} + g_{ij}^{-1}A_\mu^a(x)\frac{\lambda_{aik}}{2i}\,dx^\mu g_{kj} \tag{7.98}$$

so

$$\langle\omega, \mathbf{X}\rangle = g_{ij}^{-1}(\alpha_{ij} + \beta^\mu C_{\mu ij}) + g_{ij}^{-1}A_\mu^a(x)\frac{\lambda_{aik}}{2i}\,g_{kj}\beta^\mu \tag{7.99}$$

But if $\mathbf{X} \in H(P)$ then $\langle\omega, \mathbf{X}\rangle$ is to be zero. Thus we set $\alpha_{ij} = 0$, and equating the resultant part of (7.99) to zero gives

$$g_{ij}^{-1}\beta^\mu C_{\mu ij} + g_{ij}^{-1}A_\mu^a(x)\frac{\lambda_{aik}}{2i}\,g_{kj}\beta^\mu = 0 \tag{7.100}$$

This is valid for all β^μ so

$$g_{ij}^{-1}C_{\mu ij} + g_{ij}^{-1}A_\mu^a\frac{\lambda_{aik}}{2i}\,g_{kj} = 0$$

or, multiplying by g,

$$C_{\mu ij} = -A_\mu^a\frac{\lambda_{aik}}{2i}\,g_{kj} \tag{7.101}$$

and this determines $H(P)$. The fundamental defining requirement of a horizontal subspace given in (7.92) can now be verified immediately by multiplying g on the right by some $h \in G$ and repeating the above calculation. Thus we have acquired a bona fide $H(P)$ and can go on to uncover some of its properties.

7.11 THE CONNECTION FORM AND THE GAUGE POTENTIAL

First of all we remark that $A_\mu^a(x)$ is called the connection associated with horizontal subspace; $\mathbf{A} = A_\mu^a(x)\,(\lambda^a/2i)\,dx^\mu$ is called the connection form. Under a change of bundle coordinates the 1-form ω induces a certain transformation law for the connection form $\mathbf{A}$: let the bundle coordinates change (locally) from (x, g) to (x', g'). Further, let us only make a change of fibre coordinates i.e. $x = x'$, and also let g' be given by

$$g' = hg \tag{7.102}$$

So invariance of ω means that

$$g^{-1}dg + g^{-1}\mathbf{A}g = (g')^{-1}dg' + (g')^{-1}\mathbf{A}'g' \tag{7.103}$$

But $g' = hg$ so

$$dg' = dhg + hdg \tag{7.104}$$

The RHS of (7.103) becomes therefore

$$g^{-1}h^{-1}(dhg + hdg) + g^{-1}h^{-1}\mathbf{A}'hg \tag{7.105}$$

when we equate this with the LHS of (7.103) we deduce immediately that

$$\mathbf{A} = h^{-1}dh + h^{-1}\mathbf{A}'h \tag{7.106}$$

Using the fact that $dhh^{-1} + hdh^{-1} = 0$, we unravel from this the equation

$$\mathbf{A}' = hdh^{-1} + h\mathbf{A}h^{-1} \tag{7.107}$$

This is the transformation law of the connection form† **A**. It is what is called in electromagnetism and Yang–Mills theories, the gauge transformation law. So we see that the mathematical content of what in physics is a gauge transformation is a change in fibre coordinates of a principal bundle. Physicists also call the connection $A^a_\mu(x)$ a gauge potential.

7.12 PARALLEL TRANSPORT, COVARIANT DERIVATIVE AND CURVATURE

Next, we wish to pin down the relation between parallel transport and the connection. Recall that we have defined parallel transport by lifting any curve $\gamma(t)$ in the base space of P to a unique curve $\tilde{\gamma}(t)$ in the bundle P, a vector $\mathbf{X} \in T(P)$ is parallely transported along $\tilde{\gamma}(t)$ if it always remains in $H(P)$ as it is transported along $\tilde{\gamma}(t)$. It is simple to understand this if we write down everything in local bundle coordinates: let

$$\gamma(t) = (x_\mu(t))$$

then

$$\tilde{\gamma}(t) = (x_\mu(t), g(t)) \tag{7.108}$$

the tangents to $\tilde{\gamma}(t)$ are given by

$$\frac{\mathrm{d}}{\mathrm{d}t} = \dot{x}^\mu \frac{\partial}{\partial x^\mu} + \dot{g}\frac{\partial}{\partial g} \tag{7.109}$$

---

† This transformation law can be called affine because it both translates (by the amount hdh^{-1}) and rotates **A** (according to $h\mathbf{A}h^{-1}$). The presence of the translational quantity hdh^{-1} prevents connections from being tensors. Affine transformations of $\mathbf{R}^n$ say are given by $(n\times 1)\times(n\times 1)$ matrices of the form $\begin{pmatrix} g & a \\ 0 & 1\end{pmatrix}$, where $g \in Gl(n, \mathbf{R})$ and a is the vector in $\mathbf{R}^n$ representing the translation.

These tangents will in general lie in $T(P)$, however, we are instructed to force them to lie in $H(P) \subset T(P)$, i.e. we must demand that

$$\frac{\mathrm{d}}{\mathrm{d}t} = \dot{x}^{\mu}(t)\frac{\partial}{\partial x^{\mu}} + \dot{g}(t)\frac{\partial}{\partial g} = \beta^{\mu}\left(\frac{\partial}{\partial x^{\mu}} - A^{a}_{\mu}\frac{\lambda^{a}}{2i}g\frac{\partial}{\partial g}\right) \tag{7.110}$$

for some β^{μ}. Therefore we have at once that

$$\dot{x}^{\mu}(t) = \beta^{\mu}$$

and

$$\begin{aligned} \dot{g}(t) &= -\beta^{\mu}A^{a}_{\mu}\frac{\lambda^{a}}{2i}g \\ &= -\dot{x}^{\mu}A^{a}_{\mu}\frac{\lambda^{a}}{2i}g \end{aligned} \tag{7.111}$$

Equation (7.111) amounts essentially to the first order differential equation

$$\dot{g}(t) + \dot{x}^{\mu}(t)A^{a}_{\mu}(x)\frac{\lambda^{a}}{2i}g(t) = 0 \tag{7.112}$$

This is called the parallel transport equation, for it is only when we restrict $g(t)$ to be a solution of this equation that we will obtain parallel transport for the tangents to $\bar{\gamma}(t)$. The covariant derivative D_{μ} may now be unveiled. $\dot{x}^{\mu}D_{\mu}$ is simply the rate of change with respect to t as one moves along the horizontal lift $\bar{\gamma}(t)$, i.e.

$$D_{\mu} = \frac{\partial}{\partial x^{\mu}} - A^{a}_{\mu}\frac{\lambda^{a}}{2i}g\frac{\partial}{\partial g} \tag{7.113}$$

Thus a shorthand for the basis of $H(P)$ is simply D_{μ} and any vector $\mathbf{X}$ in $T(P) = V(P) \oplus H(P)$ is given by

$$\mathbf{X} = \alpha\frac{\partial}{\partial g} + \beta^{\mu}D_{\mu} \tag{7.114}$$

We note that, as one should expect, if the connection $A^{a}_{\mu}(x)$ is zero the covariant derivative D_{μ} becomes the ordinary partial derivative ∂_{μ}. Having obtained D_{μ} we are in a position to make explicit the curvature of the connection $A^{a}(x)$. We have referred in passing above to the fact that the curvature is a measure of the lack of commutativity of covariant differentiation. Let us therefore calculate $[D_\mu, D_\nu]$. We obtain

$$[D_\mu, D_\nu] = \left[\partial_\mu - A^a_\mu \frac{\lambda^a}{2i} g \frac{\partial}{\partial g},\quad \partial_\nu - A^b_\nu \frac{\lambda^b}{2i} g \frac{\partial}{\partial g}\right]$$

$$= -\partial_\mu A^b_\nu \frac{\lambda^b}{2i} g \frac{\partial}{\partial g} + \partial_\nu A^a_\mu g \frac{\partial}{\partial g}$$

$$+ A^a_\mu A^b_\nu \left(\frac{\lambda^a}{2i} g \frac{\partial}{\partial g}\right)\left(\frac{\lambda^b}{2i} g \frac{\partial}{\partial g}\right) - A^b_\nu A^a_\mu \left(\frac{\lambda^b}{2i} g \frac{\partial}{\partial g}\right)\left(\frac{\lambda^a}{2i} g \frac{\partial}{\partial g}\right) \tag{7.115}$$

Now the quantity $(\lambda^a/2i)g(\partial/\partial g)$ which we shall denote by R_a is actually a generator† of g, up to a sign. For

$$\frac{\lambda^a}{2i} g \frac{\partial}{\partial g} = \frac{\lambda^a_{ij}}{2i} g_{jk} \frac{\partial}{\partial g_{ki}}$$

$$= \mathrm{Tr}\left\{\lambda^a g \frac{\partial}{\partial g^T}\right\}, \qquad \mathrm{T} \equiv \text{transpose} \tag{7.116}$$

and one can then check that (using (7.94))

$$[R_a, R_b] = -f_{abc} R_c. \tag{7.117}$$

This allows (7.115) to become the equation

$$[D_\mu, D_\nu] = -F^a_{\mu\nu} R_a \tag{7.118}$$

where $F^a_{\mu\nu}$ is called the curvature tensor and is given by

$$F^a_{\mu\nu} = \partial_\mu A^a_\nu - \partial_\nu A^a_\mu + f_{abc} A^b_\mu A^c_\nu \tag{7.119}$$

---

† The Lie algebra g is often defined as the set of all left invariant vector fields on G: it is easy to see, by changing variables to $g' = hg$, that a left-invariant differential operator is given by L_a where L_a acts on some $f(g)$ as

$$L_a f = \frac{\partial f}{\partial g} f \frac{\lambda^a}{2i}$$

L_a then satisfies $[L_a, L_b] = f_{abc} L_c$, and

$$L_a = \mathrm{Tr}\left\{g \frac{\lambda^a}{2i} \frac{\partial}{\partial g^T}\right\}$$

The corresponding right-invariant operator is R_a satisfying (7.117).

The curvature tensor $F^a_{\mu\nu}$ gives rise naturally to the curvature 2-form $\mathbf{F}$ where

$$\mathbf{F} = \frac{F^a_{\mu\nu}}{2}\frac{\lambda^a}{2i}\,dx^\mu \wedge dx^\nu \qquad (7.120)$$

The affine transformation law $\mathbf{A}' = hdh^{-1} + hAh^{-1}$ of $\mathbf{A}$ produces an ordinary tensorial law

$$\mathbf{F}' = h\mathbf{F}h^{-1} \qquad (7.121)$$

for the change of $\mathbf{F}$ under gauge transformations. $\mathbf{F}$ has a simple equation relating it to $\mathbf{A}$:

$$\mathbf{F} = d\mathbf{A} + \mathbf{A} \wedge \mathbf{A} \qquad (7.122)$$

One can also check that $\mathbf{F}$ may be written in terms of g and $\boldsymbol{\omega}$:

$$\begin{aligned}\mathbf{F} &= g\,(d\boldsymbol{\omega} + \boldsymbol{\omega} \wedge \boldsymbol{\omega})g^{-1} \\ &= g\boldsymbol{\Omega}g^{-1}\end{aligned} \qquad (7.123)$$

where

$$\boldsymbol{\Omega} = d\boldsymbol{\omega} + \boldsymbol{\omega} \wedge \boldsymbol{\omega} \qquad (7.124)$$

7.13 COVARIANT EXTERIOR DERIVATIVES

We digress for a moment here on the subject of covariant exterior derivatives. Since the curvature $\mathbf{F}$ can be thought of as a 2-form obtained from a 1-form $\mathbf{A}$ by application of a certain 1st order partial differential operator D say, then D could be thought of as a kind of covariant exterior derivative, and one could write (7.122) as $D\mathbf{A} = \mathbf{F}$. However, if one is constructing covariant exterior derivatives of forms $\boldsymbol{\omega}$ which transform under the group G then one usually encounters four commonly occurring cases:

i. $\boldsymbol{\omega}$ is invariant under G; $(\boldsymbol{\omega}' = \boldsymbol{\omega})$
ii. $\boldsymbol{\omega}$ is a vector-valued 1-form under G; $(\boldsymbol{\omega}' = h^{-1}\boldsymbol{\omega})$
iii. $\boldsymbol{\omega}$ is a matrix-valued p-form under G; $(\boldsymbol{\omega}' = h^{-1}\boldsymbol{\omega}h)$
iv. $\boldsymbol{\omega} = \mathbf{A}$ the connection 1-form; $(\mathbf{A}' = h^{-1}\mathbf{A}h + h^{-1}dh)$

In each of these four cases the covariant exterior derivative must produce a form of one degree higher, with a certain specified group transformation law:

$$\begin{aligned}&\text{For i. since } \boldsymbol{\omega} \text{ is invariant under } G,\ D\boldsymbol{\omega} = d\boldsymbol{\omega};\ (D\boldsymbol{\omega})' = D\boldsymbol{\omega} \\ &\text{For ii. } D\boldsymbol{\omega} = d\boldsymbol{\omega} + \mathbf{A} \wedge \boldsymbol{\omega};\ (D\boldsymbol{\omega})' = h^{-1}D\boldsymbol{\omega} \\ &\text{For iii. } D\boldsymbol{\omega} = d\boldsymbol{\omega} + \mathbf{A} \wedge \boldsymbol{\omega} + (-1)^{p+1}\boldsymbol{\omega} \wedge \mathbf{A};\ (D\boldsymbol{\omega})' = h^{-1}(D\boldsymbol{\omega})h \\ &\text{For iv. } D\mathbf{A} = d\mathbf{A} + \mathbf{A} \wedge \mathbf{A};\ (D\mathbf{A})' = h^{-1}(D\mathbf{A})h\end{aligned} \qquad (7.125)$$

Thus the symbol D can be used in differing ways and one should try and be vigilant and remember to think of the group transformation properties of forms ω when doing calculations. For example, if $p=1$ then one may mix up formulae (iii) and (iv) in equation (7.125). Thus we may write (7.122) as

$$D\mathbf{A}=\mathbf{F} \tag{7.126}$$

7.14 THE BIANCHI IDENTITIES AND *F

For a general geometrical definition of D using the language of horizontal and vertical subspaces, cf. pp. 76, 77 of ref. [9]. The curvature satisfies an important set of tensorial identities known as the Bianchi identities, and are familiar in the context of general relativity to most physicists. Here, in this general setting, they take the form

$$D\mathbf{F}=0$$

i.e.

$$d\mathbf{F}+\mathbf{A}\wedge\mathbf{F}-\mathbf{F}\wedge\mathbf{A}=0 \tag{7.127}$$

This follows immediately on substituting $\mathbf{F}=d\mathbf{A}+\mathbf{A}\wedge\mathbf{A}$ for $\mathbf{F}$. It is instructive to look at the Bianchi identities expressed in terms of $F^a_{\mu\nu}$ rather than $\mathbf{F}$. Recall the Jacobi identity valid for nested commutators of matrices:

$$[A,[B,C]]+[C,[A,B]]+[B,[C,A]]=0 \tag{7.128}$$

Applying this to the covariant derivative D_μ we obtain the identity

$$[D_\mu,[D_\nu,D_\lambda]]+[D_\lambda,[D_\mu,D_\nu]]+[D_\nu,[D_\lambda,D_\mu]]=0 \tag{7.129}$$

Specialize for a moment to a 4-dimensional manifold M, so that Greek indices run from 1 to 4, and so that there exists the permutation symbol $\varepsilon_{\mu\nu\lambda\sigma}$ with just 4 indices. Using $\varepsilon_{\mu\nu\lambda\sigma}$, a shorthand for (7.129) is:

$$\varepsilon_{\mu\nu\lambda\sigma}[D_\nu,[D_\lambda,D_\sigma]]=0 \tag{7.130}$$

where the ε-symbol does the antisymmetric book-keeping. This we can write as

$$\varepsilon^{\mu\nu\lambda\sigma}[D_\nu,F_{\lambda\sigma}]=0 \tag{7.131}$$

(where we have defined for convenience $F_{\mu\nu}=F^a_{\mu\nu}R^a$).

It is now possible to compare (7.131) and (7.127) and see that the antisymmetric book-keeping of (7.131) is slavishly copied by (7.127) if one writes out in full the exterior differentiation and wedge products. The equations thus are equivalent. An important mathematical object occurring here is that of the dual of the curvature. If $F^a_{\mu\nu}$ is the curvature tensor then its dual is written $*F^a_{\mu\nu}$ and defined by

$$*F^a_{\mu\nu}=\tfrac{1}{2}\varepsilon_{\mu\nu\lambda\sigma}F^{a\lambda\sigma} \tag{7.132}$$

so that the Bianchi identities can be further reduced by

$$[D_\nu, *F_{\mu\nu}]=0 \tag{7.133}$$

which is to be compared with the alternative $D\mathbf{F}=\mathbf{0}$. We give these two versions of the Bianchi identities because later we shall obtain an equation of motion for the Yang–Mills field of the form $D*\mathbf{F}=0$; this equation when written in terms of D_μ becomes $[D_\mu, F_{\mu\nu}]=0$. Thus a useful point for physicists to bear in mind (since their equations are more often written in terms of D_μ rather than D), is that passage from an equation involving D_μ to one involving D requires in these two important examples an interchange of $\mathbf{F}$ and $*F_{\mu\nu}$, or of $*\mathbf{F}$ and $F_{\mu\nu}$.

A point worth mentioning in passing is that $\boldsymbol{\Omega}$ and $\boldsymbol{\omega}$ bear the same relationship to one another as do $\mathbf{F}$ and $\mathbf{A}$. However, note that $\boldsymbol{\Omega}$ and $\boldsymbol{\omega}$ are g-valued and belong to $\Lambda^2T^*(P)$ and $T^*(P)$ respectively, and $\mathbf{F}$ and $\mathbf{A}$ are g-valued and belong to $\Lambda^2T^*(M)$ and $T^*(M)$ respectively; the former are forms on the bundle, while the latter are forms on the manifold. However, by producing a section $s(x)$ of P one can go from one to the other: if we wish P to be non-trivial then $s(x)$ should only be a local section i.e.

$$s(x): U_\alpha \to P \tag{7.134}$$

where $U_\alpha \subset M$.

Given $s(x)$ then recall that the map $s^*(x)$ will pullback forms on P to forms on U_α. In fact one can readily check (and we shall do so in due course) that, the pullback $s^*\boldsymbol{\omega}$ of $\boldsymbol{\omega}$ is $\mathbf{A}$, and that the pullback $s^*\boldsymbol{\Omega}$ of $\boldsymbol{\Omega}$ is $\mathbf{F}$. If one moves to an overlapping neighbourhood U_β and wishes to repeat this process, then s changes to $s'=hs$ and the connections and the curvatures undergo a gauge transformation under the action of h. Thus gauge transformations are also characterizable as local changes of sections in the principal bundle P.

There is now enough material on connections and curvatures to enable us to discuss some of the different sorts of geometry which make use of them. We have in mind three categories (i) Riemannian or pseudo-Riemannian geometry which corresponds to the choice $G=O(n)$ or $O(n-p,p)$; (ii) Yang–Mills geometries which correspond to choosing G to be various non-Abelian Lie groups of which the prototype is $SU(2)$; (iii) the geometry of the Maxwell field which corresponds to the choice $G=U(1)$.

7.15 CONNECTION IN THE TANGENT BUNDLE

We begin with the Riemannian case. As we have seen earlier in the chapter, a Riemannian metric is provided for M in reducing the group $Gl(n, \mathbf{R})$ of the frame bundle $F(M)$ to $O(n)$. However, we shall not immediately make use of the fact the group $Gl(n, \mathbf{R})$ can be reduced to $O(n)$. The connections $\mathbf{A}$ will be called linear or affine connections, and later we shall define what we mean by a Riemannian connection.

So far, we have been dealing with connections on principal bundles P, however, most of us began our early calculations with connections for example, in general relativity, by covariantly differentiating vectors and tensors. This suggests that connections and covariant derivatives should be defined on the vector and tensor bundles $T(M)$ and $T^a_b(M)$ since this is where these vectors and tensors belong. This is indeed so and we now show how to do this. Looking back to the beginning of our discussion of connections on P, one can see that the essence of what we did was to define what we called the horizontal lift $\tilde{\gamma}(t)$ to P of a curve $\gamma(t)$ in M. Now take the bundle $T(M)$. All we have to do is define a horizontal lift $\gamma_{TM}(t)$ to $T(M)$ of a curve $\gamma(t)$ in M. There is already a natural method for doing this waiting to be discovered. Remember that $T(M)$ is an associated (vector) bundle of the principal bundle $F(M)$, i.e. its structure group is $Gl(n, \mathbf{R})$ which is the fibre of $F(M)$. One can use this fact to induce a horizontal lift $\gamma_{TM}(t)$ to TM from $\tilde{\gamma}(t)$ the horizontal lift to P. First fix a vector $\mathbf{X}$ belonging to one of the fibres $T_x(M)$ of $T(M)$, any element $O \in Gl(n, \mathbf{R})$ acts on $\mathbf{X}$ by matrix multiplication to give another vector $O\mathbf{X}$. But $\tilde{\gamma}(t)$ is a curve in P and is thus a source of such matrices O. So define the horizontal lift $\gamma_{TM}(t)$ by

$$\gamma_{TM}(t) = \tilde{\gamma}(t)\mathbf{X} \tag{7.135}$$

Thus as one goes round the horizontal lift $\tilde{\gamma}(t)$ in $P = F(M)$ one also goes round the horizontal lift $\gamma_{TM}(t)$ in $T(M)$. Next we look at

$$\frac{\mathrm{d}}{\mathrm{d}t}\gamma_{TM}(t) = \frac{\mathrm{d}}{\mathrm{d}t}(\tilde{\gamma}(t)\mathbf{X})$$

We have

$$\frac{\mathrm{d}\gamma_{TM}(t)}{\mathrm{d}t} = \dot{x}^{\mu}(t)D_{\mu}\tilde{\gamma}(t)\mathbf{X} \tag{7.136}$$

Now the curve $\gamma(t) = (x^{\mu}(t))$ is a curve in M so $\dot{x}^{\mu}(t)$ are the coordinates of a tangent vector in M. Denote this tangent vector by $\mathbf{Y}$ so that in local coordinates

$$\mathbf{Y}=\dot{x}^{\mu}(t)\frac{\partial}{\partial x^{\mu}} \tag{7.137}$$

Further, $\gamma_{TM}(t)=\tilde{\gamma}(t)\mathbf{X}$ is, for each t, a tangent vector, denote this tangent vector by $\mathbf{Z}$. Then the covariant derivative (in the bundle TM) of $\mathbf{Z}$, with respect to, or in the direction of, $\mathbf{Y}$ is written $\nabla_{\mathbf{Y}}\mathbf{Z}$ and defined by

$$\nabla_{\mathbf{Y}}\mathbf{Z}=\frac{\mathrm{d}\gamma_{TM}(t)}{\mathrm{d}t}=\dot{x}^{\mu}(t)D_{\mu}\tilde{\gamma}(t)\mathbf{X} \tag{7.138}$$

By varying both the initial point and the choice of the curve $\gamma(t)$ one can see that $\nabla_{\mathbf{X}}\mathbf{Y}$ may be calculated for any pair of vectors $\mathbf{X}$ and $\mathbf{Y}$. The following properties of $\nabla_{\mathbf{X}}\mathbf{Y}$ follow easily from the definition and are useful in calculations:

$$\begin{aligned}\nabla_{\mathbf{X}+\mathbf{Y}}(\mathbf{Z})&=\nabla_{\mathbf{X}}\mathbf{Z}+\nabla_{\mathbf{Y}}\mathbf{Z}\\ \nabla_{\mathbf{X}}(\mathbf{Y}+\mathbf{Z})&=\nabla_{\mathbf{X}}\mathbf{Y}+\nabla_{\mathbf{X}}\mathbf{Z}\\ \nabla_{\mathbf{X}}(f\mathbf{Y})&=f\nabla_{\mathbf{X}}\mathbf{Y}+(\mathbf{X}f)\mathbf{Y}\\ \nabla_{f\mathbf{X}}(\mathbf{Y})&=f\nabla_{\mathbf{X}}\mathbf{Y}\end{aligned} \tag{7.139}$$

where $f\in C^{\infty}(M)$ and $\mathbf{X}, \mathbf{Y}, \mathbf{Z}$ are vector fields on M. There is also the closely related mathematical object the covariant differential, or total covariant derivative, written $\nabla\mathbf{Y}$ and defined by:

$$\nabla_{\mathbf{X}}\mathbf{Y}=\langle\nabla\mathbf{Y},\mathbf{X}\rangle$$

or

$$\nabla_{\mathbf{Y}}=\nabla_{\partial_{\mu}}(\mathbf{Y})\otimes\mathrm{d}x^{\mu} \tag{7.140}$$

It is instructive, and essential to understanding these sophisticated (looking) bundle methods, to calculate explicitly using local coordinates $\nabla_{\mathbf{Y}}\mathbf{Z}$. Let us do this. We begin at (7.135) and we need to introduce some notation. Let

$$\mathbf{X}=a^{\mu}\frac{\partial}{\partial x^{\mu}} \tag{7.141}$$

The points $g(t)$ on the curve $\tilde{\gamma}(t)$ are group elements belonging to $Gl(n,\mathbf{R})$ so let us represent them by $n\times n$ matrices whose local coordinates we denote by the real numbers X^{μ}_{ν}; $\mu,\nu=1,\ldots,n$. The combination $A^{\alpha}_{\mu}(\lambda^{\alpha}/2i)$ is $\mathrm{gl}(n,\mathbf{R})$ valued ($\mathrm{gl}(n,\mathbf{R})$ being the Lie algebra of $Gl(n,\mathbf{R})$), and we introduce the functions $A^{\nu}_{\mu\kappa}$ to express this combination in local coordinates. With these pieces of notation the covariant derivative D_μ can be written

$$D_\mu = \frac{\partial}{\partial x^\mu} - A^a_\mu \frac{\lambda^a}{2i} g \frac{\partial}{\partial g}$$

$$= \frac{\partial}{\partial x^\mu} - A^{\nu}_{\mu\kappa} X^{\kappa}_{\lambda} \frac{\partial}{\partial X^{\lambda}_{\nu}} \tag{7.142}$$

Now note that the matrix multiplication implied in the product $\tilde{\gamma}(t)\mathbf{X}$ of (7.135) is simply the action of the matrix or group element g on the vector $\mathbf{X}$. Thus the vector $\mathbf{Z} = b^\kappa\ (\partial/\partial x^\mu)$ say is given by

$$b^\kappa\ (\partial/\partial x^\kappa) = \mathbf{Z} = \tilde{\gamma}(t)\mathbf{X} = X^{\mu}_{\nu} a^\nu\ (\partial/\partial x^\mu) \tag{7.143}$$

So

$$\nabla_{\mathbf{Y}}\mathbf{Z} = \frac{\mathrm{d}}{\mathrm{d}t}\gamma_{TM}(t) = \frac{\mathrm{d}}{\mathrm{d}t}\tilde{\gamma}(t)\mathbf{X}; \qquad \left(\mathbf{Y} = \dot{x}^\lambda \frac{\partial}{\partial x^\lambda}\right)$$

$$= \dot{x}^\lambda \left(\frac{\partial}{\partial x^\lambda} - A^{\rho}_{\lambda\sigma} X^{\sigma}_{\tau} \frac{\partial}{\partial X^{\rho}_{\tau}}\right) X^{\mu}_{\nu} a^\nu \frac{\partial}{\partial x^\mu}$$

$$= \dot{x}^\lambda \left(X^{\mu}_{\nu} \frac{\partial a^\nu}{\partial x^\lambda} \frac{\partial}{\partial x^\mu} - A^{\rho}_{\lambda\sigma} X^{\sigma}_{\tau} \delta^{\mu}_{\rho} \delta^{\tau}_{\nu} a^\nu \frac{\partial}{\partial x^\mu}\right)$$

$$= \dot{x}^\lambda X^{\mu}_{\nu} \frac{\partial a^\nu}{\partial x^\lambda} \frac{\partial}{\partial x^\mu} - \dot{x}^\lambda A^{\mu}_{\lambda\sigma} X^{\sigma}_{\nu} a^\nu \frac{\partial}{\partial x^\mu}$$

$$= \dot{x}^\lambda \frac{\partial b^\mu}{\partial x^\lambda} \frac{\partial}{\partial x^\mu} - \dot{x}^\lambda A^{\mu}_{\lambda\sigma} b^\sigma \frac{\partial}{\partial x^\mu} \tag{7.144}$$

using (7.143).

We write (7.144) in a more compact form as

$$\nabla_{\mathbf{Y}}\mathbf{Z} = C^\lambda b^{\mu}_{;\lambda} \frac{\partial}{\partial x^\mu}$$

with

$$\mathbf{Y} = C^\lambda \frac{\partial}{\partial x^\lambda}, \qquad \mathbf{Z} = b^\mu \frac{\partial}{\partial x^\mu} \tag{7.145}$$

and the $b^{\mu}_{;\lambda}$ is given by, as is the standard practice in the literature,

$$b^{\mu}_{;\lambda} = \frac{\partial b^\mu}{\partial x^\lambda} - A^{\mu}_{\lambda\sigma} b^\sigma \tag{7.146}$$

The quantity $A^{\mu}_{\lambda\sigma}$ is called a Christoffel symbol and is more usually written as $\Gamma^{\mu}_{\lambda\sigma}$; then equation (7.146) becomes the classical tensor form of the covariant derivative familiar to many physicists. It should now also be clear that if **T** is a tensor field of type (a, b) then one can compute its covariant derivative with respect to a vector **Y**. This requires one to define horizontal lifts and parallel transport in the tensor bundle $T^a_b(M)$ rather then the vector bundle $T(M)$. No new idea is needed to carry out these procedures in $T^a_b(M)$, one simply imitates exactly the procedures used for $T(M)$. The reader can verify that if

$$\mathbf{T}=T^{\mu_1,\ldots,\mu_a}_{\nu_1,\ldots,\nu_b}\,\mathrm{d}x^{\nu_1},\ldots,\mathrm{d}x^{\nu_b}\frac{\partial}{\partial x^{\mu_1}},\ldots,\frac{\partial}{\partial x^{\mu_a}}$$

and

$$\mathbf{Y}=C^{\lambda}\frac{\partial}{\partial x^{\lambda}} \tag{7.147}$$

then

$$\nabla_{\mathbf{Y}}\mathbf{T}=C^{\lambda}T^{\mu_1,\ldots,\mu_a}_{\nu_1,\ldots,\nu_b;\lambda} \tag{7.148}$$

where

$$\begin{aligned}T^{\mu_1,\ldots,\mu_a}_{\nu_1,\ldots,\nu_b;\lambda}&=\frac{\partial T^{\mu_1,\ldots,\mu_a}_{\nu_1,\ldots,\nu_b}}{\partial x^{\lambda}}+\Gamma^{\mu_1}_{\lambda\alpha}T^{\alpha\mu_2,\ldots,\mu_a}_{\nu_1,\ldots,\nu_b}+\ldots+\Gamma^{\mu_a}_{\lambda\alpha}T^{\mu_1,\ldots,\mu_{a-1}\alpha}_{\nu_1,\ldots,\nu_2}\\&\quad-\Gamma^{\alpha}_{\lambda\nu_1}\Gamma^{\mu_1,\ldots,\mu_a}_{\alpha\nu_2,\ldots,\nu_b}-\ldots-\Gamma^{\alpha}_{\lambda\nu_b}T^{\mu_1,\ldots,\mu_a}_{\nu_1,\ldots,\nu_{b-1}\alpha}\end{aligned} \tag{7.149}$$

7.16 THE TORSION TENSOR

When working on the tangent bundle $T(M)$ one encounters an important operator on $T(M)$ called the torsion. The torsion $\mathbf{T}(\mathbf{X}, \mathbf{Y})$ is defined for every pair of vector fields **X** and **Y** in $T(M)$. $\mathbf{T}(\mathbf{X}, \mathbf{Y})$ is linear and antisymmetric in **X** and **Y** and is given by

$$\mathbf{T}(\mathbf{X},\mathbf{Y})=\nabla_{\mathbf{X}}\mathbf{Y}-\nabla_{\mathbf{Y}}\mathbf{X}-[\mathbf{X},\mathbf{Y}] \tag{7.150}$$

On calculating the RHS of (7.150) one discovers that if

$$\mathbf{X}=a^{\mu}\frac{\partial}{\partial x^{\mu}},\qquad \mathbf{Y}=b^{\lambda}\frac{\partial}{\partial x^{\lambda}} \tag{7.151}$$

then

$$\begin{aligned}\mathbf{T}(\mathbf{X},\mathbf{Y})&=a^{\lambda}\Gamma^{\mu}_{\lambda\sigma}b^{\sigma}\frac{\partial}{\partial x^{\mu}}-b^{\lambda}\Gamma^{\mu}_{\lambda\sigma}a^{\sigma}\frac{\partial}{\partial x^{\mu}}\\&=(\Gamma^{\mu}_{\lambda\sigma}-\Gamma^{\mu}_{\sigma\lambda})a^{\lambda}b^{\sigma}\frac{\partial}{\partial x^{\mu}}\end{aligned} \tag{7.152}$$

Thus $\mathbf{T}(\mathbf{X}, \mathbf{Y})$ measures the antisymmetric part of the connection $\Gamma^{\mu}_{\lambda\sigma}$. The antisymmetric combination $\Gamma^{\mu}_{\lambda\sigma}-\Gamma^{\mu}_{\sigma\lambda}$ is denoted by $T^{\mu}_{\lambda\sigma}$, and $T^{\mu}_{\lambda\sigma}$ are the components with respect to these local coordinates of a tensor of type $(1, 2)$ called the torsion tensor. Turning to the curvature $F^{a}_{\mu\nu}(\lambda^{a}/2i)$ and realizing that in this case it is $gl(n, \mathbf{R})$ valued, we see that the object $F^{a}_{\mu\nu}(\lambda^{a}/2i)_{ij}$, to use our former notation, is replaced by $F^{\lambda}_{\sigma\mu\nu}$. That is to say the indices i and j now run from $1, \ldots, n$ like μ and ν instead of running over some unspecified range depending on the choice of group G for the bundle P. Also we have replaced i and j by λ and σ, the reason that the λ index is raised is that a matrix is actually a tensor of type $(1, 1)$ and it is a (mild) abuse of notation to write its entries as a_{ij} rather than a^{i}_{j}. Finally, the notation that is commonly in use dictates that instead of writing $F^{\lambda}_{\sigma\mu\nu}$ we write $R^{\lambda}_{\sigma\mu\nu}$, and this of course is the form of the components of the curvature tensor in the Riemannian case. $R^{\lambda}_{\mu\nu\sigma}$ naturally gives rise to the curvature tensor operator $\mathbf{R}(\mathbf{X}, \mathbf{Y})$ defined by

$$\mathbf{R}(\mathbf{X}, \mathbf{Y})\mathbf{Z} = \nabla_{\mathbf{X}}\nabla_{\mathbf{Y}}\mathbf{Z} - \nabla_{\mathbf{Y}}\nabla_{\mathbf{X}}\mathbf{Z} - \nabla_{[\mathbf{X},\mathbf{Y}]}\mathbf{Z} \tag{7.153}$$

where $\mathbf{X}$, $\mathbf{Y}$ and $\mathbf{Z}$ are vector fields on M. Using local coordinates one can check that the LHS of (7.153) is

$$R^{\mu}_{\nu\sigma\rho}a^{\sigma}b^{\rho}c^{\nu}\frac{\partial}{\partial x^{\mu}} \tag{7.154}$$

where $\mathbf{X} = a^{\sigma}(\partial/\partial x^{\sigma})$, $\mathbf{Y} = b^{\rho}(\partial/\partial x^{\rho})$, and $\mathbf{Z} = c^{\nu}(\partial/\partial x^{\nu})$. On the other hand, the RHS of (7.153) is found to be

$$(c^{\nu}_{;\rho\sigma} - c^{\nu}_{;\sigma\rho})a^{\sigma}b^{\rho}\frac{\partial}{\partial x^{\nu}} - (\Gamma^{\nu}_{\rho\sigma} - \Gamma^{\nu}_{\sigma\rho})a^{\sigma}b^{\rho}c^{\mu}_{;\nu}\frac{\partial}{\partial x^{\mu}}$$

$$= (c^{\nu}_{;\rho\sigma} - c^{\nu}_{;\sigma\rho})a^{\sigma}b^{\rho}\frac{\partial}{\partial x^{\nu}} - T^{\nu}_{\rho\sigma}a^{\sigma}b^{\rho}c^{\mu}_{;\nu}\frac{\partial}{\partial x^{\mu}} \tag{7.155}$$

Since $\mathbf{X}$ and $\mathbf{Y}$ are arbitrary one can then conclude that

$$c^{\nu}_{;\rho\sigma} - c^{\nu}_{;\sigma\rho} = R^{\nu}_{\mu\sigma\rho}c^{\mu} + T^{\mu}_{\rho\sigma}c^{\nu}_{;\mu}. \tag{7.156}$$

Thus $R^{\nu}_{\mu\sigma\rho}$ and $T^{\mu}_{\rho\sigma}$ express in a more intimate coordinate fashion the non-commutativity of covariant differentiation in $T(M)$. Equation (7.156) also enables one to write $R^{\nu}_{\mu\sigma\rho}$ in terms of the connection coordinates $\Gamma^{\lambda}_{\mu\nu}$. The expression that one obtains is:

$$R^{\mu}_{\nu\kappa\lambda} = \frac{\partial\Gamma^{\mu}_{\lambda\nu}}{\partial x^{\kappa}} - \frac{\partial\Gamma^{\mu}_{\kappa\nu}}{\partial x^{\lambda}} + \Gamma^{\alpha}_{\lambda\nu}\Gamma^{\mu}_{\alpha\kappa} - \Gamma^{\alpha}_{\kappa\nu}\Gamma^{\mu}_{\lambda\alpha}. \tag{7.157}$$

A quick way of obtaining the expressions for $T^{\mu}_{\nu\kappa}$ and $R^{\mu}_{\nu\kappa\lambda}$ in terms of the connection coordinates $\Gamma^{\lambda}_{\mu\nu}$ is to note that

$$\mathbf{T}\left(\frac{\partial}{\partial x^{\mu}}, \frac{\partial}{\partial x^{\nu}}\right) = T^{\lambda}_{\mu\nu}\frac{\partial}{\partial x^{\lambda}}$$

and

$$\mathbf{R}\left(\frac{\partial}{\partial x^{\mu}}, \frac{\partial}{\partial x^{\nu}}\right)\frac{\partial}{\partial x^{\kappa}} = R^{\lambda}_{\kappa\mu\nu}\frac{\partial}{\partial x^{\lambda}} \tag{7.158}$$

We have now learned how to covariantly differentiate in $F(M)$ itself and in its associated bundles $T(M)$ and $T^a_b(M)$. The bundle that we use depends on whether one wishes to differentiate vector fields or tensor fields etc. One should note that these objects are sections of the particular bundles to which they belong. So to summarize: the operation of covariant differentiation in a bundle acts on sections of that bundle (of course for the principal bundle this section will usually only be a local section unless the bundle happens to be trivial).

For the calculations in $T(M)$ and $T^a_b(M)$ with this linear connection $A^{\mu}_{\lambda\sigma}$ there is another identity, also called after Bianchi. This identity involves both the torsion and the curvature. Let $\mathbf{e}_\mu$ be a basis for $\mathbf{R}^n$ and define the vector-valued 1-form $\boldsymbol{\theta}$ by

$$\boldsymbol{\theta} = \mathbf{e}_\mu \, \mathrm{d}x^{\mu}. \tag{7.159}$$

Recall that the connection when written as $A^a_\mu(\lambda^a/2i)$ is a matrix in $gl(n, \mathbf{R})$, take as a basis for $gl(n, \mathbf{R})$ the matrices E^{μ}_{ν} where $\mu, \nu = 1, \ldots, n$ and E^{μ}_{ν} is the matrix with a 1 in the μ, νth position and zero everywhere else. Then we can write the (still matrix valued) equation

$$A^a_\mu \frac{\lambda^a}{2i} = A^{\lambda}_{\mu\sigma}E^{\sigma}_{\lambda} \tag{7.160}$$

Note from the definitions that $E^{\sigma}_{\mu}\mathbf{e}_\nu = \delta^{\sigma}_{\nu}\mathbf{e}_\mu$. Now we simply observe that

$$\begin{aligned}
D\boldsymbol{\theta} &= \mathrm{d}\boldsymbol{\theta} + \mathbf{A}\wedge\boldsymbol{\theta} \\
&= O + \mathbf{A}\wedge\boldsymbol{\theta} \\
&= A^{\lambda}_{\mu\sigma}E^{\sigma}_{\lambda}e_\nu \, \mathrm{d}x^{\mu}\wedge \mathrm{d}x^{\nu} \\
&= A^{\lambda}_{\mu\sigma}\delta^{\sigma}_{\nu}\mathbf{e}_\lambda \, \mathrm{d}x^{\mu}\wedge \mathrm{d}x^{\nu} \\
&= A^{\lambda}_{\mu\nu}\mathbf{e}_\lambda \, \mathrm{d}x^{\mu}\wedge \mathrm{d}x^{\nu} \\
&= \tfrac{1}{2}(A^{\lambda}_{\mu\nu} - A^{\lambda}_{\nu\mu})\mathbf{e}_\lambda \, \mathrm{d}x^{\mu}\wedge \mathrm{d}x^{\nu}
\end{aligned} \tag{7.161}$$

But $A^{\lambda}_{\mu\nu} - A^{\lambda}_{\nu\mu} = T^{\lambda}_{\mu\nu}$, the components of the torsion, so if we define a vector valued 2-form $\mathbf{T}$ by

$$\mathbf{T} = \tfrac{1}{2}(A^{\lambda}_{\mu\nu} - A^{\lambda}_{\nu\mu})\mathbf{e}_{\lambda}\, \mathrm{d}x^{\mu} \wedge \mathrm{d}x^{\nu} \tag{7.162}$$

Then we can write (7.161) as

$$D\boldsymbol{\theta} = \mathbf{T} \tag{7.163}$$

which can be remembered as a companion to the equation

$$D\mathbf{A} = \mathbf{F} \quad \text{or} \quad D\boldsymbol{\Gamma} = \mathbf{R} \tag{7.164}$$

for the curvature $\mathbf{F}$ or $\mathbf{R}$.

If we apply covariant exterior differentiation to (7.164) we get the Bianchi identities already derived, i.e. $D\mathbf{R} = \mathbf{0}$. The other Bianchi identities are obtained by applying covariant exterior differentiation to (7.163), the result of this is the equation

$$D\mathbf{T} = \mathbf{R} \wedge \boldsymbol{\theta} \tag{7.165}$$

We leave the straightforward verification of this to the reader. Equation (7.165) is often called Bianchi's first identity and $D\mathbf{R} = \mathbf{0}$ Bianchi's second identity.

7.17 GEODESICS

Next we shall introduce geodesics. A geodesic is a curve $\gamma(t)$ in M such that its tangent vectors are parallel along γ, i.e. parallel transport from the point t to the point t' takes the tangent vector at t into the tangent vector at t'. Since the covariant derivative $\nabla_{\mathbf{X}}\mathbf{X}$ measures the rate of change of $\mathbf{X}$ in the direction $\mathbf{X}$ under parallel transport, then an equation describing the above definition of a geodesic is simply

$$\nabla_{\mathbf{X}}\mathbf{X} = 0 \tag{7.166}$$

where $\mathbf{X}$ are the tangents to $\gamma(t)$. But $\mathbf{X} = \dot{x}^{\mu}(t)(\partial/\partial x^{\mu})$ in local coordinates so with these coordinates (7.166) becomes the equation

$$\frac{\mathrm{d}x^{\lambda}}{\mathrm{d}t}\left(\frac{\partial}{\partial x^{\lambda}}\frac{\mathrm{d}x^{\mu}}{\mathrm{d}t} + \Gamma^{\mu}_{\lambda\sigma}\frac{\mathrm{d}x^{\lambda}}{\mathrm{d}t}\frac{\mathrm{d}x^{\sigma}}{\mathrm{d}t}\right)\frac{\partial}{\partial x^{\mu}} = 0 \tag{7.167}$$

where we have used (7.145, 146). On making use of the fact that $(\mathrm{d}/\mathrm{d}t) = (\mathrm{d}x^{\lambda}/\mathrm{d}t)(\partial/\partial x^{\mu})$ this simplifies to

$$\frac{\mathrm{d}^2x^{\mu}}{\mathrm{d}t^2} + \Gamma^{\mu}_{\lambda\sigma}\frac{\mathrm{d}x^{\lambda}}{\mathrm{d}t}\frac{\mathrm{d}x^{\sigma}}{\mathrm{d}t} = 0 \tag{7.168}$$

so curves $x^\mu(t)$ satisfying (7.168) are geodesics. Note that this equation for a geodesic is insensitive to the antisymmetric part of $\Gamma^\mu_{\lambda\sigma}$. Carrying this further if we were to redefine $\Gamma^\mu_{\lambda\sigma}$ so as to have no torsion, then the geodesics would remain unchanged, i.e. if we defined $\bar{\Gamma}^\mu_{\lambda\sigma}$ by

$$\bar{\Gamma}^\mu_{\lambda\sigma} = \tfrac{1}{2}(\Gamma^\mu_{\lambda\sigma} + \Gamma^\mu_{\sigma\lambda}) \tag{7.169}$$

then $\bar{\Gamma}^\mu_{\lambda\sigma}$ is a linear connection without torsion having the same geodesics as $\Gamma^\mu_{\lambda\sigma}$; the curvatures of the two connections $\bar{\Gamma}^\mu_{\lambda\sigma}$ and $\Gamma^\mu_{\lambda\sigma}$ would, in general, be different though. From this it follows that a linear connection $\Gamma^\mu_{\lambda\sigma}$ is completely determined by knowledge of its torsion and knowledge of all the geodesics on M.

7.18 THE LEVI-CIVITA CONNECTION

Now we are ready to discuss the Levi–Civita connection or Riemannian connection of Riemannian geometry. So we invoke our right to reduce the group of the bundle $F(M)$ from $Gl(n, \mathbf{R})$ to $O(n)$, some writers use the symbol $O(M)$ for this reduced bundle to remind one that it is the bundle of orthogonal frames. In any case as we have explained earlier in this chapter, this reduction provides M with a Riemannian metric $g_{\mu\nu}$. The Levi–Civita or Riemannian connection is that unique connection Γ satisfying the following two requirements:

i. Γ has zero torsion
ii. Parallel displacement preserves scalar products (7.170)

In terms of the $\Gamma^\mu_{\lambda\sigma}$ and the $g_{\mu\nu}$ these requirements are clearly that

$$\text{i. } \Gamma^\mu_{\lambda\sigma} = \Gamma^\mu_{\sigma\lambda} \qquad \text{ii. } g_{\mu\nu;\kappa} = 0 \tag{7.171}$$

and these equations determine Γ uniquely. To actually calculate $\Gamma^\mu_{\lambda\sigma}$ in terms of the $g_{\mu\nu}$ is straightforward, we denote the inner product of any two vectors $\mathbf{X}$ and $\mathbf{Y}$ by $\mathbf{g}(\mathbf{X}, \mathbf{Y})$. Then if $\mathbf{X}$, $\mathbf{Y}$ and $\mathbf{Z}$ are any vectors the three equations below hold by virtue of (7.171. ii).

$$\begin{aligned} &\text{i. } \mathbf{X}(\mathbf{g}(\mathbf{Y}, \mathbf{Z})) = \mathbf{g}(\nabla_\mathbf{X}\mathbf{Y}, \mathbf{Z}) + \mathbf{g}(\mathbf{Y}, \nabla_\mathbf{X}\mathbf{Z}) \\ &\text{ii. } \mathbf{Y}(\mathbf{g}(\mathbf{Z}, \mathbf{X})) = \mathbf{g}(\nabla_\mathbf{Y}\mathbf{Z}, \mathbf{X}) + \mathbf{g}(\mathbf{Z}, \nabla_\mathbf{Y}\mathbf{X}) \\ &\text{iii. } \mathbf{Z}(\mathbf{g}(\mathbf{X}, \mathbf{Y})) = \mathbf{g}(\nabla_\mathbf{Z}\mathbf{X}, \mathbf{Y}) + \mathbf{g}(\mathbf{X}, \nabla_\mathbf{Z}\mathbf{Y}) \end{aligned} \tag{7.172}$$

Now if we add equations (i) and (ii) above and subtract equation (iii), and also use the zero torsion requirement in the form $\mathbf{T}(\mathbf{X}, \mathbf{Y}) = 0 =$

$\nabla_{\mathbf{X}}\mathbf{Y}-\nabla_{\mathbf{Y}}\mathbf{X}-[\mathbf{X},\mathbf{Y}]$ we obtain:

$$2g(\mathbf{Z},\nabla_X\mathbf{Y})=\mathbf{X}(\mathbf{g}(\mathbf{Y},\mathbf{Z}))+\mathbf{Y}(\mathbf{g}(\mathbf{Z},\mathbf{X}))-\mathbf{Z}(\mathbf{g}(\mathbf{X},\mathbf{Y}))+\mathbf{g}(\mathbf{Z},[\mathbf{X},\mathbf{Y}]\\ +\mathbf{g}(\mathbf{Y},[\mathbf{Z},\mathbf{X}])-\mathbf{g}(\mathbf{X},[\mathbf{Y},\mathbf{Z}]) \tag{7.173}$$

Choosing

$$\mathbf{X}=\frac{\partial}{\partial x^{\mu}},\qquad \mathbf{Y}=\frac{\partial}{\partial x^{\lambda}},\qquad \mathbf{Z}=\frac{\partial}{\partial x^{\sigma}}$$

and noting that

$$\mathbf{g}\left(\frac{\nabla_{\partial}}{\partial x^{\mu}},\frac{\nabla_{\partial}}{\partial x^{\lambda}}\frac{\partial}{\partial x^{\sigma}}\right)=\Gamma_{\lambda\mu\sigma} \tag{7.174}$$

will, on substitution into (7.173), yield the usual form for the Christoffel symbols in local coordinates namely:

$$g_{\lambda\alpha}\Gamma^{\alpha}_{\mu\sigma}=\Gamma_{\lambda\mu\sigma}=\tfrac{1}{2}\left\{\frac{\partial g_{\lambda\mu}}{\partial x^{\sigma}}+\frac{\partial g_{\lambda\sigma}}{\partial x^{\mu}}-\frac{\partial g_{\mu\sigma}}{\partial x^{\lambda}}\right\}$$

It is also useful to record the Bianchi identities in a more explicit form. Because $\mathbf{T}=\mathbf{0}$ the first Bianchi identity is simply the algebraic, rather than analytic, relation

$$O=\mathbf{R}\wedge\boldsymbol{\theta} \tag{7.175}$$

Writing this out in full gives

$$O=R^{\lambda}_{\ \sigma\mu\nu}E^{\sigma}_{\lambda}\mathbf{e}_{\alpha}\,\mathrm{d}x^{\mu}\wedge\mathrm{d}x^{\nu}\wedge\mathrm{d}x^{\alpha}\\ O=R^{\lambda}_{\ \sigma\mu\nu}\mathbf{e}_{\lambda}\,\mathrm{d}x^{\mu}\wedge\mathrm{d}x^{\nu}\wedge\mathrm{d}x^{\sigma} \tag{7.176}$$

In terms of $R^{\lambda}_{\ \mu\nu\sigma}$ this implies the usual permutation relations among the covariant indices of $R^{\lambda}_{\ \mu\nu\sigma}$:

$$R^{\lambda}_{\ \mu\nu\sigma}+R^{\lambda}_{\ \sigma\mu\nu}+R^{\lambda}_{\ \nu\sigma\mu}\\ -R^{\lambda}_{\ \nu\mu\sigma}-R^{\lambda}_{\ \mu\sigma\nu}-R^{\lambda}_{\ \sigma\nu\mu}=0 \tag{7.177}$$

But because of the definition of $R^{\lambda}_{\ \mu\nu\sigma}$ we already have $R^{\lambda}_{\ \mu\nu\sigma}=-R^{\lambda}_{\ \mu\sigma\nu}$ so (7.177) reduces to

$$R^{\lambda}_{\ \mu\nu\sigma}+R^{\lambda}_{\ \sigma\mu\nu}+R^{\lambda}_{\ \nu\sigma\mu}=0 \tag{7.178}$$

The second Bianchi identity $D\mathbf{R}=0$ in terms of the $R^{\lambda}_{\ \mu\nu\sigma}$ is:

$$R^{\lambda}_{\ \mu\nu\sigma;\alpha}+R^{\lambda}_{\ \mu\alpha\nu;\sigma}+R^{\lambda}_{\ \mu\sigma\alpha;\nu}=0 \tag{7.179}$$

The torsion has a geometrical interpretation which is worth knowing. The geometrical object that one considers is an infinitesimal parallelogram constructed by infinitesimal parallel translation. Remember that a vector $\mathbf{X}(t)$ undergoes parallel translation along a curve $\gamma(t)$ if

$$\frac{d\mathbf{X}}{dt}=0$$

or explicitly

$$\frac{da^{\mu}(t)}{dt}+\Gamma^{\mu}_{\lambda\sigma}a^{\lambda}(t)\frac{dx^{\sigma}}{dt}=0$$

where $a^{\mu}(t)$ are the components of $\mathbf{X}$. Thus for infinitesimal parallel translations the components of $\mathbf{X}$ change by da^{μ} where

$$da^{\mu}=-\Gamma^{\mu}_{\lambda\sigma}a^{\lambda}\,dx^{\sigma}$$

Now consider the triangle AOB of Fig. 7.10 with infinitesimal sides denoted by dx^{μ} and dy^{μ}

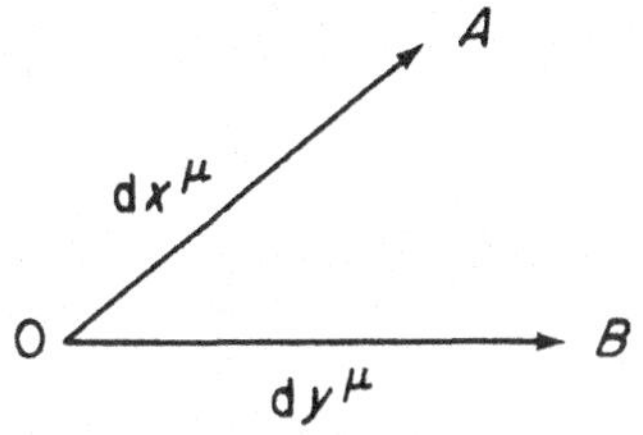

Figure 7.10

Now we infinitesimally displace the vector dx^{μ} from O to B, and similarly infinitesimally displace the vector dy^{μ} on amount dx^{μ} from O to A. We obtain therefore two new infinitesimal vectors $dx^{\mu}-\Gamma^{\mu}_{\lambda\sigma}\,dx^{\lambda}\,dy^{\sigma}$ and $dy^{\mu}-\Gamma^{\mu}_{\lambda\sigma}\,dy^{\lambda}\,dx^{\sigma}$ respectively. These four vectors make the jagged curve of Fig. 7.11.

The point now is to observe that the jagged curve of Fig. 7.11 will actually close to form a parallelogram if the vector $\mathbf{CD}=\mathbf{0}$. This requirement is simply that

$$OC-OD=0\Rightarrow\mathbf{OA}+\mathbf{AC}-O\mathbf{B}-\mathbf{BD}=0$$

$$\Rightarrow(\Gamma^{\mu}_{\lambda\sigma}-\Gamma^{\mu}_{\lambda\sigma})\,dx^{\lambda}\,dy^{\sigma}=0$$

i.e. that

$$\Gamma^{\mu}_{\lambda\sigma}=\Gamma^{\mu}_{\lambda\sigma}.$$

So zero torsion corresponds to the closing of infinitesimal parallelograms under parallel translation.

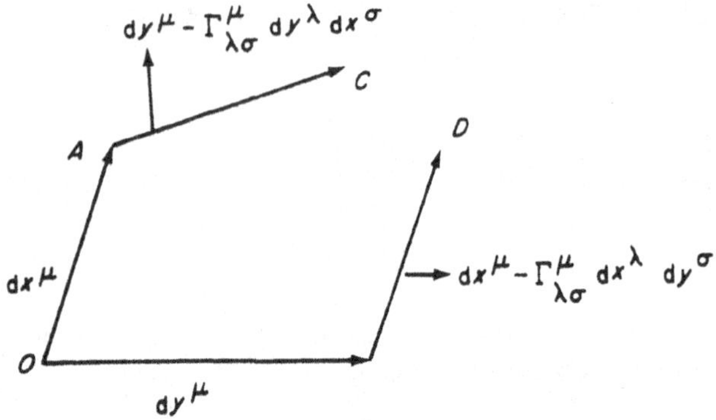

Figure 7.11

We have not dealt with pseudo-Riemannian effects here. The results that we have discussed do also hold for the pseudo-Riemannian case where the metric is no longer positive definite.

7.19 THE YANG–MILLS CONNECTION

The next category of connection to discuss is that where G is some non-Abelian Lie group. This corresponds to the connection of Yang–Mills theories. The connection $A^a_\mu(x)$ is what is called the gauge potential in physics, and the curvature $F^a_{\mu\nu}$ is called the gauge field or field tensor. Physically speaking the gauge potential is meant to reflect the existence of a certain number of massless spin one particles. There are also however, other kinds of massless particles and massive particles in the real world. These are often called matter fields by the physicist. Since we are going to discuss Yang–Mills theories in more detail later we shall not discuss them now, but we do wish to say a few words about matter fields.

Matter fields have the group theoretic property that they belong to some vector space V, of dimension n say, which is acted on by the group G. The group will act on this vector space V via a representation ρ, the representation can be a vector representation, or a tensor representation, and the representation may be by real or complex matrices. Now let us assume, just for the sake of being specific, that the representation ρ is, (i) a vector representation of dimension m, (ii) the representation is real. Then the vector space V is acted by representation matrices belonging to $Gl(m, \mathbf{R})$. Hence a bundle E containing this information can easily be constructed. E is an m-dimensional real vector bundle over M with fibre isomorphic to $\mathbf{R}^m$ and structure group G where G acts on the fibres of E by $Gl(m, \mathbf{R})$ matrices of the representation ρ. A matter field $\phi(x)$ is a section of E where $\phi(x)$ is a vector with m components:

$$\phi(x)=(\phi_1,\ldots,\phi_m) \tag{7.180}$$

In physics these m components of ϕ correspond to an m-plet of particles. The connection Γ in P induces a connection in E which we shall denote by Γ^E. The calculation of the connection 1-form Γ^E is similar to the calculation for $T(M)$. One takes a *fixed* m-dimensional vector η and regards it as a point in E. Then if $\tilde{\gamma}(t)$ is the horizontal lift in P of a curve $\gamma(t)$ in M then $\gamma_E(t)$ is the horizontal lift of $\gamma(t)$ in E with

$$\gamma_E(t)=\rho(\gamma(t))\eta \tag{7.181}$$

where $\rho(\gamma(t))$ is the matrix in $Gl(m,\mathbf{R})$ which corresponds to the group element $\gamma(t)$ under the representation ρ. Differentiation with respect to t defines the covariant derivative in E, and reference to local coordinates produces the Christoffel-like symbols $\Gamma^{a^E}_{\mu b}$ in E where $\mu=1,\ldots,n$ and $a,b=1,\ldots,m$; these calculations being similar to those already done for the vector bundle $T(M)$. Note the difference between the Greek and Latin indices. In general these fields $\phi(x)$ will have a vectorial or tensorial character corresponding to their possessing angular momentum for example. This means that $\phi(x)$ instead of taking values solely in the bundle E, takes values in the bundle $E\otimes T(M)$ or $E\otimes T^p_q(M)$. (If E and F are vector bundles over M, then $E\otimes F$ means the bundle fibre at $x\in M$ is the tensor product of the corresponding fibres of E and F at x. The transition functions $g_{\alpha\beta}(x)$ of $E\otimes F$ are then tensor products $g^E_{\alpha\beta}\otimes g^F_{\alpha\beta}(x)$ of the transition functions of E and F.) This means that the section $\phi(x)$ of E is also a section of $T(M)$ say and thus of $E\otimes T(M)$. We display this information by writing the equation

$$\phi(x)=\phi^a(x)a^\mu(x)\mathbf{e}_a\otimes\frac{\partial}{\partial x^\mu} \tag{7.182}$$

A covariant derivative $\nabla_{\mathbf{X}}\phi(x)$ now involves both the connection Γ^E (which is $gl(m,\mathbf{R})$—valued by virtue of the representation ρ of G), and the $gl(n,\mathbf{R})$—valued connection Γ in $T(M)$. One can calculate using local coordinates, that if $\mathbf{Y}=b^\mu(\partial/\partial x^\mu)$, then if we denote the covariant derivative in $E\otimes T(M)$ by $\nabla^{E\otimes T(M)}_{\mathbf{Y}}$ and those in E and $T(M)$ by $\nabla^E_{\mathbf{Y}}$ and $\nabla_{\mathbf{Y}}$ respectively, we have

$$\begin{aligned}\nabla^{E\otimes T(M)}_{\mathbf{Y}}(\phi(x))&=\nabla^{E\otimes T(M)}_{\mathbf{Y}}\left(\phi^a(x)a^\mu(x)e_a\otimes\frac{\partial}{\partial x^\mu}\right)\\&=\nabla^E_{\mathbf{Y}}(\phi^a(x)\mathbf{e}_a)\otimes\left(a^\mu\frac{\partial}{\partial x^\mu}\right)+(\phi^a(x)\mathbf{e}_a)\otimes\nabla_{\mathbf{Y}}\left(a^\mu\frac{\partial}{\partial x^\mu}\right)\end{aligned} \tag{7.183}$$

But

$$\nabla^E_{\mathbf{Y}}(\phi^a(x)\mathbf{e}_a(x)) = c^\lambda\left(\frac{\partial\phi^a(x)}{\partial x^\lambda} + \Gamma^{aE}_{\lambda b}\phi^b(x)\right)\mathbf{e}_a$$
$$= c^\lambda \phi^a_{;\lambda}\mathbf{e}_a$$

and

$$\nabla_{\mathbf{Y}}\left(a^\mu \frac{\partial}{\partial x^\mu}\right) = c^\lambda\left(\frac{\partial a^\mu}{\partial x^\lambda} + \Gamma^\mu_{\lambda\sigma}a^\sigma\right)\frac{\partial}{\partial x^\mu}$$
$$= c^\lambda a^\mu_{;\lambda}\frac{\partial}{\partial x^\mu} \tag{7.184}$$

Thus

$$\nabla^{E\otimes T(M)}_{\mathbf{Y}}\left(\phi^a a^\mu \mathbf{e}_a \otimes \frac{\partial}{\partial x^\mu}\right) = c^\lambda \phi^a_{;\lambda} a^\mu \mathbf{e}_a \otimes \frac{\partial}{\partial x^\mu}$$
$$+ c^\lambda \phi^a a^\mu_{;\lambda}\mathbf{e}_a \otimes \frac{\partial}{\partial x^\lambda} \tag{7.185}$$

with $\mathbf{Y} = c^\lambda\, \partial/\partial x^\lambda$.

The connection $\mathbf{\Gamma}^{E\otimes T(M)}$ giving rise to $\nabla^{E\otimes T(M)}_{\mathbf{Y}}$ can have some of its properties summarized by noting that

$$\mathbf{\Gamma}^{E\otimes T(M)} = \mathbf{\Gamma}^E \otimes \mathbf{I} + \mathbf{I} \otimes \mathbf{\Gamma} \tag{7.186}$$

The curvatures satisfy

$$\mathbf{R}^{E\otimes T(M)} = \mathbf{R}^E \otimes \mathbf{I} + \mathbf{I} \otimes \mathbf{R} \tag{7.187}$$

in an obvious notation. Note that $\mathbf{R}^{E\otimes T(M)} \neq \mathbf{R}^E \otimes \mathbf{R}$, this would imply that if one or other of $\mathbf{R}^E$ and $\mathbf{R}$ were zero then $\mathbf{R}^{E\otimes T(M)}$ would be zero. This is not true. For example one can have a flat space time $(\mathbf{\Gamma} = \mathbf{0})$ and a curved Yang–Mills geometry $(\mathbf{\Gamma}^E \neq \mathbf{0})$; and since this gives, from (7.186), $\mathbf{\Gamma}^{E\otimes T(M)} \neq \mathbf{0}$, then we would have $\mathbf{R}^{E\otimes T(M)} \neq O$ excluding trivial $\mathbf{\Gamma}^E$.

7.20 THE MAXWELL CONNECTION

Our third and last geometrical illustration is the geometry of the Maxwell field. The group G to be considered is the group $U(1)$. Since $U(1)$ is Abelian this simplifies enormously the geometry. In particular the curvature from $\mathbf{F}$ is given by

$$\mathbf{F} = \mathrm{d}\mathbf{A} \tag{7.188}$$

since $\mathbf{A}\wedge\mathbf{A}=\mathbf{0}$ on account of the Abelian nature of $U(1)$. If we write (7.188) in terms of $F_{\mu\nu}$ and A_μ we obtain the usual relation

$$F_{\mu\nu}=\partial_\mu A_\nu-\partial_\nu A_\mu \tag{7.189}$$

The Bianchi identities

$$D\mathbf{F}=\mathrm{d}\mathbf{F}+\mathbf{A}\wedge\mathbf{F}-\mathbf{F}\wedge\mathbf{A}=\mathbf{0}$$

become

$$\mathrm{d}\mathbf{F}=\mathbf{0} \tag{7.190}$$

consistent with $\mathbf{F}=\mathrm{d}\mathbf{A}$. The Maxwell equations are

$$D{}^*\mathbf{F}=O$$

and simplify to

$$d{}^*\mathbf{F}=\mathbf{0} \tag{7.191}$$

If we write these out in terms of components we obtain the familiar form (we leave the details as a simple exercise)

$$\partial_\mu F_{\mu\nu}=O \tag{7.192}$$

which together with the relations

$$\begin{aligned} E_i&=F_{i4;} \qquad i=1,2,3\\ B_i&=\tfrac{1}{2}\varepsilon_{ijk}F_{jk}; \qquad i,jk=1,2,3 \end{aligned} \tag{7.193}$$

imply the usual four Maxwell equations in empty space

$$\begin{aligned} \nabla\cdot\mathbf{E}&=\nabla\cdot\mathbf{B}=O\\ \nabla\times\mathbf{E}&=-\frac{\partial\mathbf{B}}{\partial t}\\ \nabla\times\mathbf{B}&=\frac{\partial\mathbf{E}}{\partial t} \end{aligned} \tag{7.194}$$

The Abelian nature of the gauge group $U(1)$ simplifies the transformation properties of $\mathbf{A}$ and $\mathbf{F}$ under gauge transformations: under gauge transformations

$$\begin{aligned} \mathbf{F}\to\mathbf{F}'&=h\mathbf{F}h^{-1}\\ &=\mathbf{F} \end{aligned} \tag{7.195}$$

The Lie algebra of $U(1)$ is simply $\mathbf{R}$ so that $\mathbf{F}$ instead of being a matrix valued form is just a real form so that $h\mathbf{F}h^{-1}=\mathbf{F}$, and $\mathbf{F}$ is invariant under gauge transformations. As for $\mathbf{A}$ we have

$$\begin{aligned}\mathbf{A}' &= hdh^{-1} + h\mathbf{A}h^{-1}\\ &= hdh^{-1} + \mathbf{A}\end{aligned} \tag{7.196}$$

or more explicitly if we write $h(x) = e^{i\wedge(x)}$ for some real function $\wedge(x)$ then (7.196), in terms of $A_\mu(x)$, is just the familiar gauge transformation of electromagnetism

$$A'_\mu(x) = A_\mu(x) + i\partial_\mu \wedge(x) \tag{7.197}$$

With regard to the bundle theoretic properties of Maxwell's equations, if the base space M of the $U(1)$ bundle is taken to be a flat manifold such as Minkowski space time, then since this manifold is topologically $\mathbf{R}^4$ and hence contractible then the bundle is trivial. This means that casting Maxwell's equations into bundle language teaches us nothing new. However, this conclusion is a little too hasty. If one tries to construct magnetic monopoles using Maxwell theory then one is led to consider the non-contractible space $\mathbf{R}^3 - \{0\}$ which is of the same homotopy type as S^2. Then one has a $U(1)$ bundle over S^2 and this bundle is not trivial. This will be dealt with when we discuss monopoles.

7.21 GENERAL REMARKS

To end this section we make a few short remarks relevant to the previous material but more conveniently given here.

The first remark is a simple definition of the sum of two vector bundles which is useful to have. If E and F are vector bundles over M then a bundle $E \oplus F$ over M called the Whitney sum bundle can be obtained by taking the fibres of $E \oplus F$ to be the direct sum of the fibres of E and F. The transition functions of $E \oplus F$ are simply the direct sum of the transition matrices g_E and g_F for E and F:

$$g_{E\oplus F} = \begin{pmatrix} g_E & 0 \\ 0 & g_F \end{pmatrix} \tag{7.198}$$

The second remark is simply a reminder that if for a linear connection $\mathbf{\Gamma}$, the torsion and curvature are both zero, then the connection $\mathbf{\Gamma}$ is called flat and there exist local coordinates in which $\mathbf{\Gamma}$ is zero also. The converse of course is also true, and if $\mathbf{\Gamma}$ is the Levi–Civita connection of relativity, which has zero torsion by definition, then one only has to demand that the curvature be zero. A general form for a flat connection is easily seen to be $\mathbf{\Gamma} = h^{-1}dh$. For principal fibre bundles a connection $\mathbf{A}$ is flat if and only if the curvature **F** is zero (there is no torsion to compute in this case), and as before a general form for a flat connection **A** is $\mathbf{A}=h^{-1}dh$. We are aware that we have not proved these important flatness properties of connections but we feel that the reader should have already met these properties at least for the Levi–Civita case. A number of different proofs can be found in reference [13].

The third and last remark is a technical loose end concerning pullbacks. We asserted above that the form $\boldsymbol{\omega}=g^{-1}dg+g^{-1}\mathbf{A}g$ which belongs to T^*P, and the form **A** which belongs to T^*M, were related by the fact that **A** is the pull-back of $\boldsymbol{\omega}$. We now show this by calculation as we promised above. As explained above one takes a section $s:U_\alpha\to P$ and then computes $s^*\boldsymbol{\omega}$, we point out that if M and N are two manifolds and $\rho=b_i\,\mathrm{d}y^i\in T^*N$, then the mechanism of the pullback of ρ to M given by a map $f:M\to N$ is just an application of the chain rule: for f is given by

$$f:M\to N$$

$$x\mapsto y=f(x)$$

so

$$\mathrm{d}y^i=\frac{\partial f^i(x)}{\partial x^\mu}\mathrm{d}x^\mu \tag{7.199}$$

and then $f^*\rho$ is the 1-form $\boldsymbol{\omega}=a_\mu\,\mathrm{d}x^\mu\in T^*M$ where

$$a_\mu=\frac{\partial f^i}{\partial x^\mu}b^i \tag{7.200}$$

(we use Greek and Latin indices to allow for the fact that in general the dimensions of M and N will be different).

Now all we have to do is to let $N=P, f=s, \rho=\boldsymbol{\omega}$ and compute. A choice of local section s is given by:

$$s:U_\alpha\to P$$

$$x^\mu\mapsto(x^\mu,g(x))=s(x) \tag{7.201}$$

Now write $\boldsymbol{\omega}$ as

$$\begin{aligned}\boldsymbol{\omega}&=a_a\,\mathrm{d}g^a+b_\mu\,\mathrm{d}x^\mu\\&=c_A\,\mathrm{d}Z^A\end{aligned} \tag{7.202}$$

In (7.202) the suffix a runs over the dimension of the group G and the suffix A runs over both a and μ; the definitions of a_a, b_μ and c_A are self-evident. Thus

$$s(x)\equiv s^A(x)=(x^\mu,g^a(x))=z^A \tag{7.203}$$

and therefore

$$dZ^A = \frac{\partial s^A(x)}{\partial x^\lambda} dx^\lambda \tag{7.204}$$

But obviously

$$\frac{\partial s^A(x)}{\partial x^\lambda} = \begin{Bmatrix} \delta_{\mu\lambda} & \text{if } A = \mu \\ \\ \dfrac{\partial g^a(x)}{\partial x^\lambda} & \text{if } A = a \end{Bmatrix} \tag{7.205}$$

Assembling these facts gives us

$$s^*\omega = \left\{ a_a \frac{g^a(x)}{\partial x^\lambda} + b_\lambda \right\} dx^\lambda \tag{7.206}$$

But this may be rewritten as

$$\begin{aligned} s^*\omega &= \{g^{-1}(x)\, \partial_\mu g(x) + g^{-1}(x) A_\mu(x) g(x)\}\, dx^\mu \\ &= A'_\mu(x)\, dx \end{aligned} \tag{7.207}$$

where $A'_\mu(x)$ is the gauge transform of $A_\mu(x)$ under the gauge change given by $g(x)$. So in general $s^*\omega$ is not quite $\mathbf{A}$ it is $\mathbf{A}$ up to a change of gauge. However, we are free to be even more specific in our choice of section s, if we make a local change of section, i.e. a gauge transformation, then we do indeed have

$$s^*\omega = \mathbf{A} \tag{7.208}$$

as promised.

7.22 CHARACTERISTIC CLASSES

In this closing topic of the chapter, we shall give a simple account of the classification of bundles. There are really two sorts of naturally arising questions that underly the desire to classify bundles. The first question is: given a total space E, a fibre F, and a base M, in how many ways can F and M be assembled to form different bundles E? The second question is: given a bundle E, can one develop a criterion, or criteria, for telling how far E is from being trivial? We shall provide a partial answer to the first question which is, in general, the more difficult of the two. The second question can be largely answered and the theory that provides the answer is called the theory of characteristic classes.

One classifies bundles by comparing them with certain special bundles called universal bundles. If one is classifying bundles, then the bundles will be either principal bundles or associated bundles. The associated bundles that we have to deal with are usually vector bundles, and these come in two varieties: real or complex according to whether their fibres are isomorphic to $\mathbf{R}^n$ or $\mathbf{C}^n$. We shall describe the classification for principal bundles P, recall that any bundle E is associated with a principal bundle and E is trivial$\Leftrightarrow P$ is trivial. Towards the end of this section we shall then add some remarks which should show the reader how to make the changes necessary to produce the classification of vector bundles that is induced by the classification of the principal bundles.

The starting point is the universal bundle and how it is connected to an arbitrary principal bundle P. To be more specific and to learn more easily what is going on, we first commit ourselves to a principal bundle with group $O(k)$ over a differentiable manifold M of dimension d; it is then very easy to proceed to the general case which shall be where P has as its fibre an arbitrary connected Lie group G. Given this principal bundle P with group $O(k)$, we compare it to a certain bundle denoted by $\xi(n-k-1, O(k))$ and called an $(n-k-1)$-universal bundle. The description of $\xi(n-k-1, O(k))$ is fairly simple and is as follows: The base space of this bundle is the (coset) space

$$O(n)/(O(k)\times O(n-k)) \tag{7.209}$$

This space is called the Grassmann manifold of k-dimensional planes† in $\mathbf{R}^n$ and is denoted by $Gr(n, k, \mathbf{R})$. So we have

$$Gr(n, k, \mathbf{R}) = O(n)/(O(k)\times O(n-k)) \tag{7.210}$$

(The RHS of (7.210) is of the form G/H where both G and H are compact Lie groups, since Lie groups are also differentiable manifolds then equation (7.210) shows that $Gr(n, k, \mathbf{R})$ is a compact manifold. The dimension of $Gr(n, k, \mathbf{R})$ is obtained by recalling that $\dim(G/H) = \dim G - \dim H$, so that since $\dim O(n) = n(n-1)/2$, we obtain $\dim Gr(n, k, \mathbf{R}) = k(n-k)$.) The total space of this bundle is

$$O(n)/O(n-k) \tag{7.211}$$

This is called a Stiefel manifold of k-frames in $\mathbf{R}^n$ and is sometimes written $V_{n,n-k}(\mathbf{R})$. The integer $n-k-1$ appearing in $\xi(n-k-1, O(k))$ is there

† A point in the manifold $Gr(n, k, \mathbf{R})$ corresponds to a k-dimensional plane through the origin of $\mathbf{R}^n$.

because $O(n)/O(n-k)$ has a string of vanishing homotopy groups, in fact [16]

$$\pi_p_{0 \le p \le n-k-1} (O(n)/O(n-k)) = 0 \tag{7.212}$$

and this is important in the details of the theory of universal bundles. A topological space X with the property $\Pi_p(X) = 0$, $0 \le p \le m$ is called m-connected, thus $O(n)/O(n-k)$ is $(n-k-1)$-connected. The real utility of displaying the integer $n-k-1$ in ξ is that $\xi(n-k-1, O(k))$ classifies principal bundles P, with group $O(k)$, over manifolds M whose dimension $d < n-k-1$. We shall amplify this remark later to demonstrate its significance.

The bundle $\xi(n-k-1, O(k))$ can be thought of more straightforwardly in a geometrical way: since a point in the base space $Gr(n, k, \mathbf{R})$ is a k-dimensional plane in $\mathbf{R}^n$, then the group $O(k)$ acts on such planes giving the fibre of $\xi(n-k-1, O(k))$ as $O(k)$. Because a trivial principal bundle with group G and base M is expressible as $M \times G$, then the non-triviality of $\xi(n-k-1, O(k))$ can be thought of as a measure of the extent to which $Gr(n, k, \mathbf{R}) \times O(k) = \{O(n)/(O(k) \times O(n-k))\} \times O(k)$ is *not* equal to $O(n)/O(n-k)$. Now we come to the classification itself, the comparison of an arbitrary principal bundle P with ξ. The comparison of P with ξ is made by means of the pullback bundle or induced bundle introduced in (7.42). The technique employed is to make use of the fact that if one takes a manifold M and maps it into $Gr(n, k, \mathbf{R})$, i.e. we have

$$f: M \to Gr(n, k, \mathbf{R}) \tag{7.213}$$

Then one can pullback the bundle ξ over $Gr(n, k, \mathbf{R})$ to the bundle $f^*\xi(n-k-1, O(k))$ over M. This is a principal bundle over M with the same group as $\xi(n-k-1, O(k))$ namely $O(k)$. Next there is a result [16] which we quote but do not prove: if P is a principal bundle over a manifold M of dimension d with group $O(k)$, then for a sufficiently large value of n there exists a map $f: M \to Gr(n, k, \mathbf{R})$ such that $f^*\xi(n-k-1, O(k)) = P$.

The content of this result is that every principal bundle with group $O(k)$ arises as a pullback bundle from $\xi(n-k-1, O(k))$. We can now amplify the remark about the significance of the integer $n-k-1$. In the above result if we fix n and k, then $f^*\xi(n-k-1, O(k))$ for various $f: M \to Gr(n, k, \mathbf{R})$ gives [16] all principal bundles P with group $O(k)$ over those manifolds whose dimension $d \le n-k-1$. Turning this statement round one can see that since usually one will be given the bundle P (and also M) first then it is d that should be thought of as fixed. Thus in choosing ξ we need to choose n sufficiently large, i.e. we should choose an n so that $n \ge d+k+1$ which is of course the same as $d \le n-k-1$.

When defining the pullback bundle in (7.42) we also had the result that if f and g were homotopic maps from $M \to Gr(n, k, \mathbf{R})$ say, then the corresponding pullback bundles $f^*\xi$ and $g^*\xi$ are equivalent. This property of $f^*\xi$ and $g^*\xi$ for $f \simeq g$ will provide us with what we referred to above as a partial answer to our first question concerning classification of bundles. Before explaining why this is so we wish to go on to show that the result $f^*\xi(n-k-1, O(k)) = P$ is not really restricted to principal bundles with group $O(k)$ but applies to any connected Lie group.

Let us first consider a principal bundle whose group is any compact Lie group G. If G is compact then G is isomorphic to a subgroup of some $O(m)$ for m large enough. To see this [6] let n be the dimension of the group G and consider a faithful representation of G over a real vector space V, take any positive definite inner product on V, and construct out of it a G-invariant inner product by averaging over the group. This will be a real, positive definite, group invariant inner product and it must therefore give a representation of G by orthogonal matrices in fact by matrices in $O(m)$ for some m. Thus G is isomorphic to a subgroup of $O(m)$, e.g. $U(n)$ is isomorphic in a natural way to a subgroup of $O(2n)$.

Now to construct the required universal bundle $\xi(n-k-1, G)$ we replace the base space $Gr(n, k, \mathbf{R}) = O(n)/(O(k) \times O(n-k))$ by $O(n)/(G \times O(n-k))$, the fibre of $\xi(n-k-1, G)$ is G by assumption, and the total space is the same as before i.e. $O(n)/O(n-k)$. Thus the bundle $\xi(n-k-1, G)$ just described is $(n-k-1)$-universal for a principal bundle with compact group G. Finally if we no longer restrict G to be compact but allow G to be an arbitrary connected Lie group then G has a maximal compact subgroup H. But we saw in (7.49) that P is then equivalent to a bundle with group H, so $\xi(n-k-1, H)$ is $(n-k-1)$-universal for a principal bundle P with arbitrary connected Lie group G and maximal compact subgroup H.

Returning then to our first classification question concerning bundles let us suppose that one wishes to know how many principal bundles P with a certain group G there are over a certain base M. Every such bundle P is a pullback bundle $f^*\xi$ for some map $f: M \to O(n)/(G \times O(n-k))$. But maps f and g which are homotopic produce equivalent bundles. This means that there is a one to one correspondence between equivalence classes of principal bundles P with group G over M and homotopy classes of maps from M to $O(n)/(G \times O(n-k))$. Thus if we know, or can calculate, how many homotopy classes of such maps there are, then one finds the number of bundles P that are possible. In general homotopy calculations are difficult to do so that the calculation will be impossible and the question left unanswered.

We are now ready to discuss the answer to the second classification question which is concerned with developing criteria which tell one how far a bundle is from being trivial. The general thinking which is used is to develop some function c say, which tells one how non-trivial a universal bundle $\xi(n-k-1, G)$ is and then somehow to pullback this function to the bundle P being studied. So proceeding along these lines, and having the advantage of hindsight, without quite knowing what sort of mathematical animal c is, we would require of c two things:

i. $c(P)=c(P')$ if P is equivalent to P'

ii. $f^*c(\xi)=c(f^*\xi)$ (7.214)

If we refer to Chapter 6 where we discussed cohomology groups then it is possible to verify that if $c(P)$ for a bundle P over M is taken to be an element of $H^p(M;\mathbf{R})$ then c will satisfy both requirements of (7.214): Requirement (i) is satisfied because equivalent bundles arise from homotopic maps f as described above so if f and g are homotopic then $P=f^*\xi$ and $P'=g^*\xi$ are equivalent. However if $c(P)$ and $c(P')$ belong to $H^p(M;\mathbf{R})$ we have, using (6.39),

$$\begin{cases} c(P)=c(f^*\xi)=f^*c(\xi) \\ c(P')=c(g^*\xi)=g^*c(\xi) \end{cases} \tag{7.215}$$

But since f and g are homotopic, then (6.44) tells us that $f^*c(\xi)=g^*c(\xi)$ and requirement (i) is satisfied. Requirement (ii) is also satisfied since we have already in (7.215) made use of the information contained in (6.39) that $c(f^*\xi)=f^*c(\xi)$. The object $c(\xi)$ is called a characteristic class and $c(\xi)\in H^p(B;\mathbf{R})$ where B is the base space of the bundle ξ. Thus the classification of bundles into categories determined by how twisted or how far they are from being non-trivial is done by first computing $c(\xi)$ for all $(n-k-1)$-universal bundles, i.e. by computing $H^p(O(n)/(G\times O(n-k));\mathbf{R})$, and then one pulls this cohomology group back to the bundle $P=f^*\xi$ via the map f so that $c(P)\in H^p(M;\mathbf{R})$.

7.23 CHERN, PONTRJAGIN AND EULER CLASSES

Having described the work of constructing characteristic classes as being the computation of the cohomology groups $H^p(O(n)/(G\times O(n-k));\mathbf{R})$, it is appropriate to point out that in practice one most frequently encounters three types of group G. These are $G=O(k)$, $G=SO(k)$, and $G=U(k)$. For principal bundles with these three types of group the corresponding characteristic classes have acquired names. For principal bundles with $G=O(k)$ the characteristic classes are called Pontrjagin classes, if $G=SO(k)$ there are still the Pontrjagin classes together with one new characteristic class called the Euler class which is only non-zero when k is even.

Finally for $G = U(k)$ all the characteristic classes are called Chern classes. Finally, there is a fourth set of characteristic classes called the Stiefel–Whitney classes, and these are defined for $O(k)$ principal bundles. They do not arise from calculating cohomology groups with real coefficients such as $H^p(M; \mathbf{R})$ but rather from cohomology groups with $\mathbf{Z}/2$ coefficients i.e. from $H^p(M; \mathbf{Z}/2)$. The Stiefel–Whitney classes are however, important for physicists since they can be used, as we shall see later, to tell whether or not a manifold M admits a spin structure. A calculational detail for principal bundles with the above three types of group is that for the latter two types, $(G = SO(k)$ or $U(k))$, rather than using the universal bundles ξ described above; it is more convenient to change the universal bundles a little. For the case $G = SO(k)$ one simply changes $O(n)$ wherever it appears to $SO(n)$ and works with the resultant (oriented) bundle denoted by $\tilde{\xi}$ say. For the case $G = U(k)$ it is convenient to replace $O(n)$ wherever it appears by $U(n)$ and work with the resultant universal bundle denoted by ξ_c say. The base space of $\tilde{\xi}$ is thus $SO(n)/(SO(k) \times SO(n-k))$ and is called the oriented Grassmann manifold and written $\tilde{G}r(n, k, \mathbf{R})$, the base space of ξ_c is similarly $U(n)/(U(k) \times U(n-k))$ and is called the complex Grassmann manifold and is written $Gr(n, k, \mathbf{C})$:

$$Gr(n, k, \mathbf{C}) = U(n)/(U(k) \times U(n-k)) \tag{7.216}$$

Incidentally from the equality of $Gr(n, k, \mathbf{C})$ with the coset spaces of the unitary groups in (7.216) we see that $Gr(n, k, \mathbf{C})$ is a *compact* differentiable manifold of real dimension $2k(n-k)$. This doubling of the dimension when passing from $Gr(n, k, \mathbf{R})$ to $Gr(n, k, \mathbf{C})$ also has the effect of making ξ_c a $(2n-2k)$-universal bundle rather than an $(n-k-1)$-universal bundle $\xi_c \equiv \xi_c(2n-2k, U(n))$. Now we take $G = O(k)$ and describe the Pontrjagin classes. Suppose one has a principal bundle P over M, ($\dim M = d$), with group $O(k)$. Then the characteristic classes for P belong, via the pullback, to $f^*H^i(Gr(n, k, \mathbf{R}); \mathbf{R})$, i.e. to $H^i(M; \mathbf{R})$ and are denoted† by $p_{i/4}(P)$. We have not the space here to compute $H^i(Gr(n, k, \mathbf{R}); \mathbf{R})$, we shall simply quote results. A very readable account of the necessary computations can be found in reference [14]. It is actually only necessary to consider $i \leqslant n-k-1$. One can easily see this because, $H^i(M; \mathbf{R})$, which is what the characteristic class belongs to, is only non-zero for $i \leqslant d$; but from above $d \leqslant n-k-1$ and so we only need those i for which $i \leqslant n-k-1$. In any case, for such i, $H^i(Gr(n, k, \mathbf{R}); \mathbf{R})$ is only non-zero [14] when i is a multiple of 4. So the possible non-zero Pontrjagin classes $p_{i/4}(P)$ of P belong to $H^i(M; \mathbf{R})$ with i a multiple of 4, and this is more conveniently written as $p_i(P) \in H^{4i}(M; \mathbf{R})$. Thus in terms of de Rham cohomology the Pontrjagin classes are all given by 4-forms, 8-forms etc. until the dimension of d or

---

† We shall see the reason for the suffix $i/4$ in a moment.

some other reason forces them to vanish. We can also take the opportunity here to define the Euler class. If k is even, say $k=2m$, then if $G=SO(k)$, the cohomology group $H^k(Gr(n, k, \mathbf{R}); \mathbf{R})$ will in general be non-zero, and so will be $H^k(M; \mathbf{R})=f^*H^k(\tilde{G}r(n, k, \mathbf{R}); \mathbf{R})$. One of the k-forms belonging to $H^k(M; \mathbf{R})$ is then called the Euler characteristic class and denoted by $e(P)$. The definition of this particular k-form is in terms of the Pontrjagin class: for because k is even, $k=2m$, the Euler class $e(P)$ is a $2m$-form so that $e(P)\wedge e(P)$ is a $4m$-form and thus belongs† to $H^{4m}(M; \mathbf{R})$. Thus if we take the $4m$-form $p_m(P)$, $m=k/2$, then the Euler class is defined to be that k-form which satisfies

$$e(P)\wedge e(P)=p_{k/2}(P) \tag{7.217}$$

The third type of group is $G=U(k)$ and then the characteristic classes for P belong to $f^*H^i(Gr(n, k, \mathbf{C}); \mathbf{R})$, i.e. to $H^i(M; \mathbf{R})$, they are denoted by $c_{i/2}(P)$ and are called Chern classes. The cohomology groups that we need to compute are $H^i(Gr(n, k, \mathbf{C}); \mathbf{R})$. As before we refer the reader to Spivak [14] for an account of the computations. The result is that the $H^i(Gr(n, k, \mathbf{C}); \mathbf{R})$ are only non-zero for even i. This means that the possible non-zero Chern classes $c_{i/2}(P)$ belong to $H^i(M; \mathbf{R})$, and again this is more conveniently written as $c_i(P)\in H^{2i}(M; \mathbf{R})$. Finally the fourth characteristic class, the Stiefel–Whitney class $w_i(P)$ is obtained using cohomology with $\mathbf{Z}/2$ coefficients, and is defined for principal bundles P over M with group $G=O(k)$. Their computation [12] requires the computation of $H^i(Gr(n, k, \mathbf{R}); \mathbf{Z}/2)$, and $w_i(P)$, which is defined for $0\leq i\leq n-1$, belongs to $f^*H^i(Gr(n, k, \mathbf{R}); \mathbf{Z}/2)$ i.e. to $H^i(M; \mathbf{Z}/2)$.

7.24 CHARACTERISTIC CLASSES IN TERMS OF CURVATURE AND INVARIANT POLYNOMIALS

The next stage in our discussion of characteristic classes is to learn how to calculate them. For the first three classes discussed, i.e. Pontrjagin, Euler and Chern, there is a remarkably satisfying and routine method of calculation. We now describe this method. So far we have said that the three characteristic classes above belong to $H^i(M; \mathbf{R})$ for various values of i. This means that they are given by certain closed, non-exact, differential forms on M. The remarkable thing is that $c_i(P)$, $p_i(P)$ and $e(P)$ are all given by various differential forms which are polynomials in $\mathbf{F}$, where $\mathbf{F}$ is a curvature 2-form for the bundle P over M. At first this claim seems highly unlikely to be true, since if one makes a different choice of connection on P then

† We know that $e(P)\wedge e(P)$ must belong to $H^{4m}(M; \mathbf{R})$ because of the cup product cf. (6.82).

the curvature form $\mathbf{F}$ will be different. Indeed this is so, but these polynomials in $\mathbf{F}$ have the marvellous property that they are actually independent (up to an exact form which is zero cohomologically speaking) of the connection $\mathbf{A}$ which determines the curvature $\mathbf{F}$. This independence is due to the fact these polynomials are specially chosen to be invariants of the Lie algebra $\mathfrak{g}$ of G. It is therefore necessary to examine these Lie algebra invariants. If $\mathfrak{g}$ is the Lie algebra of G, then the invariants of $\mathfrak{g}$ are all formed by taking polynomials in the generators $\{X_i\}$ of $\mathfrak{g}$, e.g. the Casimir operator for a compact semi-simple $\mathfrak{g}$ is $X_1^2+\ldots+X_n^2$, where n is the dimension of G. It is well known in the theory of Lie algebras (cf. ref. [14], pp. 519–523) that all these invariants are obtained by expanding, in powers of t, the $m\times m$ determinant given below in (7.218)

$$\text{Det}\,(tI+a_iX_i)=\sum_{j=0}^{m}P_{m-j}(a_i)t^j \tag{7.218}$$

Equation (7.218) defines the polynomials† $P_j(a_i)$, the (matrices) X_i have vanished on taking the determinant. The desired invariants of $\mathfrak{g}$ are formed by simply taking $P_j(a_i)$ and substituting X_i for a_i thus forming $P_j(X_i)$. The fact that these are invariant polynomials is immediate: a polynomial $P(X)$, $X\in\mathfrak{g}$, is invariant if $P(g^{-1}Xg)=P(X)$ where $g\in G$; using the fact that for a matrix T, $\text{Det}\,(g^{-1}Tg)=\text{Det}\,T$, one establishes the invariance of $P_j(a_i)$. The $P_j(a_i)$ are also homogeneous polynomials of degree j so that $P_j(\lambda a_i)=\lambda^jP_j(a_i)$, this can easily be checked. The vital introduction of curvature into the mathematics comes about by substituting the curvature form $\mathbf{F}$ for a_iX_i in (7.218), actually, to be precise, one substitutes‡ $i\mathbf{F}/2\pi$ for a_iX_i. This is reasonable since $\mathbf{F}$ is Lie-algebra-valued. The result is a form-valued version of (7.218).

$$\text{Det}\left(tI+\frac{i\mathbf{F}}{2\pi}\right)=\sum_j t^jP_{m-j}(\mathbf{F}) \tag{7.219}$$

Because $P_j(\mathbf{F})$ is homogeneous of degree j and $\mathbf{F}$ is a 2-form, then $P_j(\mathbf{F})$ is a $2j$-form. Now let $G=U(n)$. It turns out [14] that the particular $2i$-form corresponding to the Chern class $c_i(P)$ is given by $P_i(\mathbf{F})$.

$$c_i(P)=P_i(\mathbf{F}) \tag{7.220}$$

$\mathbf{F}$ is a 2-form and P_i has degree i so at least $P_i(\mathbf{F})$ is a $2i$-form, but we should also show that $P_i(\mathbf{F})$ is closed otherwise it cannot belong to $H^{2i}(M;\mathbf{R})$.

---

† Note that $P_{m-j}(a_i)$ rather than $P_j(a_i)$ is the coefficient of t^j.

‡ The i in front of the $\mathbf{F}$ is included because the X_i are complex matrices and the factor of i keeps the $P_j(a_i)$ real, although for even j they would be real anyway. The factor of $1/2\pi$ is included because the $c_i(P)$ actually determine integral cohomology classes, i.e. elements of $H^{2i}(M;\mathbf{Z})$, and they need a normalization factor of $1/2\pi$ to do this.

Here we have two important claims to establish.

$$\begin{aligned}&\text{i. } P_i(\mathbf{F}) \text{ is closed}\\ &\text{ii. } P_i(\mathbf{F}) \text{ is independent of the connection } \mathbf{A} \text{ used to compute } \mathbf{F}\end{aligned} \qquad (7.221)$$

We establish (i) first and (ii) second. To prove (i), it is convenient to think of $P_j(\mathbf{F})$ as originating from an invariant symmetric multilinear form in j variables denoted by $P(x_1, \ldots, x_j)$. The relation of $P(x_1, \ldots, x_j)$ to P_j is that when all the $x_1, \ldots, x_j$ are equal we obtain a polynomial of degree j or

$$P(x, \ldots, x) = P_j(x) \qquad (7.222)$$

Having done this, it is an exercise† to show that for any form $\mathbf{F}$ the invariance and symmetry of $P_j(\mathbf{F})$ yields the formula

$$\begin{aligned}\mathrm{d}P_j(\mathbf{F}) &= P(D\mathbf{F}, \mathbf{F}, \ldots, \mathbf{F}) + P(\mathbf{F}, D\mathbf{F}, \ldots, \mathbf{F}) + \ldots P(\mathbf{F}, \ldots, D\mathbf{F})\\ &= jP(D\mathbf{F}, \mathbf{F}, \ldots, \mathbf{F})\end{aligned} \qquad (7.223)$$

But if $\mathbf{F}$ is a curvature form $D\mathbf{F} = 0$ by the Bianchi identities (7.127), thus the RHS of (7.223) is zero and $P_j(\mathbf{F})$ is closed as claimed.

To prove (ii) we shall show that if $\mathbf{A}$ and $\mathbf{A}'$ are two connections with curvatures $\mathbf{F}$ and $\mathbf{F}'$ respectively then the difference in the polynomials P_j is exact

$$P_j(\mathbf{F}) - P_j(\mathbf{F}') = \mathrm{d}\boldsymbol{\eta} \qquad (7.224)$$

for some $\boldsymbol{\eta}$, thus the $2j$-forms $P_j(\mathbf{F})$ and $P_j(\mathbf{F}')$ belong to the same cohomology class in $H^{2i}(M; \mathbf{R})$ or

$$[P_j(\mathbf{F})] = [P_j(\mathbf{F}')] \qquad (7.225)$$

Consider then the two connections $\mathbf{A}$ and $\mathbf{A}'$ and define a family of connections $\mathbf{A}^t$ by

$$\begin{aligned}\mathbf{A}^t &= \mathbf{A} + t(\mathbf{A}' - \mathbf{A})\\ &= \mathbf{A} + t\mathbf{a}\end{aligned} \qquad (7.226)$$

where $\mathbf{a} = \mathbf{A}' - \mathbf{A}$. Let $\mathbf{A}^t$ have curvature $\mathbf{F}^t$, then as t varies from 0 to 1, $\mathbf{A}^t$ varies from $\mathbf{A}$ to $\mathbf{A}'$ and $\mathbf{F}^t$ from $\mathbf{F}$ to $\mathbf{F}'$. Note that we may write

$$P_j(\mathbf{F}) - P_j(\mathbf{F}') = \int_0^1 \frac{\mathrm{d}}{\mathrm{d}t} P_j(\mathbf{F}^t)\, \mathrm{d}t \qquad (7.227)$$

Actually we show below that $\mathrm{d}P_j/\mathrm{d}t\ (\mathbf{F}^t)$ is exact for $0 \leqslant t \leqslant 1$,

$$\frac{\mathrm{d}P_j}{\mathrm{d}t}(\mathbf{F}^t) = \mathrm{d}\boldsymbol{\theta}(t), \qquad 0 \leqslant t \leqslant 1 \qquad (7.228)$$

† e.g. one can start by using the definition of covariant derivative as $\mathrm{d}/\mathrm{d}t$ along $\gamma(t)$.

Thus

$$P_j(\mathbf{F})-P_j(\mathbf{F}')=\int_0^1 \mathrm{d}\boldsymbol{\theta}(t)\,\mathrm{d}t$$
$$=d\int_0^1 \boldsymbol{\theta}(t)\,\mathrm{d}t$$
$$=\mathrm{d}\boldsymbol{\eta} \tag{7.229}$$

where $\boldsymbol{\eta}=\int_0^1 \boldsymbol{\theta}(t)\,\mathrm{d}t$, and hence we have our required result. Now we have to show that $\mathrm{d}P_j/\mathrm{d}t\;(\mathbf{F}^t)$ is exact for $0\leqslant t\leqslant 1$. Well if $\mathrm{d}P_j/\mathrm{d}t\;(\mathbf{F}^t)$ is exact for $t=0$, then it is also exact for any subsequent value of t, t' say. For one could always replace the interval [0, 1] by the interval $[t, 1]$ and consider instead of $\mathbf{A}$ and $\mathbf{A}'$ the connections $\mathbf{A}^t$ and $\mathbf{A}'$. Thus it is sufficient to prove that $\mathrm{d}P_j/\mathrm{d}t\;(\mathbf{F}^t)|_{t=0}$ is exact. This follows because,

$$\frac{\mathrm{d}}{\mathrm{d}t}P_j(\mathbf{F}^t)=\frac{\mathrm{d}}{\mathrm{d}t}P(\mathbf{F}^t,\mathbf{F}^t,\ldots,\mathbf{F}^t)$$
$$=P\left(\frac{\mathrm{d}F^t}{\mathrm{d}t},\mathbf{F}^t,\ldots,\mathbf{F}^t\right)+P\left(\mathbf{F}^t,\frac{\mathrm{d}\mathbf{F}^t}{\mathrm{d}t},\ldots,F\right)$$
$$+\ldots P\left(\mathbf{F},\mathbf{F},\ldots,\frac{\mathrm{d}\mathbf{F}^t}{\mathrm{d}t}\right) \tag{7.230}$$

But

$$\mathbf{F}^t=\mathrm{d}\mathbf{A}^t+\mathbf{A}^t\wedge\mathbf{A}^t$$
$$=\mathrm{d}(\mathbf{A}+t\mathbf{a})+(\mathbf{A}+t\mathbf{a})\wedge(A+t\mathbf{a})$$
$$=\mathbf{F}+t\,(\mathrm{d}\mathbf{a}+\mathbf{A}\wedge\mathbf{a}+\mathbf{a}\wedge\mathbf{A})+t^2\mathbf{a}\wedge\mathbf{a} \tag{7.231}$$

and therefore†

$$\left.\frac{\mathrm{d}\mathbf{F}^t}{\mathrm{d}t}\right|_{t=0}=D\mathbf{a} \tag{7.232}$$

Combining (7.230–232) we obtain

$$\left.\frac{\mathrm{d}}{\mathrm{d}t}P_j(\mathbf{F}^t)\right|_{t=0}=P(D\mathbf{a},\mathbf{F},\ldots,\mathbf{F})+P(\mathbf{F},D\mathbf{a},\ldots,\mathbf{F})+\ldots P(\mathbf{F},\ldots,\mathbf{F},D\mathbf{a})$$
$$=jP(D\mathbf{a},\mathbf{F},\ldots,\mathbf{F})$$
$$=\mathrm{d}(jP(\mathbf{a},\mathbf{F},\ldots,\mathbf{F})) \tag{7.233}$$

† Note that $D\mathbf{a}=\mathrm{d}\mathbf{a}+\mathbf{A}\wedge\mathbf{a}+\mathbf{a}\wedge\mathbf{A}$ because $\mathbf{a}$, being the *difference* of two connections, transforms as a matrix-valued 1-form cf. discussion before equation (7.125).

where in the last line we have used a very similar formula to (7.223) and the fact that $D\mathbf{F}=\mathbf{0}$ causes most of the terms on the RHS of (7.233) to vanish. Finally, we can summarize (7.233) using the notation of (7.228) as

$$\frac{\mathrm{d}}{\mathrm{d}t}P_j(\mathbf{F}^t)\bigg|_{t=0}=\mathrm{d}\boldsymbol{\theta}(o) \tag{7.234}$$

where the $(2j-1)$-form $\boldsymbol{\theta}(0)$ is given by

$$\boldsymbol{\theta}(0)=jP(\mathbf{a},\mathbf{F},\ldots,\mathbf{F}) \tag{7.235}$$

We can go on to give the formulae relating the Pontrjagin and Euler classes to curvatures. The Pontrjagin classes are given by substituting $-\mathbf{F}/2\pi$ for a_iX_i in (7.218), the group G has now been chosen to be $O(n)$ so that $\mathbf{F}$ is $o(n)$-valued ($o(n)$ is the Lie algebra of $O(n)$). We obtain

$$\operatorname{Det}\left(tI-\frac{\mathbf{F}}{2\pi}\right)=\sum_j t^jP_{m-j}(\mathbf{F}) \tag{7.236}$$

and the fact that $\mathbf{F}$ is $o(n)$-valued causes $P_j(\mathbf{F})=0$ for j odd. To see this we simply observe that because $\mathbf{F}\in o(n)$, $\mathbf{F}^T=-\mathbf{F}$ ($T\equiv$ transpose) so we have

$$\begin{aligned}\operatorname{Det}\left(tI-\frac{\mathbf{F}}{2\pi}\right)&=\operatorname{Det}\left(tI-\frac{\mathbf{F}}{2\pi}\right)^T\\&=\operatorname{Det}\left(tI+\frac{\mathbf{F}}{2\pi}\right)\end{aligned} \tag{7.237}$$

and from this we obtain in terms of the $P_j(\mathbf{F})$ the equality

$$\sum_j t^jP_{m-j}(\mathbf{F})=(-1)^m\sum_j(-t)^jP_{m-j}(\mathbf{F}) \tag{7.238}$$

which implies that P_j is zero for odd j (consider the cases m even and m odd separately). Now because of this we naturally obtain the result that $P_j(P)\in H^{4j}(M;\mathbf{R})$, for $P_j(\mathbf{F})$ is a $2j$-form and if j is even, say $j=2i$, then $P_{2i}(\mathbf{F})$ is a $4i$-form. So the Pontrjagin class $p_i(P)$ is given by

$$p_i(P)=P_{2i}(\mathbf{F}) \tag{7.239}$$

As for the Euler class we choose $G=SO(k)$ where $k=2m$, the Lie algebra of $SO(k)$ is still $o(k)$. Then the Euler class $e(P)$ is given by (7.217) as

$$e(P)\wedge e(P)=p_m(P), \tag{7.240}$$

but

$$p_m(P)=P_{2m}(\mathbf{F}) \tag{7.241}$$

so $e(P)$ is given by that $2m$-form $e(P)$ which satisfies

$$e(P) \wedge e(P) = P_{2m}(\mathbf{F}) \tag{7.242}$$

Notice that so far we have not defined $e(P)$ uniquely, for example its sign is not determined. The existence of the Euler class has to do with the intriguing property that the determinant of an antisymmetric $2m \times 2m$ matrix is the square of a polynomial in its entries. Further, this polynomial is $SO(k)$ invariant. This polynomial is called the Pfaffian of the matrix concerned. To see how the Pfaffian is connected with the Euler class $e(P)$, one displays the $O(k)$ matrix content of $\mathbf{F}$ by writing $\mathbf{F}$ as $\mathbf{F}^{\alpha}_{\beta}$; $\alpha, \beta = 1, \ldots, k = 2m$, then we define $e(P)$ by the expression

$$\begin{aligned} e(P) &= \frac{(-1)^m}{2^{2m}\Pi^m m!} \varepsilon_{\alpha_1\alpha_2\ldots\alpha_{2m}} \mathbf{F}^{\alpha_1}_{\alpha_2} \wedge \mathbf{F}^{\alpha_3}_{\alpha_4} \wedge \ldots \wedge \mathbf{F}^{\alpha_{2m-1}}_{\alpha_{2m}} \\ &= \frac{1}{(2\Pi)^m} Pf(\mathbf{F}) \end{aligned} \tag{7.243}$$

where $Pf(\mathbf{F})$ is called the Pfaffian of the matrix $\mathbf{F}^{\alpha}_{\beta}$. $Pf(\mathbf{F})$ is only $SO(k)$ invariant [14] rather than $O(k)$ invariant i.e. $Pf(g^{-1}\mathbf{F}g) = Pf(\mathbf{F})$, $g \in SO(k)$; thus the Euler class $e(P)$ is an $SO(k)$ invariant and is the extra characteristic class that one obtains if $k = 2m$ and one goes from the group $O(k)$ to the group $SO(k)$.

7.25 CLASSIFICATION OF BUNDLES

Next we wish to discuss briefly how the classification of principal bundles that we have given extends to the classification of their associated vector bundles. We consider real and complex vector bundles E over a manifold M. The structure group of these bundles will be taken to be $O(k)$ and $U(k)$, respectively. These bundles E are also classified by universal bundles which we shall denote by ξ_E.

While the fibre of the universal principal bundles $\xi(n-k-1, G)$ is the group G, the fibre of the universal bundles ξ_E will be a real or complex vector space of real or complex dimension k. The ξ_E are associated vector bundles of the ξ and arise in a natural way from them. First of all the base spaces of the ξ_E and the ξ are the same, it is only the fibres that are different.

Let E be a real vector bundle over M of rank k, with structure group $Gl(k, \mathbf{R})$ which without loss of generality we can reduce to $O(k)$, cf. (7.45). Let M have dimension d. Then the bundle $\xi_E(n-k-1, \mathbf{R}^k)$ is defined as the real vector bundle of rank k over the base $Gr(n, k, \mathbf{R})$ whose fibre above the point $x \in Gr(n, k, \mathbf{R})$ is simply the k-plane represented by the point x, i.e. the fibre at the point x is x itself but x is a real vector space of dimension k and can therefore be the fibre of a vector bundle. $\xi_E(n-k-1, \mathbf{R}^k)$ classifies real vector bundles E of rank k in exactly the way one would expect: if $d \leq n-k-1$ and $f: M \to Gr(n, k, \mathbf{R})$ then every E arises as $f^*\xi_E(n-k-1, \mathbf{R})$ for some such map f, and homotopic maps f and g give the same pullback bundle. Now if we let E be a complex vector bundle of rank k then its group is reducible to $U(k)$ and an exactly similar classification goes through with $Gr(n, k, \mathbf{R})$ being replaced by $Gr(n, k, \mathbf{C})$. The universal bundle is written as $\xi_E(2n-2k, \mathbf{C}^k)$ and is defined in exactly the same way as $\xi_E(n-k-1, \mathbf{R}^k)$. So $\xi_E(2n-2k, \mathbf{C}^k)$ classifies complex vector bundles E of rank over M for $d \leq 2n-2k$.

The characteristic Chern, Pontrjagin, and Euler classes may all be calculated for vector bundles E and are given in terms of curvatures $\mathbf{F}$ in exactly the same way as before.

7.26 THE STIEFEL–WHITNEY CLASS

Next we wish to discuss the fourth characteristic class that we have mentioned, the Stiefel–Whitney class. The Stiefel–Whitney class unlike the other three characteristic classes is not given in terms of a curvature $\mathbf{F}$. The Stiefel–Whitney classes are defined for real vector bundles with group $O(k)$ or principal bundles with group $O(k)$ and are denoted by $w_i(E)$ and $w_i(P)$, respectively. As we pointed out above their calculation requires the computation [12] of $H^i(Gr(n, k, \mathbf{R}); \mathbf{Z}/2)$. We simply wish to state a few facts about the w_i. If the dimension of d of M is even then $w_n(E)$ actually coincides with the Euler class $e(E)$. For $i = 1$ and 2 the values of w_i are important. If we take $E = TM$, then

$$w_1(TM) = 0 \Leftrightarrow M \text{ is orientable.} \tag{7.244}$$

If

$$w_2(TM) \neq 0, M \text{ has no spin structure.} \tag{7.245}$$

What happens in the second example above is that if one attempts to define spinors on manifold M for which $w_2(TM)$ is non-zero, then one finds that one cannot define parallel transport of these spinors on M, a sign ambiguity always arises. Note that if $w_2(TM)$ is zero one also needs M to be orientable, i.e. $w_1(TM) = 0$, for spinors to be definable on M. This is an important result for physics since it imposes a restriction on the spectrum, or kinds of particle, allowed on M in terms of the topology of M.

7.27 CALCULATION OF CHARACTERISTIC CLASSES

The next thing to do is to calculate characteristic classes for some bundles. We begin with the Chern class. Let P be a principal bundle over M with group $G = SU(2)$. The curvature $\mathbf{F}$ is given by

$$\mathbf{F} = \mathbf{F}^a \frac{\sigma^a}{2i} \tag{7.246}$$

where $\sigma^a = 1, 2, 3$ are the Pauli matrices:

$$\sigma^1 = \begin{pmatrix} 0 & 1 \\ 1 & 0 \end{pmatrix} \qquad \sigma^2 = \begin{pmatrix} 0 & -i \\ i & 0 \end{pmatrix}$$
$$\sigma^3 = \begin{pmatrix} 1 & 0 \\ 0 & -1 \end{pmatrix} \tag{7.247}$$

So

$$\begin{aligned} &\mathrm{Det}\left(tI + \frac{i}{2\pi}\mathbf{F}\right) \\ &= \mathrm{Det}\left(tI + \frac{\mathbf{F}^a\sigma^a}{4\pi}\right) \\ &= t^2 + \frac{t}{2\pi}\, tr(\mathbf{F}^a\sigma^a) + \frac{1}{32\pi^2}\{tr(\mathbf{F}^a\sigma^a \wedge \mathbf{F}^b\sigma^b) - tr(\mathbf{F}^a\sigma^a) \wedge tr(\mathbf{F}^b\sigma^b)\} \end{aligned}$$

But since $\sigma^a \in su(2)$, $tr\,\sigma^a = 0$. Hence

$$\mathrm{Det}\left(tI + \frac{i\mathbf{F}}{2\pi}\right) = t^2 - \frac{\mathbf{F}^a \wedge \mathbf{F}^a}{16\pi^2} \tag{7.248}$$

where we have used $tr(\sigma^a\sigma^b) = 2\delta^{ab}$. The result in terms of Chern classes is that

$$\begin{aligned} c_1(P) &= 0 \\ c_2(P) &= -\frac{1}{16\pi^2}\mathbf{F}^a \wedge \mathbf{F}^a \end{aligned} \tag{7.249}$$

Note that to obtain formulae such as (7.248) one merely needs to exploit the definition of a determinant as a product of eigenvalues. To see this, let T be a diagonal $n \times n$ matrix with eigenvalues $\lambda_1 \ldots \lambda_n$. Then

$$\begin{aligned}\operatorname{Det}\left(tI + \frac{i}{2\pi} T\right) &= \left(t + \frac{i\lambda_1}{2\pi}\right)\left(t + \frac{i\lambda_2}{2\pi}\right) \ldots \left(t + \frac{i\lambda_n}{2\pi}\right) \\ &= t^n + t^{n-1} \frac{i}{2\pi} (\Sigma\lambda_i) + t^{n-2} \left(\frac{i}{2\pi}\right)^2 \left\{\frac{(\Sigma\lambda_i)^2 - \Sigma\lambda_i^2}{2}\right\} \\ &\quad + \ldots \left(\frac{i}{2\pi}\right)^n (\lambda_1\lambda_2 \ldots \lambda_n) \end{aligned} \tag{7.250}$$

Thus in terms of Chern classes (7.250) contains the information that

$$\begin{aligned} c_0(P) &= 1 \\ c_1(P) &= \frac{i}{2\pi} Tr(\mathbf{F}) \\ c_2(P) &= \left(\frac{i}{2\pi}\right)^2 \left\{\frac{(Tr\mathbf{F}) \wedge Tr(\mathbf{F}) - Tr(\mathbf{F} \wedge \mathbf{F})}{2}\right\} \\ &\vdots \\ c_n(P) &= \left(\frac{i}{2\pi}\right)^n \operatorname{Det}(\mathbf{F}) \end{aligned} \tag{7.251}$$

For our example $n = 2$ so that it should be true that

$$\begin{aligned} c_2(P) &= \left(\frac{i}{2\pi}\right)^2 \operatorname{Det}(\mathbf{F}) \\ &= \left(\frac{i}{2\pi}\right)^2 \frac{\{Tr(\mathbf{F}) \wedge Tr(\mathbf{F}) - Tr(\mathbf{F} \wedge \mathbf{F})\}}{2} \end{aligned} \tag{7.252}$$

It is clear that for 2×2 matrices $\operatorname{Det} A = \frac{1}{2}\{(Tr(A))^2 - Tr(A^2)\}$ since $\lambda_1\lambda_2 = \frac{1}{2}\{(\lambda_1 + \lambda_2)^2 - (\lambda_1^2 + \lambda_2^2)\}$. But to obtain familiarity with calculations involving matrix-valued forms we shall go through the calculation of Det (**F**). Let us work with $su(2)$ so that we are required to calculate $c_2(P) = (i/2)^2 \operatorname{Det}(\mathbf{F})$ and show that it is equal to $-(1/16\Pi^2)\mathbf{F}^a \wedge \mathbf{F}^a$ in agreement with (7.247). Using the Pauli matrices as given in (7.247), we obtain

$$\mathbf{F} = \mathbf{F}^a \frac{\sigma^a}{2i} = \frac{1}{2i} (\mathbf{F}^1\sigma^1 + \mathbf{F}^2\sigma^2 + \mathbf{F}^3\sigma^3) \tag{7.253}$$

$$\begin{aligned} &= \frac{\mathbf{F}^1}{2i}\begin{pmatrix} 0 & 1 \\ 1 & 0 \end{pmatrix} + \frac{\mathbf{F}^2}{2i}\begin{pmatrix} 0 & -i \\ i & 0 \end{pmatrix} + \frac{\mathbf{F}^3}{2i}\begin{pmatrix} 1 & 0 \\ 0 & -1 \end{pmatrix} \\ &= \frac{1}{2i}\begin{pmatrix} \mathbf{F}^3 & \mathbf{F}^1 - i\mathbf{F}^2 \\ \mathbf{F}^1 + i\mathbf{F}^2 & -\mathbf{F}^3 \end{pmatrix} \end{aligned} \tag{7.254}$$

Thus

$$\left(\frac{i}{2\pi}\right)^2 \text{Det}\,\mathbf{F} = \left(\frac{1}{2i}\right)^2 \left(\frac{i}{2\pi}\right)^2 \{-\mathbf{F}^3 \wedge \mathbf{F}^3 - (\mathbf{F}^1 + i\mathbf{F}^2) \wedge (\mathbf{F}^1 - i\mathbf{F}^2)\}$$

$$= \tfrac{1}{4}\left(\frac{i}{2}\right)^2 \{\mathbf{F}^3 \wedge \mathbf{F}^3 + \mathbf{F}^2 \wedge \mathbf{F}^2 + \mathbf{F}^1 \wedge \mathbf{F}^1 + i\mathbf{F}^2 \wedge \mathbf{F}^1 - i\mathbf{F}^1 \wedge \mathbf{F}^2\} \tag{7.255}$$

But since the $\mathbf{F}^a$ are 2-forms $\mathbf{F}^1 \wedge \mathbf{F}^2 = \mathbf{F}^2 \wedge \mathbf{F}^1$ and we have

$$c_2(P) = \left(\frac{i}{2\pi}\right)^2 \text{Det}\,\mathbf{F} = \left(\frac{i}{2\pi}\right) \frac{\mathbf{F}^a \wedge \mathbf{F}^a}{4} \tag{7.256}$$

in agreement with (7.250).

Next we take an example which calculates Pontrjagin classes. The bundle is a vector bundle namely the tangent bundle to a manifold M. Further, we shall be specific and choose M to be a Riemannian manifold with dimensions 2 and 4, respectively. In treating these examples, we shall also, since we have chosen n to be even in both cases, be able to calculate the Euler class of TM. The first case is where $\dim M = 2$. Now since $p_i(TM)$ is a $4i$-form, then the $p_i(TM)$ are only non-zero for $4i \leq 2$, hence the only non-zero $p_i(TM)$ is $p_0(TM)$ which is of course trivially equal to 1 (cf. 7.219, 7.239). It is then tempting to conclude without further calculation that since $e(TM)$ satisfies, $e(TM) \wedge e(TM) = p_1(TM)$, then $e(TM)$ is zero also. This is too hasty and indeed incorrect: it is true that if $\mathbf{F}$ is $o(n)$-valued, then (7.251) gives simply:

$$\text{Det}\left(tI - \frac{\mathbf{F}}{2\pi}\right) = t^2 \tag{7.257}$$

so that $p_0(TM) = 1$ and $p_1(TM) = 0$. However, the definition (2.243) of se(TM) gives

$$e(TM) = \frac{1}{(2\pi)} Pf(\mathbf{F})$$

$$= \frac{-1}{2^2\pi} \varepsilon_{\alpha\beta} \mathbf{F}^\alpha_\beta \tag{7.258}$$

and we can see that $e(TM) \neq 0$. In fact, since α and β only range over the values 1 and 2, we have the simple expression

$$e(TM) = \frac{-1}{2\pi} \mathbf{F}^1_2 \tag{7.259}$$

We can also see that because $\dim M=2$, $e(TM)\wedge e(TM)=0$, for

$$\mathbf{F}^1_2 = F^1_{2\mu\nu}\,\mathrm{d}x^\mu \wedge \mathrm{d}x^\nu \tag{7.260}$$

But since μ and ν are limited to the 1 and 2, and because $F^1_{2\mu\nu}$ is antisymmetric in μ and ν, we find

$$\mathbf{F}^1_2 = 2F^1_{212}\,\mathrm{d}x^1 \wedge \mathrm{d}x^2$$

So

$$\begin{aligned} p_1(TM) = e(TM)\wedge e(TM) &= \frac{4}{(2\pi)^2}(F^1_{212})^2\,\mathrm{d}x^1\wedge \mathrm{d}x^2 \wedge \mathrm{d}x^1 \wedge \mathrm{d}x^2 \\ &= 0 \end{aligned} \tag{7.261}$$

This of course was all evident from the beginning but it is instructive to check the details explicitly. The second case is where $\dim M=4$. This time we expect $p_1(TM)$ to have a chance of being non-zero and our calculation confirms this expectation. However, recall that $\dim M=4$, the group of TM is $O(4)$ and according to the definition $e(TM)$ is a 4-form, thus we shall again have $p_1(TM)=e(TM)\wedge e(TM)=0$, and we shall again find nevertheless, that $e(TM)\neq 0$. Because $\mathbf{F}$ is $o(4)$-valued $\mathrm{Det}\,[tI-(\mathbf{F}/2\pi)]$ contains only even powers of t:

$$\mathrm{Det}\left(tI-\frac{\mathbf{F}}{2\pi}\right) = t^4 + t^2 p_1(TM) + p_2(TM) \tag{7.262}$$

We know already that $p_2(TM)$ is zero, to calculate $p_1(TM)$ we simply use the calculation in (7.251) and obtain

$$p_1(TM) = \frac{-1}{(2\pi)^2}\frac{Tr(\mathbf{F}\wedge\mathbf{F})}{2} \tag{7.263}$$

We could also write this as $p_1(TM)=(-1/8\pi^2)Tr(\mathbf{R}\wedge\mathbf{R})$ where the $\mathbf{R}$ would serve to remind us that the geometry is Riemannian. The Euler class is given by

$$\begin{aligned} e(TM) &= \frac{(-1)^2}{2^4\pi^2 2!}\varepsilon_{\alpha\beta\mu\nu}\mathbf{R}^\alpha_\beta \wedge \mathbf{R}^\mu_\nu \\ &= \frac{1}{32\pi^2}\varepsilon_{\alpha\beta\mu\nu}\mathbf{R}^\alpha_\beta \wedge \mathbf{R}^\mu_\nu \end{aligned} \tag{7.264}$$

As an example of bundle where e and $p_m = e\wedge e$ are both non-zero we consider a principal bundle P with group $SO(4)$ over an 8-dimensional manifold M. In this example p_0, p_1, p_2 and e are all non-zero. We have already obtained the expressions for p_0, p_1 and e above so that

$$p_0(P) = 1$$

$$p_1(P) = -\frac{1}{8\pi^2} Tr\{\mathbf{F} \wedge \mathbf{F})$$

$$e(P) = \frac{1}{(2\pi)^2} Pf(\mathbf{F}) = \frac{1}{32\pi^2} \varepsilon_{ijkl} \mathbf{F}^i_j \wedge \mathbf{F}^k_l \tag{7.265}$$

(the Latin indices range from 1 to 4). Since M is 8-dimensional the curvature $\mathbf{F}^i_j$ is expressible as (the Greek indices range from 1 to 8)

$$\mathbf{F}^i_j = F^i_{j\mu\nu}\, dx^\mu \wedge dx^\nu \tag{7.266}$$

in local coordinates, hence

$$\begin{aligned} p_2(P) &= e(P) \wedge e(P) \\ &= \frac{1}{2^{10}\pi^4} \varepsilon_{ijkl}\varepsilon_{abcd} F^i_{j\mu\nu} F^k_{l\lambda\sigma} F^a_{b\alpha\beta} F^c_{d\gamma\delta} \\ &\quad \times dx^\mu \wedge dx^\nu \wedge dx^\lambda \wedge dx^\sigma \wedge dx^\alpha \wedge dx^\beta \wedge dx^\gamma \wedge dx^\delta \end{aligned} \tag{7.267}$$

For examples of Stiefel–Whitney classes we can take complex projective space $\mathbf{C}P^n$, (incidentally $\mathbf{C}P^n$ is the Grassmannian manifold $Gr(n+1, k, \mathbf{C})$ for $k = 1$ because then $Gr(n+1, 1, \mathbf{C})$ is the space of complex lines in $\mathbf{C}^{n+1}$) and consider its tangent bundle $T\mathbf{C}P^n$. We simply quote [12] the result that

$$w_1(T\mathbf{C}P^n) = 0 \text{ for all } n$$

$$w_2(T\mathbf{C}P^n) = \begin{cases} 0 & \text{if } n \text{ is odd} \\ x & \text{if } n \text{ is even} \end{cases} \tag{7.268}$$

where x is the generator of the cohomology group $H^2(\mathbf{C}P^n; \mathbf{Z}/2)$. This means that $\mathbf{C}P^{2n}$, $n = 1, 2, \ldots$ etc. do not admit spin structures, and that $\mathbf{C}P^{2n+1}$, $n = 1, 2, \ldots$ etc. do admit spin structures.

7.28 GENERAL REMARKS

Some general informative remarks including points of nomenclature and notation are due now, these will all relate (except where explicit reference to the contrary is given) to the first three characteristic classes treated, i.e. those of Chern, Pontrjagin and Euler.

We have seen that these characteristic classes are all obtained by Lie algebra invariants and converting them into cohomology classes by appropriate substitution of curvature forms. In more precise mathematical terms this substitution can be thought of as a homomorphism called the Weil homomorphism. The Weil homomorphism w is a map from the set of all invariant Lie algebra polynomials, denoted by $I(G)$, to all the cohomology classes, denoted as in (6.84) by $H^*(M; \mathbf{R})$: (for details of w see [10])

$$w: I(G) \to H^*(M; \mathbf{R}) \tag{7.269}$$

Both $I(G)$ and $H^*(M; \mathbf{R})$ are rings and w is called a homomorphism because it is a homomorphism of the ring structures of $I(G)$ and $H^*(M; \mathbf{R})$.

Our notation for characteristic classes of a bundle P or E was to write $c_j(P)$, $c_j(E)$, $p_j(P)$ etc. We now want to *change* this. Instead of writing the above quantities we shall write $c_j(\mathbf{F})$, $p_j(\mathbf{F})$ etc. where $\mathbf{F}$ are the curvatures used in the respective bundle calculations. We shall also call these $c_j(\mathbf{F})$, $p_j(\mathbf{F})$, etc. characteristic forms because they are indeed forms. Each of these forms may be integrated either, over the manifold M if $c_j(\mathbf{F})$, $p_j(\mathbf{F})$ is an n-form, or over the appropriate lower dimensional object a chain if $c_j(\mathbf{F})$, $p_j(\mathbf{F})$ are forms of lower degree than n. When such integrations are carried out the result is a number. These numbers are called Chern numbers, Pontrjagin numbers etc., and are denoted by $C_j(P)$, $C_j(E)$, $P_j(P)$, $P_j(E)$ etc. For example

$$C_j(P) = \int_c c_j(\mathbf{F}) \tag{7.270}$$

where c is a j-chain, and in particular if $j = n = \dim M$

$$C_{n/2}(P) = \int_M c_{n/2}(\mathbf{F}) \tag{7.271}$$

with similar formulae for the other two characteristic classes. Since strictly speaking a form such as $c_j(\mathbf{F})$ is only a representative of the cohomology class $[c_j(\mathbf{F})]$ rather than being the class itself, we shall now use $c_j(P)$ to denote the cohomology class itself:

$$c_j(P) = [c_j(\mathbf{F})] \tag{7.272}$$

and $c_j(P)$ is called the jth Chern class. Another piece of terminology is the *total* characteristic class, this is given by the sum of all the individual characteristic classes and is denoted by, $c(P)$, $c(E)$, $p(P)$ etc. The total Chern class and the total Pontrjagin class are given by

$$\begin{aligned} c(P) &= 1 + c_1(P) + c_2(P) + \dots \\ p(P) &= 1 + p_1(P) + p_2(P) + \dots \end{aligned} \tag{7.273}$$

This gives rise to a total Chern form $c(\mathbf{F})$ and a total Pontrjagin form $p(\mathbf{F})$ via equation (7.272), we have therefore in analogy with (7.272),

$$c(P) = [c(\mathbf{F})]$$
$$p(P) = [p(\mathbf{F})] \tag{7.274}$$

The total Chern and Pontrjagin forms are really given by using $\text{Det}[tI + (i\mathbf{F}/2\pi)]$ with $t = 1$, for when $t = 1$ we obtain

$$\begin{aligned} \text{Det}\left(I + \frac{i\mathbf{F}}{2\pi}\right) &= 1 + c_1(\mathbf{F}) + c_2(\mathbf{F}) + \ldots = c(\mathbf{F}) \\ \text{Det}\left(I - \frac{\mathbf{F}}{2\pi}\right) &= 1 + p_1(\mathbf{F}) + p_2(\mathbf{F}) + \ldots = p(\mathbf{F}) \end{aligned} \tag{7.275}$$

Among the examples of fibre bundle for which we have computed characteristic classes is TM the tangent bundle of M. Instead of writing $c_i(TM)$, $p_i(TM)$ etc, it is common to find the slightly confusing usage of writing $c_i(M)$ for $c_i(TM)$ and $p_i(M)$ for $p_i(TM)$ etc. Also the characteristic classes are then called characteristic classes of the manifold M instead of the more precise characteristic classes of TM. If $\dim M = n$, then the integrals over M of n-forms made from wedge products of $c_i(\mathbf{F})$ and $p_i(\mathbf{F})$ are also called characteristic numbers of M.

7.29 FORMULAE OBEYED BY CHARACTERISTIC CLASSES

When bundles are expressible as sums or products of other bundles it is natural to be interested in whether or not this phenomenon induces similar formulae among their characteristic classes. For example, if we take a vector bundle which is a Whitney sum $E \oplus F$, then are there simple relations between $c(E \oplus F)$ and $c(E)$ and $c(F)$ for E and F complex vector bundles, or between $p(E \oplus F)$ and $p(E)$ and $p(F)$ for E and F real vector bundles? The answer is yes. For the complex case one can prove that for the Chern classes

$$c(E \oplus F) = c(E)c(F) \tag{7.276}$$

For the real case the same formula holds for the corresponding Pontrjagin quantities (see reference [12], Theorem (15.3) for the most general formula for $p(E \oplus F)$):

$$p(\mathbf{F}_{E\oplus F}) = p(\mathbf{F}_E) \wedge p(\mathbf{F}_F)$$
$$p(E \oplus F) = p(E)p(F) \tag{7.277}$$

For the real even dimensional orientable case, for which the Euler class is defined, then we also have

$$e(E\oplus F)=e(E)e(F) \tag{7.278}$$

However, if a vector bundle occurs as a tensor product then the simplest relations between the characteristic classes are not given by considering $c(E\oplus F)$ and $p(E\oplus F)$. The simplest relations between tensor product bundles are given by considering what are called characteristic polynomials. The first characteristic polynomial to consider is called the Chern character and is written $ch(E)$. The polynomial $ch(T)$, $T\in \mathfrak{g}$ is defined by

$$ch(T)=tr\exp(T) \tag{7.279}$$

and if we substitute $T=i\mathbf{F}/2\Pi$, then

$$ch(\mathbf{F})=tr\exp\left[\frac{i\mathbf{F}}{2\pi}\right] \tag{7.280}$$

The cohomology class determined by $ch(\mathbf{F})$ is the Chern character $ch(E)_n$ of the complex vector bundle E. The property of $ch(E)$ when evaluated on tensor products is that it is multiplicative

$$ch(E\otimes F)=ch(E)ch(F) \tag{7.281}$$

It is also nicely behaved for sums

$$ch(E\oplus F)=ch(E)+ch(F) \tag{7.282}$$

Thus the Chern character has the advantage that it behaves well for sums and products. There are further characteristic polynomials which are of use when studying the Atiyah–Singer index theorem, cf. reference [7].

Returning to the distinction between real and complex vector bundles E, and realizing that given a real vector bundle E one can complexify it to obtain a complex vector bundle E_c (E_c is constructed by replacing the fibres $\mathbf{R}^k$ of E by $\mathbf{C}^k$), there is then a relation between the Chern classes of E_c and the Pontrjagin classes of E. This relation is

$$p_i(E)=(-1)^i c_{2i}(E_c) \tag{7.283}$$

and is sometimes used as a method of definition of Pontrjagin classes [4, 10, 18]. Alternatively if one has a complex vector bundle E of rank k one can think of it as a real vector bundle E_R of rank $2k$. The relation that results in terms of $p_i(E_R)$ and $c_i(E)$ is a little more complicated;

$$(-1)^i p_i(E_R)=\sum_{j=0}^{2i}(-1)^j c_j(E)c_{2i-j}(E) \tag{7.284}$$

e.g. if $i=1$, we obtain

$$-p_1(E_R)=c_0(E)c_2(E)-c_1(E)c_1(E)+c_2(E)c_0(E)$$

or since $c_0(E)=1$,

$$p_1(E_R)=c_1^2(E)-2c_2(E) \tag{7.285}$$

Sometimes [7] there is a natural bundle E_c associated with a real bundle E, or a natural bundle E_R associated with a complex bundle E. Given either of those situations one may be able to compute the characteristic classes of the bundle one is interested in by use of the relations (7.283, 4).

A final item of terminology is that if $c(P)$ is some characteristic class of a bundle P, then we saw that $c(P)=c(f^*\xi)=f^*c(\xi)$ for some universal bundle ξ. The $c(\xi)$ are called universal characteristic classes, and thus one may speak of the universal Stiefel–Whitney class, the universal Pontrjagin class and so on.

7.30 GLOBAL INVARIANTS AND LOCAL GEOMETRY

We discuss in this short section the connections that exist between the local geometry of a manifold M and its global invariants. These connections will be provided by characteristic classes. In particular, they will be provided by appropriate characteristic classes $c(TM)$ of the bundle TM. As we pointed out above it is common to denote these by $c(M)$. We wish to distinguish three types of global invariant;

i. invariants of the topological structure of M.
ii. invariants of the differential structure of M.
iii. invariants of the complex structure of M.

A word of explanation is helpfull about each of these. A topological invariant is one which is unchanged if one changes M for a homeomorphic manifold M'. An invariant of the differential structure of M is one which is unchanged if one changes M for a diffeomorphic manifold M'. An invariant of the complex structure of M is one which is unchanged if one changes M for a holomorphic manifold M'. In terms of a map $f:M\to M'$ we have,

$$f:M\to M'$$

$$f^{-1}:M'\to M$$

in

i. f and f^{-1} are continuous (f is called a homeomorphism).
ii. f and f^{-1} are C^∞ (f is called a diffeomorphism)

and in

iii. f and f^{-1} are holomorphic (f is called biholomorphic). (7.286)

Note that holomorphic$\Rightarrow C^{\infty}\Rightarrow$continuity but not the other way round. For complex manifolds M the Chern classes $c_i(M)$ are invariants of the complex structure of M, for real manifolds M the Pontrjagin classes $p_i(M)$ are invariants of the differential structure of M. The Stiefel–Whitney classes are topological invariants of M.

To develop this further consider the manifold S^7. In Chapter 2 (cf. equation (2.70)) we first pointed out the existence of distinct differentiable structures on S^7. To summarize the discussion in Chapter 2 is to say that not all manifolds homeomorphic to S^7 are also diffeomorphic to S^7. In fact, if we realize that the property of being diffeomorphic for a pair of manifolds is an equivalence relation on the collection of all manifolds, just as is the property of being homeomorphic, then we have the conclusion that the equivalence class of all topological spaces homeomorphic to S^7 is made up of distinct diffeomorphism classes, 28 in all. For spaces M and M' within the same diffeomorphism class we have $p_i(M)=p_i(M')$, but this will be false in general if M and M' are homeomorphic but in different diffeomorphism classes. Actually for S^7, $H^p(S^7;\mathbf{R})$ is only non-zero for $p=7$ so that the possible non-trivial Pontrjagin class p_i is zero whatever differentiable structure is put on S^7. For an example, where the Pontrjagin classes differ on two homeomorphic but not diffeomorphic manifolds, cf. reference [11].

Similar general remarks may be made for the Chern classes $c_i(M)$ of complex manifolds M. We use the adjective global in front of the word invariant when we wish to emphasize that the particular invariance being discussed is an invariant of the whole manifold M. Local properties are in general different from global ones. For example the tangent space $T_p(M)$ at the point p is a local property, the forms $\mathbf{F}$ and $\mathbf{F}\wedge\mathbf{F}$ being made up of local differential operators give local information about M, the curvature k of a surface in $\mathbf{R}^3$ may be constant locally but the only such compact closed surface with constant curvature globally is the sphere S^2. The property of being a geodesic is a local property for a curve on M, however, the number of geodesics between a pair of fixed points on M is a global invariant of M. The characteristic classes are global invariants of a bundle P which measure the deviation of P from being a product bundle. Except for the Stiefel–Whitney classes, the characteristic classes are determined by curvature, and if a flat connection exists then the characteristic classes vanish; on the other hand if one computes using some connection a characteristic class and finds it be non-zero, then M does not admit a flat connection†. We can see here therefore a source for links between local

† A technical point here is that the connection A flat or otherwise must be $\mathfrak{g}$-valued where $\mathfrak{g}$ is the Lie algebra of the group G of the bundle P. For example there can be flat connections A for a bundle P over M, but $A\in\mathfrak{gl}(k,\mathbf{R})\supset\mathfrak{g}$, $A\notin\mathfrak{g}$.

geometry and global invariants of M. The characteristic numbers obtained by integrating the characteristic forms over M provide concrete numerical invariant data about M. A celebrated result obtained by integrating the Euler form $e(\mathbf{F})$ over M gives a topological invariant the Euler number or Euler–Poincaré characteristic $\chi(M)$ of M. This result was first obtained in 2 dimensions and is called the Gauss–Bonnet theorem. We now use this result and part of a discussion due to Chern [5], to illustrate our remarks above. We begin in 2 dimensions with a manifold M with a boundary ∂M. ∂M is piecewise smooth, i.e. it may have vertices where it is not differentiable. The Gauss–Bonnet theorem states that

$$\sum_{i=1}^{n} (\pi-\alpha_i)+\int_M k_g \, \mathrm{d}s+\int_M k \, \mathrm{d}A = 2\pi\chi(M) \tag{7.287}$$

In (7.287) the $\alpha_1, \ldots, \alpha_n$ are the n interior angles at the vertices of the boundary ∂M, k_g is called the geodesic curvature of the curve ∂M, and k is the ordinary Gaussian curvature of M while $\mathrm{d}s$ and $\mathrm{d}A$ are elements of arc length and area, respectively. In (7.287) we have examples of three kinds of curvature; point curvature or 0-dimensional curvature given by the abrupt α_i change in direction of ∂M by an amount α_i at a vertex, and line curvature or 1-dimensional curvature given by k_g, and surface curvature or 2-dimensional curvature given by k. The content of the Gauss–Bonnet theorem is that if one sums over all the curvatures of M then one gets 2π times Euler number $\chi(M)$ of M. Recall that the much older Euler formula for polyhedra P

$$V-E+F=2(=\chi(P)) \tag{7.288}$$

also has terms of increasing dimension on the LHS, further the + and − signs alternate, (admittedly only once), on the LHS. This alternation is present in the higher dimensional definition of the Euler characteristic

$$\chi(M)=\sum_{0}^{n}(-1)^i \dim H^i(M;\mathbf{R}) \tag{7.289}$$

where M is n-dimensional. If we take M to be simply connected and use Poincaré's lemma $\chi(M)=1$. Consider also a slightly degenerate case where the surface lies in the plane giving $k=0$, then we obtain

$$\sum_i (\pi-\alpha_i)+\int_{\partial M} k_g \, \mathrm{d}s = 2\pi \tag{7.290}$$

Now the Gauss–Bonnet theorem is reduced to a statement about piecewise-smooth curves in the plane, namely that the rotation index of such curves is 2π. (The rotation index of a piecewise-smooth curve is defined as the angle through which the tangent turns as it moves along the curve, at a vertex the tangent is defined to rotate an amount equal to the exterior angle i.e. $\pi-\alpha_i$.)

If we further simplify things by making the pieces of ∂M straight instead of curved then $k_g=0$, and M is some sort of irregular polygon, and we have

$$\sum_i (\pi-\alpha_i)=2\pi \tag{7.291}$$

But $(\Pi-\alpha_i)$ are the exterior angles of the polygon and their sum is always 2Π. These are nice examples of a 2 dimensional kind and fortunately the Gauss–Bonnet theorem can be generalized to higher dimensions. The generalization is to higher even dimensions for orientable compact manifolds M. If M has dimension $2n$ is compact and orientable then the bundle TM has rank $2n$ and structure group $SO(2n)$ so that the Euler form $e(\mathbf{F})$ is defined, the generalized Gauss–Bonnet theorem states that

$$\int_M e(\mathbf{F})=\chi(M) \tag{7.292}$$

the LHS containing local geometry in the form of $e(\mathbf{F})$ and the RHS being the topological invariant $\chi(M)$. For complex manifolds of complex dimension n, this can also be stated in terms of Chern classes. If TM is a complex vector bundle of rank n and TM_R its real counterpart of rank $2n$, then we find that (use (7.284) with $i=n$ the summation reduces to one term given by $j=n$ because for TM, $c_j(M)=0$ if $j>n$)

$$p_n(TM_R)=c_n^2(TM) \tag{7.293}$$

But $e^2(TM_R)=p_n(TM_R)$ so $e(TM_R)=c_n(TM)$ taking the positive square root which we do not justify here. So in terms of Chern classes for complex n-dimensional manifolds M we have

$$\int_M c_n(\mathbf{F})=\chi(M) \tag{7.294}$$

Of course complex manifolds are necessarily even-dimensional and orientable as we pointed out when discussing G-structures earlier in the chapter. Also, if M is odd-dimensional then the Gauss–Bonnet theorem yields nothing since both $e(TM)$ and $\chi(M)$ are then zero.

Finally, as an example of the theorem at work take the manifold M to be compact closed $(\partial M=\phi)$ orientable and 2-dimensional. The Gauss–Bonnet theorem then gives

$$\int_M e(\mathbf{F})=\chi(M) \tag{7.295}$$

with $e(\mathbf{F}) = (1/2\pi)\mathbf{F}_1^2$ cf. (7.259), but $\mathbf{F}_1^2$ is $k\ \mathrm{d}A$, so

$$\int_M k\ \mathrm{d}A = 2\pi\chi(M) \tag{7.296}$$

Now it is known that all compact closed orientable 2-dimensionable manifolds are given by spheres with g handles which we denote by M_g^2, g is called the genus, clearly $M_0^2 = S^2$, $M_1^2 = T^2$ (the torus) etc. Then we can say:

i. $k < 0$ everywhere on $M \Rightarrow \chi(M) < 0 \Rightarrow M$ diffeomorphic to some M_g^2, $g \geqslant 2$;

ii. $k = 0$ everywhere on $M \Rightarrow \chi(M) = 0 \Rightarrow M$ diffeomorphic to T^2;

iii. $k > 0$ everywhere on $M \Rightarrow \chi(M) = 2 \Rightarrow M$ diffeomorphic to S^2.

(7.297)

All this follows from realizing that

$$\begin{aligned}\chi(M_g^2) &= \dim H^0(M_g^2; \mathbf{R}) - \dim H^1(M_g^2; \mathbf{R}) + \dim H^2(M_g; \mathbf{R}) \\ &= 1 - 2g + 1 = 2 - 2g \end{aligned} \tag{7.298}$$

Thus $\chi(M_g^2)$ has only one positive value $\chi = 2$, and $\chi = 0$ can only be obtained by taking $g = 1$. A torus with $k = 0$ everywhere is unconventional but can be obtained by identifying opposite ends of a parallelogram, this surface then lies in $\mathbf{R}^4$ rather than $\mathbf{R}^3$ (cf. Chapter 2). The conventional curved torus which lies in $\mathbf{R}^3$ has $k > 0$ and $k < 0$ on its surface in such a way as to satisfy the Gauss–Bonnet theorem. The conventional torus is of course diffeomorphic to the flat torus.

This is an appropriate point to close this chapter.

REFERENCES

1. ADAMS, J. F., On the non-existence of elements of Hopf invariant one. *Ann. Math.* **72**, 20 (1960).
2. ATIYAH, M. F., "K-Theory". W. A. Benjamin Inc., 1967.
3. BOREL, A. and SERRE, J. P., "Détermination des p-puissances réduites de Steenrod dans la cohomologie des groupes classiques". Applications, C.R. Acad. Sci. Paris, 233,680, 1951.
4. CHERN, S. S., "Complex Manifolds Without Potential Theory". Van Nostrand, 1967.
5. CHERN, S., *Am. Math. Monthly* p. 339 (1979).
6. CHEVALLEY, C., "Theory of Lie Groups". Princeton University Press, 1946.
7. EGUCHI, T., GILKEY, P. B. and HANSON, A. J., Gravitation, Gauge theories and Differential Geometry. *Phys. Rep.* **66**, No. 6 (1980).
8. IWASAWA, K., *Ann. Math.* **50**, 507 (1949).
9. KOBAYASHI, S. and NOMIZU, K., "Foundations of Differential Geometry". Vol. I, Interscience, 1969.

10. KOBAYASHI, S. and NOMIZU, K., "Foundations of Differential Geometry". Vol. II, Interscience, 1969.
11. MILNOR, J. W., *Topology* **3** (Suppl. 1), 53 (1964).
12. MILNOR, J. W. and STASHEFF, J. D., "Characteristic Classes". Princeton University Press, 1974.
13. SPIVAK, M., "Differential Geometry". Vol. 2, Publish or Perish, 1975.
14. SPIVAK, M., "Differential Geometry". Vol. 5, Publish or Perish, 1975.
15. SPIVAK, M., "Differential Geometry". Vol. 5, p. 423. Publish or Perish, 1975. The form given in this reference is not quite that of (7.51); to complete the calculation the basis elements should be permuted by the permutation

$$\begin{pmatrix} 1 & 2 & 3 & 4 & \dots & 2m \\ 1 & (m+1) & 2 & (m+2) & \dots & (2m) \end{pmatrix}$$

16. STEENROD, N., "The Topology of Fibre Bundles". Princeton University Press, 1970.
17. SYNGE, J. L. and GRIFFITH, B. A., "Principles of Mechanics". McGraw-Hill, 1959.
18. WELLS, R. O., "Differential Analysis on Complex Manifolds". Springer-Verlag, 1979.

CHAPTER 8

Morse Theory

8.1 MORSE INEQUALITIES

In this chapter we will prove the Morse inequalities for a compact manifold and briefly discuss two applications of these inequalities in theoretical physics.

Let us start by explaining what the Morse inequalities involve. Suppose M is a compact differential manifold of dimension n. Suppose further that f represents a smooth real valued function on M i.e.

$$f: M \to \mathbf{R} \tag{8.1}$$

The Morse inequalities place restrictions on number of critical points (i.e. extrema) that the function f can have due to the topology of M. In order to state the nature of these restrictions a few definitions have to be introduced.

Definition

In terms of a local co-ordinate system a point $a^* \equiv (a_1^*, a_2^*, \ldots, a_n^*)$ $a^* \in M$ is a critical point of f if:

$$\left| \frac{\partial f}{\partial x_i}(x_1, \ldots, x_n) \right|_{x=a^*} = 0, \text{ for } \forall i \tag{8.2}$$

Definition

A critical point a^* is called non-degenerate if and only if the matrix $|\partial^2 f/\partial x_i \partial x_j|_{x=a}{}^*$, called the Hessian, is non-singular, i.e. Determinant $|\partial^2 f/\partial x_i \partial x_j|_{x=a^*} \neq 0$.

Furthermore $f(a^*)$ is called the critical value of f. (8.3)

In our discussions we will only consider non-degenerate critical points.

There are a number of reasons for this. First of all it is possible to show that the notion of non-degeneracy is co-ordinate independent (see reference [5] for the proof). Secondly, as we will see, a non-degenerate critical point is necessarily isolated, i.e. there is a neighbourhood of the critical point in which no other critical points of f are present. Thirdly in a neighbourhood U of a non-degenerate critical point a^* the function f can be represented as:

$$f(y) = f(a^*) - y_1^2 - y_2^2 \ldots - y_k^2 + y_{k+1}^2 + \ldots + y_n^2 \qquad (8.4)$$

The number of negative terms present is called the index of the critical point a^*. In expression (8.4) the index is k. We will prove this result shortly. Finally requiring f to have only non-degenerate critical points is not unduly restrictive. Indeed the following theorem can be proved (see reference [5] for proof).

Theorem

Any bounded smooth function $f: M \to \mathbf{R}$ can be uniformly approximated by a smooth function $g: M \to \mathbf{R}$ which has only non-degenerate critical points. Furthermore g can be chosen so that the ith derivatives of g on a compact set $K \subset M$ uniformly approximate the corresponding derivatives of f for all $i \leqslant p$ (p arbitrary but finite). (8.5)

The restrictions on the number of critical points of f due to the topology of M can now be precisely stated in the form of the following theorem.

Theorem (the Morse inequalities)

Let C_k denote the number of non-degenerate critical points with index k of a smooth real valued function f on a compact differentiable manifold M. Then

$$R_k(M) - R_{k-1}(M) + \ldots \pm R_0(M) \leqslant C_k - C_{k-1} + \ldots \pm C_0$$

with equality for $k = n$, where $R_k(M)$ denotes the kth Betti number, i.e. $R_k(M) = \text{Dim}\,[H_k(M)]$ (8.6)

The set of inequalities immediately lead to the Corollary.

Corollary

Under the conditions of Theorem (8.6) it follows that:

$$C_k \geqslant R_k, \forall k \qquad (8.7)$$

This result is obtained by adding the inequalities for k and $k-1$ given in (8.6).

The Morse inequalities imply that if a compact manifold M has $R_k \neq 0$ for some k then *any* smooth function f defined on M must have at *least* R_k critical points with index k. This is clearly a useful result.

Now for the proofs. We start by establishing (8.4) in the form of a Lemma.

8.2 MORSE LEMMA

Let a^* be a non-degenerate critical point of a smooth real valued function f defined on a compact differentiable manifold M of dimension n. Then there is a local co-ordinate system $(y_1, \ldots, y_n)$ in a neighbourhood U of $a^* \in M$ with $y_i(a^*) = 0$ for all i such that

$$f = f(a^*) - (y_1)^2 - (y_2)^2 \ldots - (y_k)^2 + (y_{k+1})^2 + \ldots + (y_n)^2 \tag{8.8}$$

holds throughout U. k is said to be the index of f at a^*.

Proof

Let us choose $a^* = 0$ and $f(0) = (0)$. We can then write

$$f(x_1, \ldots, x_n) = \int_0^1 \mathrm{d}t \frac{\mathrm{d}}{\mathrm{d}t} f(tx_1, \ldots, tx_n), \text{ since } f(0) = 0$$

Thus

$$f(x_1, \ldots, x_n) = \int_0^1 \sum_{j=1}^n x_j \frac{\partial f}{\partial x_j}(tx_1, \ldots, tx_n)\, \mathrm{d}t$$

$$= \sum_{j=1}^n x_j g_j(x_1, \ldots, x_n) \tag{8.9}$$

where $g_j(x_1, \ldots, x_n) \equiv \int_0^1 \mathrm{d}t(\partial/\partial x_j) f(tx_1, \ldots, tx_n)$. Since the point 0 is assumed to be a critical point of f by definition:

$$g_j(0) = \frac{\partial}{\partial x_j} \begin{matrix} f(tx_j) = 0 \\ tx = 0 \end{matrix}$$

So that repeating the argument just made it follows that

$$g_j(x_1, \ldots, x_n) = \sum_{i=1}^n x_i h_{ij}(x_1, \ldots, x_n) \tag{8.10}$$

Hence in the neighbourhood of a critical point 0 where $f(0)$ is chosen to be zero the function f can always be represented as:

$$f(x_1, \ldots, x_n) = \sum_{i,j=1}^{n} x_i x_j h_{ij}(x_1, \ldots, x_n) \tag{8.11}$$

We can assume without any loss of generality that $h_{ij} = h_{ji}$, since we can always replace h_{ij} in (8.11) by $\{\frac{1}{2}(h_{ij} + h_{ji}) + \frac{1}{2}(h_{ij} - h_{ji})\}$ and get

$$f(x_1, \ldots, x_n) = \sum_{ij=1}^{n} x_i x_j \tfrac{1}{2}(h_{ij} + h_{ji}) \tag{8.12}$$

Note also that $h_{ij}(0) = (\partial^2 f/\partial x_i \partial x_j)_{x=0}$ and hence is non-singular by the definition of a non-degenerate critical point. We now claim that there is a non-singular transformation of the co-ordinate functions which gives us the desired expression for f in a perhaps smaller neighbourhood of 0. To see this we proceed as follows. Suppose $h_{11}(0) \neq 0$. Then

$$f(x_1, \ldots, x_n) = [(x_1)^2(\sqrt{h_{11}})^2 + \ldots] \tag{8.13}$$

If we now introduce co-ordinates:

$$\begin{aligned} y_1 &= \sqrt{h_{11}}\left[x_1 + \sum_{r\neq 1} \frac{h_{1r}}{h_{11}} x_r\right] \\ y_r &= x_r, \qquad r \neq 1 \end{aligned} \tag{8.14}$$

Then in terms of these variables the function f is diagonal in the variable y_1. Also by the inverse function theorem the variables $y_1, \ldots y_n$ can serve as smooth co-ordinate functions possibly in a smaller neighbourhood of 0 than the set $x_1, \ldots, x_n$. By repeated application of this procedure the Lemma is established.

An immediate consequence of the Lemma is the corollary.

Corollary

Non-degenerate critical points are isolated. (8.15)

This is because in the neighbourhood of 0 in which the representation of f established in Lemma (8.8) holds f does not have any other critical points.

To proceed further we need to know how the topology of a manifold changes in the neighbourhood of a critical point. For any real number c let M_c be the set of all x in M with $f(x) < c$ where c is a real number. If there is no critical value in the interval $[b, c]$ then $f^{-1}[b, c]$ and $f^{-1}(b) \times [b, c]$ have the same homotopy type. A geometrical way of understanding this result is to note that since there is no critical value of the function f in $[b, c]$ the direction of the gradient of f can be defined using a metric. Using these directions orthogonal trajectories to the surface $f(x)=b$ can be constructed and used to deform the region $f(x)\leqslant b$ to the region $f(x)\leqslant c$. A proof of this result can be found in reference [5]. Now suppose that there is exactly one critical point of f with index k and value c, i.e. $f(a^*)=c$ in the interval $(c+\varepsilon, c-\varepsilon)$. How is the topology of $M_{c+\varepsilon}$ related to that of $M_{c-\varepsilon}$ for small enough ε? We have the following theorem.

Theorem

Let $f:M\rightarrow \mathbf{R}$ be a smooth function and let a^* be a non-degenerate critical point with index k. Suppose $f(a^*)=c$ and $f^{-1}[c-\varepsilon, c+\varepsilon]$ contains no critical point of f other than a^* for some $\varepsilon>0$. Then for all sufficiently small ε the set $M_{c+\varepsilon}$ has the homotopy type of $M_{c-\varepsilon}$ with a k-cell, e_k attached. Where $M_c=\{x|f(x)\leqslant c\}$. (8.16)

We will not prove this fundamental theorem but only explain its geometrical content. First we must explain what attaching a k-cell e_k means. A k-cell is a k-dimensional ball i.e.

$$e_k = \text{Set } (x_1, \ldots, x_k) \text{ such that } x_1^2+x_2^2+\ldots+x_k^2<1 \tag{8.17}$$

Attaching such a k-cell to $M_{c-\varepsilon}$ means that we first take the point set union of $M_{c-\varepsilon}$ and e_k and then glue the two spaces along the boundary $\dot{e}_k$ of e_k. Where

$$\dot{e}_k = \text{Set } (x_1, \ldots, x_k) \text{ such that } x_1^2+\ldots+x_k^2=1 \tag{8.18}$$

using a smooth function

$$g:\dot{e}_k\rightarrow M_{c-\varepsilon} \tag{8.19}$$

This means that the points x and $g(x)$ are identified in the point set union $e_k U_g M_{c-\varepsilon}$. Let us now examine the theorem. We know that in a neighbourhood U of a non-degenerate critical point a^* with index k the function f can be represented as:

$$f=c-y_1^2-y_2^2\ldots-y_k^2+y_{k+1}^2+\ldots+y_n^2$$

Observe that in this neighbourhood there are k directions in which the function f decreases in value from c, namely, in the directions $y_1\ldots y_k$. The theorem tells us that as far as the homotopy type of the manifolds $M_{c-\varepsilon}$, $M_{c+\varepsilon}$ is concerned, $M_{c+\varepsilon}$ can be obtained from $M_{c-\varepsilon}$ by 'filling in' this k-dimensional depression by a k-cell e_k (see Fig. 8.1). We now proceed to prove the Morse inequalities (Theorem 8.6). It is convenient to introduce the following definition.

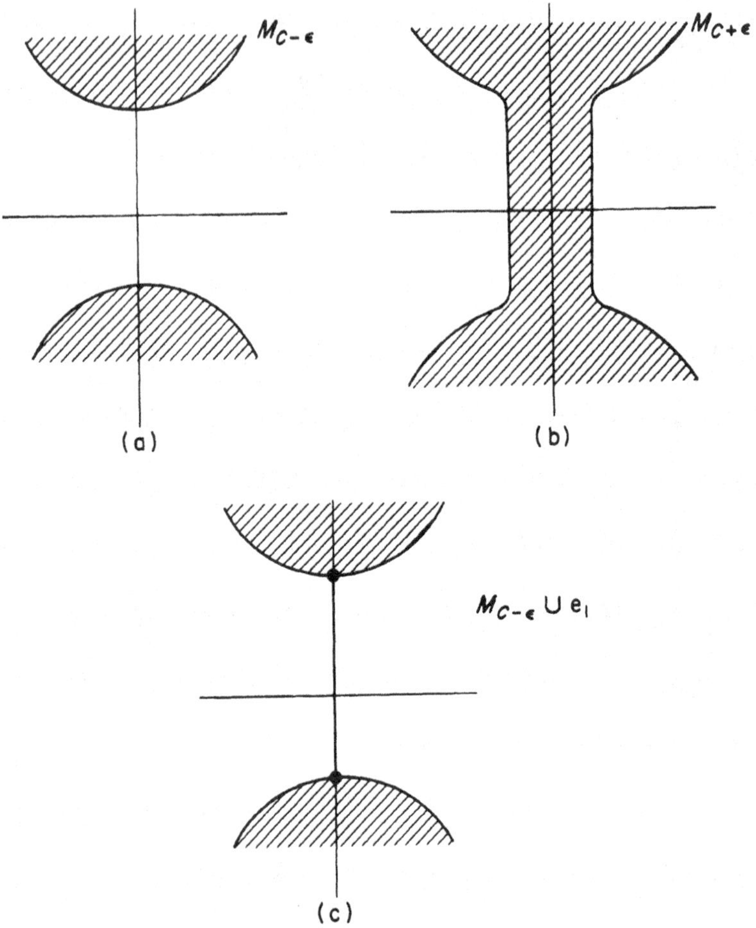

Figure 8.1

Definition

An integer valued function $S(X, Y)$ of a pair of spaces X, Y is said to be sub-additive if for three spaces $X \supset Y \supset Z$ we have

$$S(X, Z) \leqslant S(X, Y) + S(Y, Z) \tag{8.20}$$

If the equality sign holds then the function is said to be additive.

An immediate consequence of this definition is the Lemma.

Lemma

If $X_0 \subset X_1 \subset X_2 \ldots X_N \subset X_{N+1}$, then

$$S(X_{N+1}, X_0) \leqslant \sum_{i=1}^{N} S(X_i, X_{i-1}) \tag{8.21}$$

Proof

The result is true trivially for $N = 1$ and is the definition of sub addivity for $N = 2$. Suppose the result is true for $(N - 1)$ i.e.

$$S(X_N, X_0) \leqslant \sum_{i=1}^{N-1} S(X_i, X_{i-1})$$

Since S is a sub-additive function if we consider the space $X_0 \subset X_N \subset X_{N+1}$ we must have

$$\begin{aligned} S(X_{N+1}, X_0) &\leqslant S(X_{N+1}, X_N) + S(X_N, X_0) \\ &\leqslant S(X_{N+1}, X_N) + \sum_{i=1}^{N-1} S(X_i, X_{i-1}) \\ &\leqslant \sum_{i=1}^{N} S(X_i, X_{i-1}) \end{aligned}$$

Which establishes the Lemma.

We now make the following claim.

Claim

For $X \supset Y$ the function

$$R_k(X, Y) - R_{k-1}(X, Y) + R_{k-2}(X, Y) - \ldots \pm R_0(X, Y)$$

is a sub-additive function. Here $R_k(X, Y)$ denotes the dimension of the relative homology group $H_k(X, Y)$. (8.22)

Proof

We prove the claim by examining the exact homology sequence associated with a triple of spaces X, Y, Z. Generalizing the results of Chapter 4 slightly it is possible to show that the following is an exact sequence:

$$\begin{aligned} &\to H_{k+1}(X, Y) \xrightarrow{\partial_k} \overset{*}{H}_k(Y, Z) \xrightarrow{l_k} H_k(X, Z) \\ &\qquad \xrightarrow{i_k} H_k(X, Y) \xrightarrow{\partial_{k-1}} \ldots H_0(X, Y) \xrightarrow{\partial_0} 0 \end{aligned} \tag{8.23}$$

The exactness of the sequence of relative homology groups means that the following relationships hold:

$$\text{Rank } \partial_k = \text{Kernel } i_k$$

$$\text{Rank } i_k = \text{Kernel } j_k$$

$$\text{Rank } j_k = \text{Kernel } \partial_{k\ 1}, \qquad n = n, \ldots, 0$$

where

$$\text{Rank } \partial_k \equiv \text{Dimension of Image of } \partial_k$$

$$\text{Kernel } i_k \equiv \text{Dimension of Kernel of } i_k, \text{ etc.}$$

On the other hand from the rank and nullity theorem of linear algebra it follows that

$$\begin{aligned} R_k(Y, Z) &\equiv \text{Dimension of } H_k(Y, Z) \\ &= \text{Rank } i_k + \text{Kernel } i_k \\ &= \text{Rank } i_k + \text{Rank } \partial_k, \text{ from (8.24)} \end{aligned}$$

Therefore

$$\text{Rank } \partial_k = R_k(Y, Z) - \text{Rank } i_k \tag{8.25}$$

Again

$$\begin{aligned} R_k(X, Z) &= \text{Rank } j_k + \text{Kernel } j_k \\ &= \text{Rank } j_k + \text{Rank } i_k \end{aligned}$$

Therefore

$$\text{Rank } i_k = R_k(X, Z) - \text{Rank } j_k \tag{8.26}$$

Similarly

$$\text{Rank } j_k = R_k(X, Y) - \text{Rank } \partial_{k-1}$$

Thus from (8.25) and (8.26) we get

$$\text{Rank } \partial_k = R_k(Y, Z) - R_k(X, Z) + R_k(X, Y) - \text{Rank } \partial_{k-1} \tag{8.27}$$

By repeated use of (8.27) and noting that Rank $\partial_0 = 0$, we get

$$\begin{aligned} \text{Rank } \partial_k = R_k(Y, Z) - R_k(X, Z) + R_k(X, Y) - R_{k-1}(Y, Z) \\ + R_{k-1}(X, Z) - R_{k-1}(X, Y) + \ldots \end{aligned} \tag{8.28}$$

Now Rank $\partial_k > 0$, except when $k = n$, when rank $\partial_{n=0}$, since $H_{n+1}(Y, Z) = \{0\}$.

Thus if we write

$$S_k(X, Y) = R_k(X, Y) - R_{k-1}(X, Y) + R_{k-2}(X, Y) - \dots \pm R_0(X, Y) \quad (8.29)$$

Then equation (8.28) can be written as

$$S_k(X, Z) \leqslant S_k(X, Y) + S_k(Y, Z)$$

where the equality sign holds only for $k = n$. Thus the claim is established.

Let us now consider a compact differentiable manifold M on which a smooth real valued function f having only non-degenerate critical points is defined. Let us arrange the critical points of f in order $c_1 < c_2 \dots < c_k$ such that M_{a_i} contains exactly i critical points and $M_{a_k} = M$. Thus we have the spaces

$$\phi \subset M_{c_1} \subset M_{c_2} \subset \dots M_{c_k} = M$$

where ϕ denotes the empty space. Applying the sub-additive function S_k to these spaces we get

$$S_k(M) \equiv S_k(M, \phi) \leqslant \sum_{i=1}^{k} S_k(M_{c_i}, M_{c_{i-1}}) \quad (8.30)$$

where by definition

$$R_k(M_{c_i}, M_{c_{i-1}}) = \text{Dimension of } H_k(M_{c_i}, M_{c_{i-1}}) \quad (8.31)$$

From Theorem (8.18) we know that M_{c_i} has the same homotopy type as $M_{c_{i-1}} U_g e_p$ if the critical point contained in $(M_{c_i} - M_{c_{i-1}})$ has index p. Since homology groups are the same for spaces of the same homotopy type it follows that

$$H_k(M_{c_i}, M_{c_{i-1}}) = H_k(M_{c_{i-1}} U_g e_p, M_{c_{i-1}}) \quad (8.32)$$

By using the excision theorem of Chapter 4 we can write

$$H_k(M_{c_{i-1}} U_g e_p, M_{c_{i-1}}) = H_k(e_p, \dot{e}_p) \quad (8.33)$$

where the interior of $M_{c_{i-1}} - M_{c_{i-1}} \cap e_p$ has been excised. In Chapter 4 we showed that

$$H_k(e_p, \dot{e}_p) = \begin{cases} Z & \text{if } k = p \\ \{0\}, & \text{otherwise} \end{cases} \quad (8.34)$$

(remember $\dot{e}_p = S^{P-1}$, $e_p = B^P$). So that

$$\text{Dimension } H_k(e_p, \dot{e}_p) = \delta_{kp} \quad (8.35)$$

Hence (8.30) together with (8.29) and (8.35) imply:

$$R_k(M) - R_{k-1}(M) + \dots \pm R_0(M) \leqslant C_k - C_{k-1} \dots \pm C_0$$

with equality holding only for $k = n$. These are the Morse inequalities of Theorem (8.6).

8.3 SYMMETRY BREAKING SELECTION RULES IN CRYSTALS

A crystal can be described as a state of matter in which the density function $\rho(\vec{x})$ is invariant under transformations $T_1, T_2, \ldots T_k$ belonging to a finite group G, i.e.

$$\rho(\vec{x}) = \rho(T_1\vec{x}) = \ldots = \rho(T_k\vec{x}) \tag{8.38}$$

The group G is called the symmetry group of this crystal. It is an experimental fact that as the external conditions of the crystal are changed the crystal and its density function change continuously but the symmetry group of the crystal can change abruptly from G to H where H is usually a subgroup of G. It is also an experimental fact that not all subgroups H_i of G are realized, i.e. selection rules seem to operate in these symmetry breaking transitions.

Recently Michel and Mozrzymas [4] have applied Morse theory to the problem of determining these symmetry breaking selection rules. We will briefly explain how this is done. The framework for discussing the changes in symmetry of a crystal used in reference [4] was Landau's theory of phase transitions [3]. The equilibrium properties of a crystal are described by its density function $\rho(\vec{x})$, temperature T and pressure P. The equilibrium density $\rho_E(\vec{x})$ of the crystal is determined, according to classical thermodynamics, by minimizing an appropriate real valued thermodynamic function $F(\rho; T; P)$. F is a functional of ρ which depends on the parameters P and T. If F were known then the equilibrium density $\rho_E(\vec{x})$ could be determined as the function for which F takes its absolute minimum value.

Let us consider a crystal with symmetry group G for, say $T > T_0$, and symmetry group H for $T < T_0$ where H is a subgroup of G. We want to determine whether any subgroup H_i of G is possible as the 'broken' symmetry group H or if selection rules operate. In view of our assumptions we write

$$\rho(\vec{x}) = \rho_0(\vec{x}) + \delta\rho(\vec{x}) \tag{8.39}$$

where $\rho_0(\vec{x})$ is invariant under G and $\delta\rho(\vec{x})$ is invariant under H. Thus if $\delta\rho = 0$, the symmetry group of $\rho(\vec{x})$ is G. From group theory [2] we know that if G contains $n(G)$ elements $g_1, \ldots g_{n(G)}$ then a representation of G can be constructed by regarding the group elements as linear transformations on a set of $n(G)$ linearly independent functions, $\Phi_1(\vec{x}), \ldots,$

$\phi_{n(G)}(\vec{x})$, i.e.

$$g\phi_k(\vec{x}) = \sum_{j=1}^{n(G)} D_{kj}[g]\phi_j(\vec{x}),\ g \in G \tag{8.40}$$

This suggests that as far as the action of the group elements $g_1, g_2, \ldots,$ is concerned any arbitrary function $\psi(\vec{x})$ can be expanded in terms of the complete set of functions $\phi_1(\vec{x}), \ldots, \phi_K(\vec{x}) \ldots \phi_{n(G)}(\vec{x})$. In particular

$$\delta\rho(\vec{x}) = \sum_{i=1}^{n(G)} C_i\phi_i(\vec{x}) \tag{8.41}$$

Note that once the complete set of functions $\phi_i(\vec{x})$ is selected different density functions $\delta\rho(\vec{x})$ are described by the set of constants $(C_1, C_2, \ldots, C_{n(G)})$.

Furthermore since

$$\begin{aligned}\delta\rho'(\vec{x}) = g\delta\rho(\vec{x}) &= \sum_{i=1}^{n(G)} C_i g\phi_i(\vec{x}) = \sum_{i=1}^{n(G)} C_i\phi'_i(\vec{x}) \\ &= \sum_{i=1}^{n(G)}\sum_{j=1}^{n(G)} C_i D_{ik}[g]\phi_k(\vec{x}) \\ &= \sum_{i=1}^{n(G)} C'_i\phi_i(\vec{x})\end{aligned}$$

where

$$C'_i = \sum_{k=1}^{n(G)} C_k D_{ki}[g],\ g \in G \tag{8.42}$$

we can regard the action of the group G on $\delta\rho$ in two ways, namely $g\delta\rho(\vec{x}) = \delta\rho'(\vec{x})$ because $C'_i = \sum_{k=1}^{n(G)} C_k D_{ki}[g]$ or because $\phi'_i = g\phi_i$. The co-ordinates $(C_1, \ldots, C_{n(G)})$ of $\delta\rho$ can thus be regarded as elements of an $n(G)$ dimensional vector space on which the representation matrices of G, $D[G]$ act. For the representation of a finite group G the following theorems hold [2].

Theorem

Every representation of a finite group G is equivalent to a unitary representation, i.e. the vectors $x \equiv (x_1, \ldots, x_{n(G)})$, $y \equiv (y_1, \ldots, y_{n(G)})$ on which G acts $x \to x' = D(g)x$; $y \to y' = D(g)y$, $g \in G$ can be given a scalar product structure $(\vec{x}, \vec{y})$ and the representation of G, i.e. the matrices $D[g]$, keeps this scalar product invariant, i.e.

$$(\vec{x}, \vec{y}) = (\vec{x}', \vec{y}') \tag{8.43}$$

An immediate Corollary of this Theorem is:

Corollary

Every real representation of a finite group is equivalent to an orthogonal representation. (8.44)

These theorems tell us that if we consider only real representations of G then the space of co-efficients $(C_1, \ldots, C_{n(G)})$ can be chosen to be an $n(G)$ dimensional euclidean space. The restriction to only real representations is necessary since $\rho(\vec{x})$, the density function of the crystal, is necessarily real.

After this group theoretical digression we can return to our problem. In the neighbourhood of the transition temperature T_0 we require the symmetry group to change abruptly from G to H. This will happen if $\delta\rho(\vec{x})$ changes from a zero value to a non-zero value as the temperature T crosses T_0. In this neighbourhood, then, $\delta\rho(\vec{x})$ is small, i.e. the co-ordinates $(C_1, \ldots, C_{n(G)})$ are small. On the basis of this simple observation and using physical arguments Landau [3] supposed that, for $T \sim T_0$, the thermodynamic potential could be represented as polynomial in the vector C containing only second and fourth degree terms with co-efficients that varied with T and P. This polynomial $F(C)$ we will call the 'Landau polynomial'. Since the thermodynamic potential is a co-ordinate independent object it also follows that $F(C)$ has to be G invariant, i.e.

$$F[D(g)C] = F[C] \tag{8.45}$$

The equilibrium density distribution of the system is to be determined by finding vectors C_0 such that

$$F'[C_0] = 0 \tag{8.46}$$

If the point at infinity is added on to the $n(G)$ dimensional space in which the vector C_0 lives then the real function F becomes a mapping from S^k to $\mathbf{R}$ where $(k = n(G))$ and the equation $F'(C_0) = 0$ corresponds to a critical point of this mapping. If we suppose that the critical points of F where $F: S^k \to \mathbf{R}$ are non-degenerate (this, in fact can be proved) then Morse theory is applicable as S^k is a differential manifold. We can then write down the set of Morse inequalities appropriate for S^k namely.

$$n_0 \geqslant 1$$

$$n_1 - n_0 \geqslant R_1 - R_0 = -1$$

$$n_2 - n_1 - n_0 \geqslant R_2 - R_1 + R_0 = +1, \text{ etc}$$

where n_i denotes the number of critical points of $F[c_0]$, of index i and we have used the fact that

$$R_i(S^k) = \begin{cases} 1, & (\text{for } i = 0 \text{ or } i = k) \\ 0, & \text{otherwise} \end{cases}$$

We also observe that the equation $F'(C_0) = 0$ is cubic in the k-dimensional vector C_0 so that the total number of roots, real and complex, of this system (counting the point at infinity) is $3^k + 1$. The solutions of interest to us are only the real ones. Each *real* solution corresponds to a critical point of F with an index ranging from 0 to k. We thus have the inequality,

$$\begin{aligned} n_0 + n_1 + \ldots + n_k &= \text{Total number of real solutions} \\ &\leqslant 3^k + 1 \end{aligned} \tag{8.48}$$

Note n_0 describes the number of critical points of F with index zero, i.e. the number of minima. The numbers n_i are related to the subgroup H_i of G and thus if determined contain in them information regarding the nature of symmetry breaking. We proceed to find this relation. Suppose C_0 is a critical point of F with index p. Suppose further that $D[H_j]C_0 = C_0$,

i.e.

$$D(h)C_0 = C_0, \; h \in H_j \subset G \tag{8.49a}$$

Thus H_j is the invariance group of c_0. This equation also implies that

$$H_j \delta\rho = \delta\rho \tag{8.49b}$$

since $\delta\rho = \sum_i c_0^i \phi_i$. Thus H_j satisfying (8.49a) is a possible symmetry group of the crystal when the critical point C_0 has index zero and thus corresponds to a minima of F. In general once H_j, a subgroup of G, is chosen the group G can be partitioned into disjoint classes in terms of the cosets of H_j, i.e. we can write

$$G = H_j + g_1 H_j + g_2 H_j + \ldots + g_{m-1} H_j \tag{8.50}$$

where $g_1 H$ represents the coset generated by the element g_1 and contains in it all elements f which can be written as $g_1 h$, $h \in H_j$.

If $n(G)$ represent the order of the group G and $n(H_j)$ the order of the subgroup H_j then Lagranges Theorem [2] tells us that $n(G)/n(H_j)$ is an integer and is indeed equal to m of (8,50). Also since $F[C_0]$ is, by construction, G invariant it follows that if C_0 is a critical point of F then so is $D[g]C_0$, where $g \in G$. The number n_p of distinct critical points that can be generated in this way is equal to $n(G)/n(H)$. Explicitly these critical points are C_0, $D[g_1]C_0$, $D[g_2]C_0, \ldots, D[g_{m-1}]C_0$ where $g_1, g_2, \ldots, g_{m-1}$ represent the elements of G defined in (8.50). Any other element of G, f say must belong to one of the cosets of H_j, gH_j say and can thus be written as

$f=g_1h$, $h \in H_j$ so that $D[f]C_0=D[g_1h]C_0=D[g_1]C_0$, since $D[h]c_0=C_0$ by the definition of H_j. Thus the connection between the subgroups H_j of G and the numbers n_p can be stated as follows. If the invariance group of C_0, a critical point of F of index p, is H_j then $n_p=n(G)/n(H_j)$. Let us now look at a specific example and see how the ideas outlined are actually used to determine the structure of a symmetry breaking transition. Let G, the symmetry group of the crystal, be the group O_h, which contains 48 elements, i.e. $n(G)=48$ (see reference [2] for details). An analysis of this group [4] leads to the following possibilities for its symmetry breaking subgroups:

$$n(H)=8 \text{ or } 6 \text{ or } 4 \text{ or } 2 \tag{8.51}$$

Thus the ratio $n(G)/n(H)$ can be 6 or 8 or 12 or 24. A few remarks regarding the dimension k of the representation space to be used are now necessary. In general, as we have stated earlier, the number k of functions $\phi_i(\vec{x})$ needed to form a complete set is $n(G)$ the order of the group G. For the group O_h, $n(G)$ is 48. However the representation of G constructed in this way is not an irreducible one and can be split into the sum of the different irreducible representations of G. We have the following theorem [2].

Theorem

If $n(G)$ is the order of the finite group G and if $n_1, n_2, \ldots, n_s$ are the dimensions of the irreducible representations of G then

$$n_1^2+n_2^2+\ldots+n_s^2=n(G) \tag{8.52}$$

Thus the set $\phi_1(\vec{x}), \ldots, \phi_{n(G)}(\vec{x})$ can be written as ϕ_l^n, where n is the irreducible representation label and l the 'partner' functions which belong to the irreducible representation n. On the basis of physical arguments Landau suggested that in the neighbourhood of a symmetry breaking transition $\delta\rho(\vec{x})$ could be expressed in terms of the functions belonging to only *one* of the irreducible representation of G, although these functions on their own *do not* form a complete set. For the group O_h the irreducible representation are 1, 2 and 3 dimensional. The representation we will analyse is the three dimensional one. In this case the symmetry breaking vector C_0 is an element of 3 dimensional rather than 48 dimensional space. Thus $k=3$ and (8.48) gives the constraint

$$n_0+n_1+n_2+n_3 \leqslant 3^3+1=28 \tag{8.53}$$

Since $n(G)/n(H) \geqslant 6$, it follows that n_0, n_1, n_2 must all be greater than or equal to six. While n_3, which counts the number of maxima of F need not be so constrained. This is because a maximum point of F cannot, by definition, represent an equilibrium configuration of the crystal hence it is not necessary to require such a point to break the symmetry. The Landau polynomial in the region where we want the crystal to have H rather than G as its symmetry group is constructed to have a maximum at the origin and at infinity. The maximum at the origin ensures that the solution $C_0 = 0$ with symmetry G is not thermodynamically preferred to a solution with $C_0 \neq 0$, and symmetry H. While the maximum at infinity is introduced to make sure that the crystal is stable. The constraint on n_3 is thus $n_3 \geqslant 2$. Collecting all our constraints we have:

$$\begin{aligned} n_3+n_2+n_1+n_0 &\leqslant 28 \quad \text{(i)} \\ n_3-n_2+n_1-n_0 &= 0 \quad \text{(ii)} \\ n_2-n_1+n_0 &> 1 \quad \text{(iii)} \\ n_1-n_0 &> -1 \quad \text{(iv)} \end{aligned} \tag{8.54}$$

These inequalities immediately give:

$$\begin{aligned} n_3+n_1 &\leqslant 14 \quad \text{(i)} \\ n_2+n_0 &\leqslant 14 \quad \text{(ii)} \\ n_1 &> n_0-1 \quad \text{(iii)} \end{aligned} \tag{8.55}$$

The possible values of n_0 are, as we saw, 24, 12, 8 or 6. Let us examine these in turn.

a. $n_0 = 24$. From (8.55.iii) it follows that $n_1 > 23$. But this is not allowed in view of (8.54.i)
b. $n_0 = 12$. Again from (8.55) it follows that $n_1 > 11$, so that $n_1 = 12$ or 24. Again this possibility is not allowed in view of (8.54.i).
c. $n_0 = 8$. Arguing as before we find that $n_1 = 8$ or 12 or 24. On the other hand since $n_0 + n_2 \leqslant 14$, and $n_2 \geqslant 6$ it follows that $n_2 = 6$. Also $n_3 = 2$ or 7, or higher values and from (8.55.i) it follows that only $n_3 = 2$ is allowed (8.54.ii) then determines n_1 uniquely to be 12. Thus we get

$$n_0 = 8, \qquad n_1 = 12, \qquad n_2 = 6, \qquad n_3 = 2 \tag{8.56}$$

as the structure of a possible symmetry breaking solution.
d. Finally the case $n_0 = 6$ has to be considered. A similar analysis yields the unique structure:

$$n_0 = 6, \qquad n_1 = 12, \qquad n_2 = 8, \qquad n_3 = 2 \tag{8.57}$$

The symmetry group corresponding to $n_0 = 8$ is c_{3v} and for $n_0 = 6$ is c_{4v} [4]. Thus we get the following symmetry breaking selection rule: a crystal with symmetry group O_h, which contains 48 elements, can change its symmetry group only to its subgroup C_{4v} (containing 8 elements) or C_{3v} (containing 6 elements).

In this example the selection rules could be determined using group theory and the Morse inequalities without ever having to consider the detailed structure of the Landau polynomial! In our discussion we deliberately glossed over various group theoretical details so as to highlight the simple way in which the Morse inequalities are used in the problem. For the group theoretical details left out and further applications reference [4] may be consulted.

8.4 ESTIMATING EQUILIBRIUM POSITIONS

Consider N point particles having masses $m_i (i = 1, \ldots, N)$ which influence each other by a force law derivable from a potential function. Suppose that each m_i is constrained to move on some compact manifold M_i and that the potential function V depends only on the position of the N particles. Thus V, a real valued function, is defined on the manifold M where

$$M = M_1 \times M_2 \times \ldots M_N \tag{8.58}$$

An equilibrium configuration of the system can be defined to be a critical point of V. Assuming these critical points of V are non-degenerate it immediately follows that the number of equilibrium positions C by:

$$C = \sum_{k=0}^{\text{Dim } M} C_k > \sum_{k=0}^{\text{Dim } M} R_k(M) \tag{8.59}$$

If, for instance, all the M_i are homeomorphic to S^1, a simple application of the Kunneth formula of Chapter 4 tells us that

$$H_k(M) = H_k(\underbrace{S^1 \times S^1_x \ldots S^1}_{N\text{-terms}}) = \underbrace{Z + Z \ldots + Z}_{p\text{-terms}}$$

where

$$p = {}^N C_k = \frac{N!}{k!(N-k)^1}$$

Hence

$$R_k(M) = \dim (H_k(M) = {}^N C_k$$

So that

$$C = \sum_{k=0}^{\text{Dim } M} C_k > \sum_{k=0}^{\text{Dim } M} {}^N C_k = (1+1)^{\text{Dim } M} = 2^N$$

Since Dim $M = N$. Thus

$$C \geqslant 2^N \tag{8.60}$$

For a discussion of this problem using a different topological inequality reference [1] may be consulted. A further non-trivial application of this nature is the work of Palmore [7] on new relative equilibrium positions in the 4 body problem.

REFERENCES

1. EL'SGOL'C, L. E., "Qualitative Methods in Mathematical Analysis". Amer. Math. Soc., Rhode Island, 1964.
2. HAMMERMESH, M., "Group Theory". Addison Wesley, 1962.
3. LANDAU, L. D. and LIFSHITZ, E. M., "Statistical Physics". Pergamon Press, 1959.
4. MICHEL, L. and MOZRZYMAS, B., Group theoretical methods in physics. *In* "Lecture Notes in Physics" (Ed, Tubingen), Vol. 79. p. 447. Springer-Verlag, 1963.
5. MILNOR, J., "Morse Theory". Princeton University Press, 1963.
6. MORSE, M., "The Calculus of Variation in the Large". Amer. Math. Soc., Rhode Island, 1964.
7. PALMORE, J., *Bull. Amer. Math. Soc.* **79**, 904–908 (1973).

CHAPTER 9

Defects, Textures and Homotopy theory

In this chapter we will briefly explain how homotopy theory has been used to classify defects and textures in an ordered medium. We start by looking at a simple example.

9.1 PLANAR SPIN IN TWO DIMENSIONS

The ordered medium in this case is a region of $\mathbf{R}^2$ with the property that at every point there is a spin vector $\vec{S}$. The vector $\vec{S}$ has fixed length (which can be taken to be unity), is free to point in any direction in $\mathbf{R}^2$ and is a continuous function of $\vec{r} \in \mathbf{R}^2$, except possibly for a subset $\Sigma \subset \mathbf{R}^2$ (Σ, the defect set, could, of course, be empty). We can thus write, for $r \in \mathbf{R}^2 - \Sigma$

$$\vec{S}(\vec{r}) = \hat{u} \cos \phi(\vec{r}) + \hat{v} \sin \phi(\vec{r}) \tag{9.1}$$

where $\hat{u}$, $\hat{v}$ are a fixed pair of orthonormal vectors. The ordered medium will be said to be in a uniform state if $\vec{S}(\vec{r})$ is independent of $\vec{r}$. Such a uniform spin system could, for example, describe a ferromagnet.

We will now demonstrate, using essentially homotopy theory, that the defect set Σ for the planar spin system need not be empty. More precisely we will show that topologically stable point defects are possible for this system. The argument is simple. Suppose that an isolated point defect of the medium is located at P. We then surround P by a contour C. By hypothesis $\vec{S}(\vec{r})$ is well defined on C. Since $\vec{S}(\vec{r})$ is a unit vector in 2-dimension it can be represented as a point θ on S^1 as is apparent from Fig. 9.1. $\vec{S}(\vec{r})$ thus represents a mapping from points on C, a closed curve in physical space, to a point on S^1, the space which describes the directions of the spin vector $\vec{S}$. Since $S(\vec{r})$ is a continuous function of position, the angle through which $\vec{S}$ turns as the contour C is traversed must be an integral multiple of 2π, $2\pi n$ say (see Fig. 9.2). Furthermore again since $S(\vec{r})$ is a continuous function while n is a discrete quantity changing the size or the shape of the contour C cannot change the value of n as long as $\vec{S}(\vec{r})$ remains continuous throughout these changes. Hence if C is shrunk

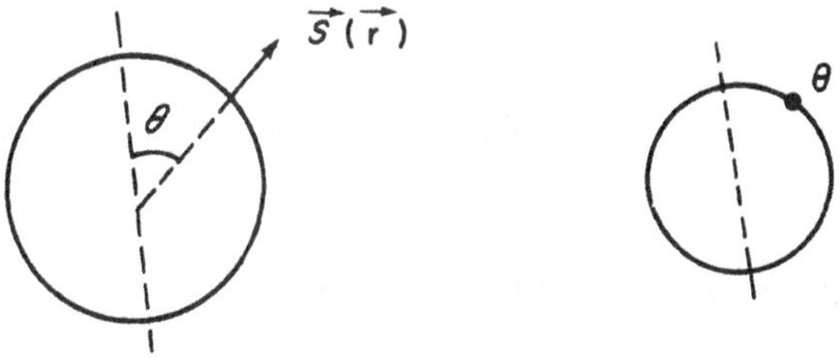

Figure 9.1

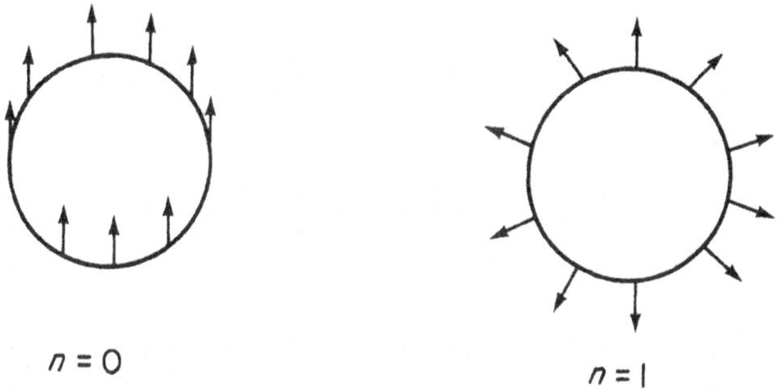

Figure 9.2

(in a continuous manner) to a circular shape of radius ε round P we will continue to get $(2\pi n)$ for the total change in angle through which $\vec{S}$ turns, $\vec{S}$ will then certainly be singular at P if $n \neq 0$. A singularity at the point P with $n \neq 0$ is said to be 'topologically stable' and may be classified by its n-value. Figure 9.2 shows two configurations of $\vec{S}$ corresponding to $n = 0$ and $n = 1$.

From our discussions of homotopy theory it should be apparent that the integer n is related to $\pi_1(S^1)$ and represents the winding number of the map $\vec{S}(\vec{r})$. We now proceed to give a more precise description of an ordered medium. We have the following definitions.

9.2 DEFINITION OF AN ORDERED MEDIUM

The physical space M^3 will be taken to be a smooth 3-dimensional manifold. It will be assumed that M^3 is connected, orientable and compact with boundary ∂M^3. (9.2)

Definition

There will also be given in the description of an ordered medium a topological space V called the *order parameter space*. (For our simple example V was S^1.) (9.3)

Finally we have,

Definition

In M^3 there is a subset $\Sigma \subset M^3$ which will be called the *set of defects*. Outside Σ a continuous map ϕ from M^3 to V is given. Thus

$$\phi : M^3 - \Sigma \to V$$

ϕ is called the order parameter field. (9.4)

The following theorem can now be established.

9.3 STABILITY OF DEFECTS THEOREM

There are no topologically stable

i. Point defects if $\pi_2(V) = 0$
ii. Line defects if $\pi_1(V) = 0$ (9.5)
iii. Wall defects if $\pi_0(V) = 0$

Proof

We first prove (ii). It is convenient for the proof to replace the possible line defect Σ_1 by a cylindrical region C (Fig. 9.3) of cross-section D. The order parameter ϕ is now well defined on the boundary of D with Σ_1 inside C. For a closed loop γ on boundary of C we have:

$$\phi(\gamma) : S^1 \to V \tag{9.6}$$

This map defines an element of $\pi_1(V)$ up to, as it is possible to show, conjugacy. If $\pi_1(V) = 0$ then any closed loop $\phi(\gamma)$ in V is homotopic to the constant map. We can then make use of the following theorem.

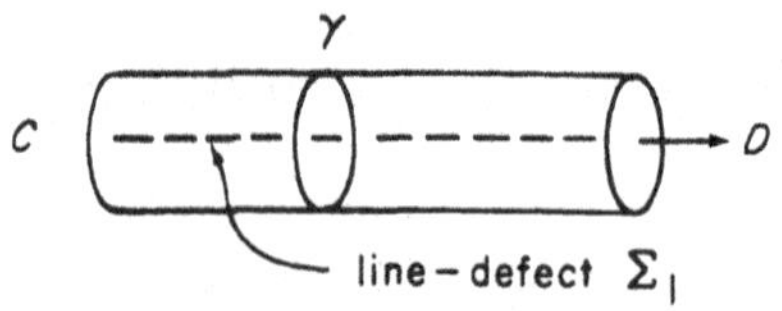

Figure 9.3

Theorem

A continuous map $f: S^n \to X$ is homotopic to the constant map if and only if it can be extended to a map $F: D^{n+1} \to X$ where F restricted to the boundary S^n of D^{n+1} is f. (9.7)

Thus when $\pi_1(V)=0$ and $\phi(r)$ is homotopic to the constant map the order parameter function can be extended, in a continuous manner, over D which means that a topologically stable line defect is not possible and (ii) is established. Using essentially the same procedure, namely replacing a possible point defect Σ_0 at P by a spherical region containing Σ_0 at P (Fig. 9.4) so that, as in the previous case, the order parameter ϕ is well defined on the boundary S^2 of the three-dimensional disc D^3. Since the map ϕ

$$\phi: S^2 \to V \tag{9.8}$$

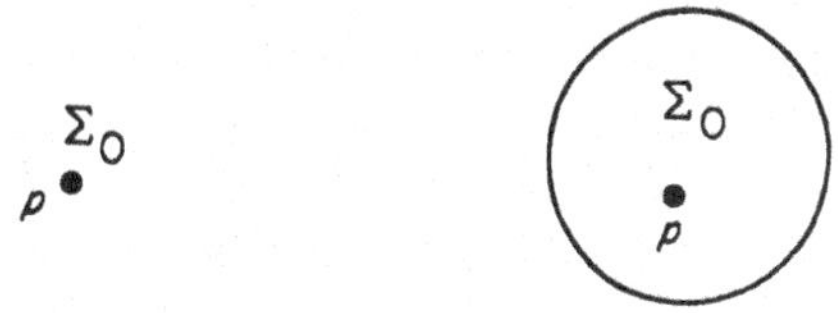

Figure 9.4

is related to the group $\pi_2(V)$ we can conclude that when $\pi_2(V)=0$, $\phi(S^2)$ is homotopic to the constant map. An application of Theorem (9.7) then again leads to the result that $\phi(S^2)$ can be extended in a continuous manner throughout D^3 which means that a topologically stable point defect is not possible if $\pi_2(V)=0$ and establishes (i).

Finally to prove (iii) we replace the two dimensional wall defect by a three dimensional wall on whose boundary the order parameter ϕ is well defined. We then note that $\pi_0(V)=0$ means that the space V is path connected. Thus a path from a point p, say on one side of the wall to a point q, say on the other side along which ϕ is well defined must exist. This path effectively represents a hole in the wall which can then be made to disappear by continuity as demonstrated in Fig. 9.5. Thus when $\pi_0(V)=0$ wall defects are topologically unstable and (iii) is established.

To complete our proof of Theorem (9.6) we need to prove Theorem (9.7) which was stated without proof. The proof is straightforward.

Suppose that $f: S^n \to X$ is homotopic to a constant map. This means that there is map (homotopy) $h_t: S^n \to X$, $t \in [0, 1]$ $h_0 = f$ and $h_1 = c$, a constant

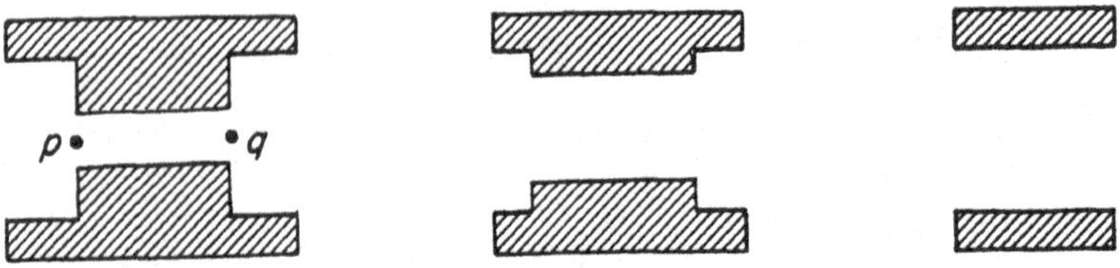

Figure 9.5

map. Now we note that any point $y \in D^{n+1} - 0$, be written uniquely as $y = tx$ where $t \in [0, 1]$ and $x \in S^n$. Thus we can define: $F: D^{n+1} \to X$ by $F(tx) = h_{1-t}(x)$, for $x \in S^n$, $t \in [0, 1]$ and $F(0) = h_1(S^1)$. By construction F is a continuous extension of f.

Conversely if we suppose $f: S^n \to X$ extends to a map $F: D^{n+1} \to X$ then we can define $f_t: S^n \to X$, $t \in [0, 1]$ by $f_t(X) = F(tx)$. Then f_t is a homotopy between f and a constant map and Theorem (9.8) is established.

Theorem (9.6) tells us that the existence of topologically stable defects has to do with the non-vanishing of certain homotopy groups of the order parameter space V. Our next goal will be to try and relate specific wall, line or point defects to more or less well defined elements of specific homotopy groups of V. (Our discussions will only deal with line defects. For a thorough discussion of the analogous problem for point or wall defects references [2, 3] may be consulted.) We start then by examining an ordered medium for which $\pi_1(V) \neq 0$ so that topologically stable line defects are possible in the medium. We test the nature of these line defects by examining the order parameter ϕ along a closed circular curve S^1 in the medium surrounding the defect. We suppose that ϕ is well defined on S^1. Thus we have the loop map:

$$\phi: S^1 \to V \tag{9.9}$$

The homotopy class of such loops when a point $a = \phi(x_0) \in V$ is kept fixed generates, as we saw in Chapter 4, the homotopy group $\pi_1(V; a)$. As far as the classification problem of line defects is concerned, however, there is no reason why two loops, both of which surround the line defect and on both of which ϕ is well defined, should have any point in common. Thus a natural way of classifying a line defect would be in terms of the free homotopy class of loops surrounding the line defect in question. Our problem then is to find a relation between the notion of a class of freely homotopic loops $\alpha_F \in V$ and elements $[\alpha]$, $[\beta]$ of $\pi_1(V; a)$ the fundamental group of order parameter space. The basic theorem which we state without proof is [3]:

Theorem

Two loops in order parameter space V are freely homotopic if and only if they belong to the same conjugacy class of $\pi_1(V; a)$. (9.10)

A few remarks are in order. First of all a loop $\alpha(t)$ will be said to be freely homotopic to a loop $\beta(t)$ if there exist a continuous map $H(s, t)$ with the property $H(0, t) = \alpha(t)$, $H(1, t) = \beta(t)$. We no longer require $H(s, 0) = H(s, 1) = x_0$ the common point of α, β. We recall that we argued that a line defect was to be classified not by an element of $\pi_1(V)$ but by the free homotopy class of loops surrounding the defect. Theorem (9.10) tells us that such loops describe conjugacy classes of $\pi_1(V)$. For a group G the conjugacy class of an element g_i of G is defined to be the set of elements $g_K g_i g_K^{-1}$, $g_K \in G$. Note that if $\pi_1(V)$ is an abelian group then the conjugacy classes are the group elements of $\pi_1(V)$ since $g_K g_i g_K^{-1} = g_i g_K g_K^{-1} = g_i$ themselves. So that for such a system each group element of $\pi_1(V)$ represents a topologically distinct type of line defect. A classification scheme of defects in terms of conjugacy classes of $\pi_1(V)$ rather than the elements of $\pi_1(V)$ themselves also has interesting implications when the merger of two line defects is considered. Suppose there are two non intersecting line defects P and Q in a medium with $\pi_1(V) \neq 0$. If the two defects are surrounded by a real space contour S^1 then from Fig. 9.6 it is apparent S^1 gives rise to a loop in V that is freely homotopic to the product of loops in V determined by contours that surround each line defect separately.

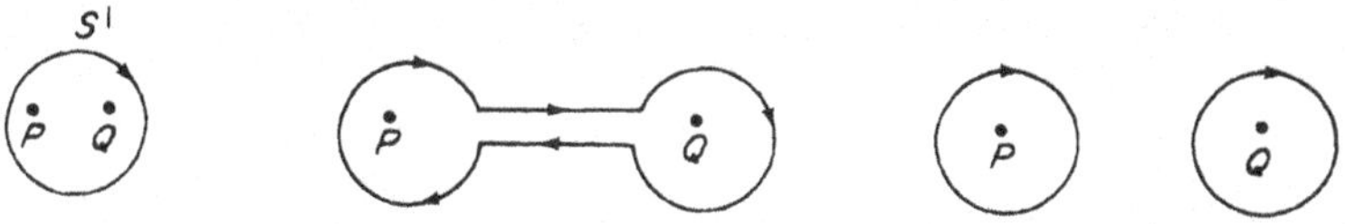

Figure 9.6

If we try to combine the two line defects P and Q into a single one R, say, subject only to the condition that at some sufficiently large distance from the original pair the order parameter remains unchanged throughout the process then R can only be described by one of the conjugacy classes of $\pi_1(V)$ which is contained in the product of the conjugacy classes characterizing the original two defects P and Q.

There are two cases to examine. If $\pi_1(V)$ is abelian then P, Q are described by elements $[\alpha]$, $[\beta]$ of $\pi_1(V)$ and when these two defects merge into a defect R, then R, is described by the unique element $[\alpha]\circ[\beta]$ of $\pi_1(V)$. Thus the combination rule for line defects, in this case, reflects the group multiplication rule for $\pi_1(V)$. When $\pi_1(V)$ is not abelians things are more interesting. If C_1, C_2 are the conjugary classes of $\pi_1(V)$ which describe the defects P and Q then the defect R is described by one of the conjugary classes contained in the product $C_1 \cdot C_2$. This is because in general $C_1 \cdot C_2$ is not itself a conjugary class but can be written as a sum of conjugary classes. If we suppose, for example, that $C_1 \cdot C_2 = C_3 + C_4$, say where C_3, C_4 represent conjugary classes of $\pi_1(V)$ different from C_1 and C_2 then by combining the defects P and Q a defect R_1 described by the conjugacy class C_3 or a defect R_2 described by the conjugacy class C_4 may result. In other words if $\pi_1(V)$ is non-abelian then there could be more than one outcome when two line defects in such a medium combine. Physically this could be taken to mean that the precise path followed in merging the two defects is important when $\pi_1(V)$ is non-abelian.

A few examples illustrating the power of the topological classification scheme might be useful.

9.4 EXAMPLES

Example 1. Nematic liquid crystal

This is a medium consisting of long molecules with the symmetry of ellipsoids of revolution. The order parameter ϕ in this case is a direction (the direction of the axis of revolution). The space of directions is then the order parameter space V. V can be described geometrically as a sphere S^2 with antipodal points identified. Thus V in this case is $\mathbf{R}P^2$ (the real projective plane). From Chapter 4, Example 6 we know $\pi_1(\mathbf{R}P^2) = \mathbf{Z}/2$, the group of integers modulo 2. There is thus precisely one class of line defects in the medium which can be characterized by the number ± 1. Furthermore when two defects combine they annihilate one another since $1 \pm 1 = 0$, modulo 2.

Example 2. Three dimensional spin

In this case the order parameter space V represents the space of possible directions in 3-dimensional and can be taken to be S^2, the surface of a 2-sphere. From Example 3 of Chapter 4 we know that $\pi_1(S^2) = 0$. Thus topologically stable line defects are not possible for this system.

Example 3

This is a mathematical example first considered by Toulouse and Poenaru [3], where the order parameter space V is the Klein bottle considered in

Example 7 of Chapter 4. The fundamental group $\pi_1(V)$ thus consists of elements of the form $g^n K^m$ with the relation $gK = Kg^{-1}$ between the generators g and K. Thus $\pi_1(V)$ is not abelian hence such a medium will exhibit the theoretical ambiguities discussed in this chapter. A physical model for a space with the Klein bottle as its order parameter space was provided in reference [3] which we now describe. Suppose there is a two dimensional medium whose local properties (for example density) deviate from uniformity by an amount proportional to $\cos 2\pi/d(\vec{K}\cdot\vec{r}-\phi)$. We can represent the direction of $\vec{K}$ in 2-dimensions by an angle α. Since $\cos 2\pi/d(\vec{K}\cdot\vec{r}-\phi)$ is unchanged if $\alpha \to \alpha+\pi, \phi \to (-)\phi$ the order parameter can be taken to be the rectangle $0 \leqslant \alpha \leqslant \pi, -\pi \leqslant \phi \leqslant \pi$ provided opposite sides are identified as shown in Fig. 9.7.

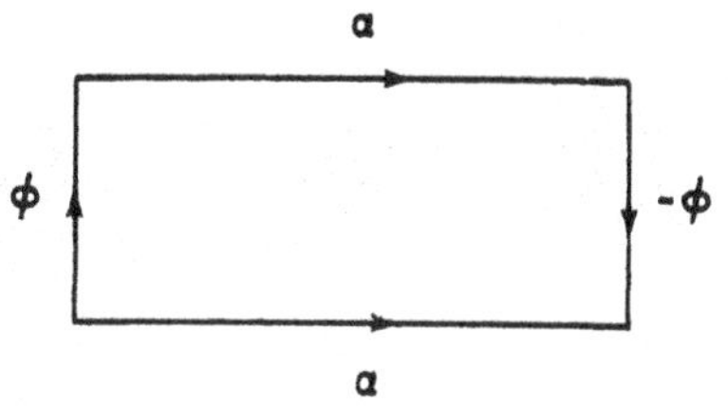

Figure 9.7

A rectangular region identified in this manner represents the Klein bottle. A configuration of this ordered medium is represented by drawing lines of equal phase $\phi = 2\pi n$, $n \in \mathbf{Z}$. In Fig. 9.8 two defects A and B in this medium are combined and it is clear from the pictures that the outcome depends on the medium between A and B. This is, of course, expected from our discussions on the outcome of merging two line defects in a medium with a non-abelian fundamental group for its order parameter space.

9.5 GENERAL REMARKS ON CROSSING OF DEFECTS, TEXTURES AND $\pi_3(S^2)$

Although this theoretical example seems quite useful it should not be taken seriously. The main criticism made about the physical model is the following: the physical model comes equipped with a certain euclidean invariant structure, namely, the curves of constant ϕ in the medium should remain a distance d apart. A general continuous distortion of the medium which is allowed as far as homotopy theory is concerned would not keep

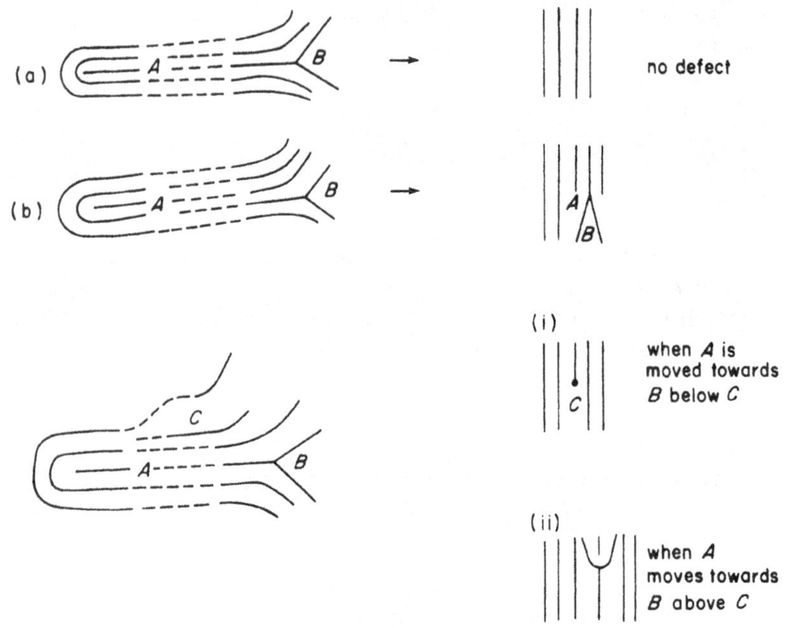

Figure 9.8

the euclidean distance d invariant and hence a classification scheme which ignores this important feature of the medium might be physically quite irrelevant. For further discussions on this point Mermin's article [2] or Poenaru's paper [4] where an attempt to face up to euclidean constraints mathematically is made, may be consulted. Before turning to the final topic of this chapter, textures, we would like to briefly mention the intriguing behaviour that can arise when two line defects are made to cross in a medium with a non vanishing fundamental group. Poenaru and Toulouse [5] showed that a line defect α could move through a line defect β like a ghost if and only if $\alpha\beta \sim \beta\alpha$. The symbols α, β are explained in Fig. 9.9.

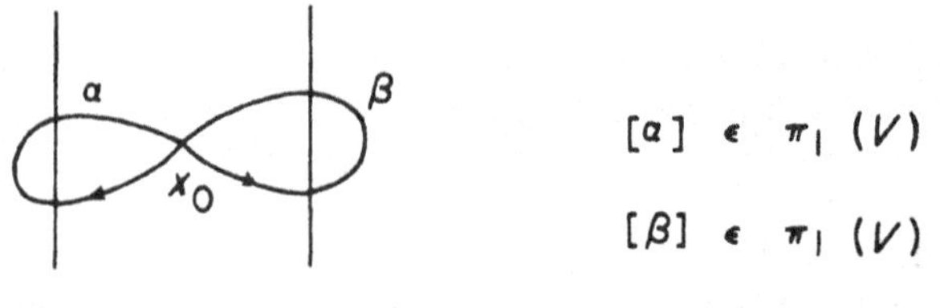

Figure 9.9

An intuitive argument for this theorem is contained in reference [2]. Here we give a plausibility argument. Suppose C_1, C_2 represent two contours in the medium as shown in Fig. 9.10. We fix once and for all the order in which these contours are to be traversed. If the line defect α is

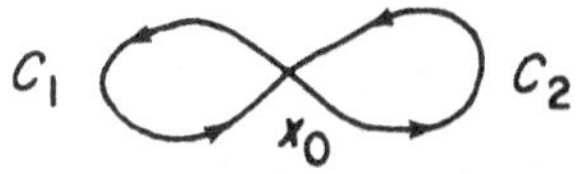

Figure 9.10

placed inside C_1 and the line defect β inside C_2 we get the element $(\alpha \circ \beta)$ of $\pi_1(V)$ is generated. If the line defects are interchanged the element $(\beta \circ \alpha)$ of $\pi_1(V)$ is generated. If these two line defect configurations are to be continuously interchangeable into each other we must have $\alpha \circ \beta \sim \beta \circ \alpha$ which is the result of Poenaru and Toulouse. Thus if $\alpha\beta \not\sim \beta\alpha$ or in other words, $\pi_1(V)$ is non-abelian then one line defect cannot go smoothly through another line defect. The mobility of line defects in such a medium is impaired. Now let us turn to textures. Consider a 3-dimensional medium with order parameter V and suppose that the medium is uniform at points far from the origin. A configuration of this medium can be regarded as a mapping from a 3-dimensional cubical box I^3: in which the entire surface of the box maps into a single point in V, v_0 say which represents the constant uniform behaviour of the system at large distance. But a map such as this corresponds to an element of $\pi_3(V)$. If $\pi_3(V)=0$ then any two configurations of the median are topologically identical as can be shown using essentially the arguments of Theorem (9.8). If, however, $\pi_3(V) \neq 0$ then we have the interesting possibility of topologically inequivalent non-singular configuration in a medium. These are called textures. To illustrate what is possible let us consider a medium with order parameter space S^2, e.g. a three dimensional spin system. We will now show that $\pi_3(S^2)=\mathbf{Z}$, i.e. textures are possible in this medium. What we will show is that the space S^3, may be regarded as a fibre bundle with fibre S^1 and base S^2. Once this geometrical picture has been established a calculation of $\pi_3(S^2)$ proceeds from the fact that for a fibre bundle E with base B and fibre F the following sequence of homotopy groups forms an exact sequence:

$$\to \pi_{n+1}(F) \to \pi_n(E) \to \pi_n(B) \to \pi_n(F) \to \pi_{n-1}(E) \to \dots \qquad (9.11)$$

For a proof of this important result Steenrod's book [6] may be consulted. Armed with this result our homotopy group of interest can be determined. Replacing E by S^3, B by S^2 and F by S^1 in (9.12) we get:

$$\to \pi_3(S^1) \to \pi_3(S^3) \to \pi_3(S^2) \to \pi_2(S^1) \to \dots \qquad (9.12)$$

We next remark that $\pi_n(S^1)=0$, for $n \geqslant 2$. This follows from the following important theorem which we state without proof: [1] and the remark that R^1 is a covering space for S^1 with projection $p: e^{2\pi i\theta} \to \theta$.

Theorem

If B is a covering space of X and $p: B \to X$ is the projection map then the induced homomorphism: $p_*: \pi_n(B; b) \to \pi_n(X, x)$ is an isomorphism onto for $n \geqslant 2$ (where $p(b)=x$). (9.13)

We recall that a covering space is a fibre bundle with discrete fibre. From the remark preceeding Theorem (9.13) and the theorem itself it follows that $\pi_n(S^1)=\pi_n(R^1)$ for $n \geqslant 2$. But R^1 is a contractible space, hence $\pi_n(S^1)=0$, for $n \geqslant 2$ as claimed. Using this result in (9.12) we immediately infer that

$$\pi_3(S^3)=\pi_3(S^2) \tag{9.14}$$

So that from Corollary (5.21) it follows that

$$\pi_3(S^2)=\mathbf{Z} \tag{9.15}$$

as claimed. It remains for us to show that S^3 can indeed be regarded as a fibre bundle with base space S^2 and fibre S^1. To do this we proceed as follows: we regard S^3 as a unit sphere in the space C^2 of two complex variables, i.e. S^3 is regarded as the space of a pair of complex points (z_1, z_2) subject to the constraint:

$$z_1\bar{z}_1+z_2\bar{z}_2=1 \tag{9.16}$$

where $\bar{z}$ represents the complex conjugate of the complex number z. In terms of these variables S^2 is represented as the complex projective line, that is to say, as pairs of complex numbers (z_1, z_2) not both zero with equivalence relation:

$$(z_1, z_2) \sim (\lambda z_1, \lambda z_2),\ \lambda \neq 0 \tag{9.17}$$

The equivalence class thus generated is an element of S^2 and will be written as $[z_1, z_2]$. The projection ρ from S^3 to S^2, called the Hopf map after H. Hopf who introduced the map in 1931, can now be defined as:

$$\rho:(z_1, z_2) \to [z_1, z_2] \tag{9.19}$$

i.e. ρ maps a pair of complex numbers into the equivalence class defined by (9.17) to which it belongs. The continuity of ρ is straightforward, also ρ maps S^3 onto S^2 since any pair in $[z_1, z_2]$ can be normalized by dividing by $(z_1\bar{z}_1+z_2\bar{z}_2)^{1/2}$.

To show that S^3 is a bundle space over S^2 relative to the projection ρ we note that points in S^1 can be represented by the set of complex numbers λ with $|\lambda| = 1$. Now let $n = [1, 0]$ and $s = [0, 1]$ be the north polar and south polar points of S^2 and we cover S^2 with the open sets

$$U = S^2 - n$$
$$V = S^2 - s \tag{9.19}$$

Now every point in U can be represented by a pair $[z, 1]$ and we have the map:

$$\phi_u : U \times S^1 \to S^3$$

defined by:

$$\phi_u([z, 1], \lambda) = \frac{\lambda z}{\sqrt{(z\bar{z}+1)}}, \frac{\lambda}{\sqrt{(z\bar{z}+1)}} \tag{9.20}$$

for each $[z, 1] \in U$ and $\lambda \in S^1$. It is possibly to check that ϕ_u maps $U \times S^1$ homeomorphically onto $\rho^{-1}(U)$ and that $\rho\phi_u(u, d) = u$ for each $u \in U$ and $d \in D$. Similarly a function $\phi_v : V \times S^1 \to S^3$ can be constructed. Thus S^3 is a bundle space over S^2 relative to the Hopf map ρ.

It remains to show that the fibre in this example is S^1. This is immediate since $\rho^{-1}[z_1, z_2]$ consists of all points $(\lambda z_1, \lambda z_2)$ with $|\lambda| = 1$, i.e. points on S^1.

We end this chapter with a cautionary note. All the arguments of this chapter ignored energetic problems. Two configurations A and B may be topologically equivalent but the twisting and bending required to change A to B might involve enormous amounts of energy. Thus a purely topologically classification must be regarded only as the first step in the analysis of a problem.

REFERENCES

1. HOCKING, J. G. and YOUNG, G. S., 'Introduction to Topology'. Addison Wesley, 1961.
2. MERMIN, N. S., *Rev. Mod. Phys.* **51**, 591 (1979).
3. POENARU, V., *In* "Les Houches Summer School Proceedings, 1978". North-Holland, 1979.
4. POENARU, V., *Commun. Math. Phys.* **80**, 127 (1981).
5. POENARU, V. and TOULOUSE, G., *Journal de Physique* **8**, 887 (1977).
6. STEENROD, N., "The Topology of Fibre Bundles". Princeton University Press, 1970.

CHAPTER 10

Yang–Mills Theories: Instantons and Monopoles

10.1 INTRODUCTION

In this chapter we study a particular class of solutions to the Euler-Lagrange equations for a non-Abelian gauge theory.

We employ the notation of Chapter 7 for connections and curvatures. As a preliminary to obtaining Euler-Lagrange equations we need a Lagrangian. The Lagrangian for a gauge theory with group $U(n)$ say is given by

$$L = \frac{-1}{2} Tr(F_{\mu\nu}F^{\mu\nu}) \tag{10.1}$$

(If one wishes to replace $U(n)$ by an arbitrary Lie group G then one replaces the trace in (10.1) by the Killing form of G; i.e. $Tr(F_{\mu\nu}F^{\mu\nu})$ is replaced by

$$I_{ab}F^{a}_{\mu\nu}F^{b\mu\nu} \tag{10.2}$$

where I_{ab} is the metric tensor of the bilinear Killing form, $a, b = 1, \ldots, d$, d being the dimension of $\mathfrak{g}$, and $F^{a}_{\mu\nu}$ are the components of $F_{\mu\nu}$ relative to a basis of $\mathfrak{g}$.) We specialize straightaway to the choice $G = SU(n)$ and we choose M, the manifold on which $F_{\mu\nu}$ is defined, to be 4-dimensional. Recall that if M is 4-dimensional, then the $\varepsilon_{\mu\nu\alpha\beta}$ symbol can be used to define the 2-form $^*F_{\mu\nu}$, c.f. (7.132) where we had

$$^*F_{\mu\nu} = \tfrac{1}{2}\varepsilon_{\mu\nu\alpha\beta}F^{\alpha\beta} \tag{10.3}$$

The action S for this Lagrangian L is obtained by integrating L over M so that

$$S = \int_M \frac{-1}{2} tr(F_{\mu\nu}F^{\mu\nu})\, dv \tag{10.4}$$

where dv is the volume element on M. S is more conveniently expressed in terms of the 2-forms $\mathbf{F}$ and $^*\mathbf{F}$ as

$$S = -\int_M tr(\mathbf{F} \wedge {}^*\mathbf{F}) \tag{10.5}$$

This is a formula that is trivial to verify. This form of S is the most useful for our purposes, it also displays properly the invariance properties of S. The Euler–Lagrange equations for the extrema of S are

$$[D_\mu, F_{\mu\nu}] = 0 \tag{10.6}$$

or in terms of forms they are

$$D^*\mathbf{F} = \mathbf{0} \tag{10.7}$$

a form which is equivalent to (10.6) (compare also the Bianchi identities in equations (7.127) (7.133)). These equations are to be solved for the connection $\mathbf{A}$ after suitable boundary conditions have been given. They are however non-linear coupled partial differential equations containing quadratic and cubic terms in $\mathbf{A}$ and are thus not soluble by any elementary methods. The situation does not look quite so bleak if one observes that if one can find $\mathbf{A}$ such that $\mathbf{F}$ is proportional to $^*\mathbf{F}$

$$\mathbf{F} = \lambda\, ^*\mathbf{F} \tag{10.8}$$

for some λ, then the Euler–Lagrange equations are automatically satisfied. This is so because the Bianchi identities

$$D\mathbf{F} = \mathbf{0} \tag{10.9}$$

then also imply

$$D^*\mathbf{F} = \mathbf{0} \tag{10.10}$$

The subject of instantons is in fact the construction of solutions to equation (10.8) with $\lambda = \mp 1$. We now need to make some observations on the possible values that λ can take. The first point to realise is that the expressions for the action S in equation (10.4) or (10.5) both require a metric $g_{\mu\nu}$ to be defined on the 4-dimensional manifold M. The duality operation $*$ is only defined relative to a metric or inner product. In fact given a metric or inner product on M denoted by $\langle , \rangle$ then we define $^*\boldsymbol{\omega}$ for any 2-form $\boldsymbol{\omega}$ by

$$\langle \boldsymbol{\eta}, \boldsymbol{\omega} \rangle = \int_M \boldsymbol{\eta} \wedge {}^*\boldsymbol{\omega} \tag{10.11}$$

where $\boldsymbol{\eta}$ is an arbitrary 2-form. Note that $\boldsymbol{\eta} \wedge {}^*\boldsymbol{\omega}$ is a 4-form and is therefore integrable over M. Given this definition it is easy to see that for $M = \mathbf{R}^n$

$$^*\omega_{\mu\nu} = \tfrac{1}{2}\varepsilon_{\mu\nu\alpha\beta}\omega^{\alpha\beta} \tag{10.12}$$

in agreement with (10.3). Now recall that if

$$g = \mathrm{Det}\,(g_{\mu\nu}) \tag{10.13}$$

then $\varepsilon_{\mu\nu\alpha\beta}$ and $\varepsilon^{\mu\nu\alpha\beta}$ are related by

$$\varepsilon_{\mu\nu\alpha\beta} = g\varepsilon^{\mu\nu\alpha\beta} \tag{10.14}$$

Then if we perform the $*$ operation twice on a 2-form, we obtain the identity

$$**\omega = g^{-1}\omega \tag{10.15}$$

for an arbitrary 2-form ω. Hence if we choose $M = \mathbf{R}^4$ with the usual Euclidean metric so that $g = 1$, then

$$**\omega = \omega \tag{10.16}$$

However, if we choose M to be Minkowski space, i.e. $\mathbf{R}^4$ with the metric $g_{\mu\nu} = \text{diag}\,(+++-)$ so that $g = -1$, then

$$**\omega = -\omega \tag{10.17}$$

If we use (10.16, 17) and take the $*$ of equation (10.8) we obtain

$$\begin{aligned} &{}^*\mathbf{F} = \lambda\,{}^{**}\mathbf{F} \\ &\Rightarrow \mathbf{F} = \lambda^{2**}\mathbf{F} \\ &\Rightarrow \mathbf{F} = \begin{cases} \lambda^2\mathbf{F} & \text{for a Euclidean metric} \\ -\lambda^2\mathbf{F} & \text{for a Lorentz metric} \end{cases} \end{aligned} \tag{10.18}$$

Hence, if we solve $\mathbf{F} = \lambda{}^*\mathbf{F}$ in Euclidean space, then $\lambda = \mp 1$, but if we solve $\mathbf{F} = \lambda{}^*\mathbf{F}$ in Minkowski space then $\lambda = \mp i$, and these are the only values of λ that one can have. This result has important consequences for the choice of the gauge group G. To see this recall that $\mathbf{F}$ is $\mathfrak{g}$-valued, so therefore is ${}^*\mathbf{F}$. Then if M has a Lorentz metric so that we are interested in solving

$${}^*\mathbf{F} = \mp i\mathbf{F} \tag{10.19}$$

then, we must have $i\mathfrak{g} = \mathfrak{g}$ in an obvious notation. It is elementary to check that this latter condition is not satisfied for the Lie algebras of any compact Lie groups G. One must choose non-compact G such as $SL(n, \mathbf{c})$ or $Gl(n, \mathbf{c})$ say. This is a serious restriction since in physics the gauge groups chosen are usually compact. However, there is no such restriction if one works with Euclidean metric since then one is solving

$${}^*\mathbf{F} = \mp\mathbf{F} \tag{10.20}$$

We shall therefore only look at this equation with $G = U(n)$. We shall return later in the chapter to remark on the results obtainable if one works with the Minkowski space equations (10.19). We also reserve until later any comments on the physical meaning and interpretation of solving equations such as (10.20) in Euclidean space. On a point of nomenclature since $^*\mathbf{F}$ is called the dual of $\mathbf{F}$, then an $\mathbf{F}$ such that $\mathbf{F}={}^*\mathbf{F}$ is called self-dual, and an $\mathbf{F}$ such that $\mathbf{F}=-{}^*\mathbf{F}$ is called anti-self-dual.

10.2 INSTANTONS

The next thing to establish is the boundary conditions under which we shall solve our self, anti-self-dual equations. These involve some topological considerations concerning the homotopy of the group G. Our boundary conditions are actually asymptotic conditions which describe how the connection $\mathbf{A}$ behaves at infinity in $\mathbf{R}^4$. They are determined by the physical requirement that the Euclidean action S should be finite. In terms of $F_{\mu\nu}$ this is just an $L^2(\mathbf{R}^4)$ requirement—the requirement that the curvature should be square integrable. Thus we must have

$$\begin{gathered} F_{\mu\nu}(x)\to 0 \\ |x|\to\infty \end{gathered} \tag{10.21}$$

We have not yet mentioned the rate at which $F_{\mu\nu}$ should decay at infinity. Before doing that we wish to understand the implication of the decay of $F_{\mu\nu}$ at infinity for A_μ. It is not necessary for A_μ to decay as fast as $F_{\mu\nu}$. All that is necessary is that $A_\mu(x)$ should decay to a gauge transformation from the zero connection. More precisely one must have

$$\begin{gathered} A_\mu(x)\to g^{-1}(x)\partial_\mu g(x) \\ |x|\to\infty \end{gathered} \tag{10.22}$$

where again we do not specify the rate of decay. The rate of decay is determined by choosing to solve the equations on a compactification of $\mathbf{R}^4$, namely the conformal compactification obtained by adding the point at infinity. This was described in Chapter 6 where we showed $\mathbf{R}^n\cup\{\infty\}\simeq S^n$, in our case we shall make use of the fact that $\mathbf{R}^4\cup\{\infty\}\simeq S^4$. Next, we simply restrict the rate of decay of $A_\mu(x)$ by the requirement that $A_\mu(x)$ should be smoothly extendable from $\mathbf{R}^4$ to S^4. This transfers our problem from $\mathbf{R}^4$ to S^4. Since the stereographic projection from $\mathbf{R}^4\cup\{\infty\}$ to S^4 is a conformal (i.e. angle preserving) map, then we need to know the behaviour of the self-dual equation under conformal transformations. In fact both the Euler–Lagrange equation

$$D^*\mathbf{F}=0 \tag{10.23}$$

and the self-dual or anti-self-dual equations

$$\mathbf{F}=\mp{}^*\mathbf{F} \tag{10.24}$$

are invariant under conformal transformations. To see this one shows that the action S is a conformal invariant, so then are its extrema which are given by (10.23, 24). The action S is a conformal invariant because $\mathbf{F}\wedge{}^*\mathbf{F}$ is a conformal invariant. The only thing to check here is that the $*$ operator is conformally invariant, the 2-form $\mathbf{F}$ is the same whether one uses Cartesian type coordinates on $\mathbf{R}^4\cup\{\infty\}$ or the conformally projected spherical coordinates on S^4. Finally, the derived invariance of $*$ is verified by using the definition (10.11) and confirming that if two conformally related metrics $g_{\mu\nu}(x)$ and $g'_{\mu\nu}(x)$ are used to define $*$ (i.e. $g_{\mu\nu}(x)=f(x)g'_{\mu\nu}(x), f>0$) then, the $*$ operation is unchanged. In summary we can say that equations (10.23, 24) depend only on the conformal structure of the 4-dimensional manifold on which they are defined. Further on S^4 the integral giving the action S always converges since S^4 is compact and $A_\mu(x)$ is smooth on the whole of S^4. We see that we have eliminated our boundary condition by encoding the problem into another manifold without boundary.

Another point conveniently disposed of here is the difference between the self-dual and the anti-self-dual equations. There is no essential difference between these equations. This is because the definition of $*$ requires S^4 to be given an orientation and since, for example, $\varepsilon_{\mu\nu\alpha\beta}$ changes sign under a change of orientation, then a change of orientation interchanges self-dual and anti-self-dual connections. We shall see later simple explicit ways of passing from one type of solution to the other.

10.3 TOPOLOGY AND BOUNDARY CONDITIONS

The topological considerations concerning the boundary conditions can now be described. If we return to the asymptotic condition (10.22) which was

$$\begin{gathered} A_\mu(x)\to g^{-1}(x)\partial_\mu g(x) \\ |x|\to\infty \end{gathered} \tag{10.25}$$

Then $g(x)$ can be thought of as being defined by a sphere at infinity in $\mathbf{R}^4$, i.e. on S^3. Now choose $G=SU(2)$, then for each x, $g(x)\in SU(2)$. Thus $g(x)$ is a continuous map

$$g:S^3\to SU(2) \tag{10.26}$$

We know that such maps fall into homotopy classes and are simply the elements of $\pi_3(SU(2))$. But $SU(2)$ is topologically an S^3. So

$$\begin{aligned} \pi_3(SU(2)) &= \pi_3(S^3) \\ &= \mathbf{Z} \end{aligned} \tag{10.27}$$

This result means that every g is labelled by an integer k, k is called the degree of g. This integer $k \in \mathbf{Z}$ labels the equivalence class of $\pi_3(S^3)$ to which g belongs, and if g and h are homotopic maps then they have the same degree, if not, then their degrees are different. If instead of $G = SU(2)$ we choose $G = U(n)$, $n \geq 2$ then we still have the same situation since for such G [15]

$$\pi_3(G) = \mathbf{Z} \tag{10.28}$$

From now on we set $G = SU(2)$ except where explicit reference to the contrary is made.

The integer k just introduced actually classifies principal bundles with group $SU(2)$ over S^4. Let us see how this arises by considering an arbitrary $SU(2)$-bundle over S^4 and examining what makes it non-trivial. We need to remember an important result obtained in Chapter 7; this was that a bundle over a contractible base is trivial. The sphere S^4 is not contractible but it can be thought of as being made up of two contractible pieces, c.f. Fig. 10.1. In Fig. 10.1, we depict S^4 as being split into two pieces denoted by A and B. A and B are separately contractible but their union S^4 is not. A and B are intended to overlap so that their intersection is a circular strip-like region in the neighbourhood of the equator of S^4. The equator

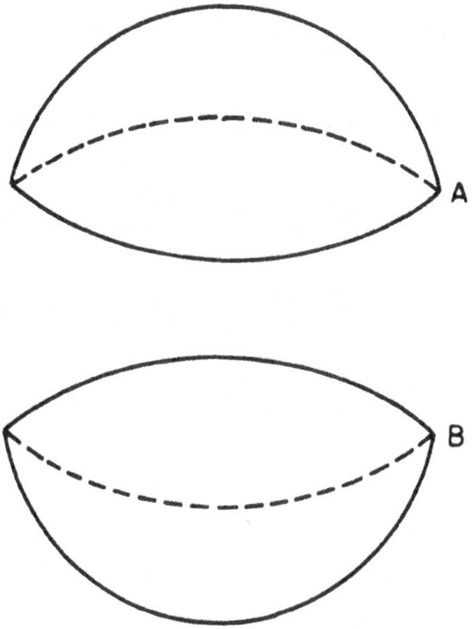

Figure 10.1

of S^4 is an S^3 and topologically $A \cap B$ is therefore a cylinder of the form $L \times S^3$ where L is a line segment. Thus we have

$$\begin{aligned} A \cup B &= S^4 \\ A \cap B &= L \times S^3 \end{aligned} \tag{10.29}$$

Now since A and B are separately contractable the bundle P when restricted to either A of B is trivial. It is therefore in the joining together of the coordinate patches A and B that the non-triviality of P resides. More specifically if we use coordinates (x_A, g_A) for the part of P over A and coordinates (x_B, g_B) for the part of P over B, then over the intersection region $A \cap B$ one changes coordinates by use of a transition function $g_{AB}(x)$. It is in this choice of transition function $g_{AB}(x)$ that the non-triviality of P resides.

The transition function determines a map from $A \cap B$ to $G = SU(2)$

$$\begin{aligned} g_{AB} : A \cap B &\to SU(2) \\ x &\to g_{AB}(x) \end{aligned} \tag{10.30}$$

But $A \cap B$ is homeomorphic to S^3 thus

$$g_{AB} : S^3 \to SU(2) \tag{10.31}$$

which says that

$$g_{AB} \in \pi_3(SU(2)) \tag{10.32}$$

so the homotopy group $\pi_3(SU(2))$ classifies $SU(2)$ bundles over S^4. (More generally one can also see that $\pi_{n-1}(G)$ classifies G-bundles over S^n.) There is then an $SU(2)$-bundle over S^4 for every element of $\pi_3(SU(2))$, and since $\pi_3(SU(2)) = \mathbf{Z}$, then we attach to each bundle an integer k. The transition function g_{AB} determines this k, and two homotopic transition functions g_{AB} and g'_{AB} correspond to the same value of k, and correspond also to the same bundle P. This non-uniqueness of g_{AB} was already uncovered in Chapter 7, c.f. (7.13) where this non-uniqueness was expressed as

$$g'_{AB}(x) = \lambda_A^{-1}(x) g_{AB}(x) \lambda_B(x), \; x \in A \cap B \tag{10.33}$$

It is perhaps not immediately clear that the integer k with which we have labelled g_{AB}, is the same as the integer k with which we labelled $g(x)$ where $g(x)$ gives the behaviour of the connection $A_\mu(x)$ at infinity. These integers are the same. In other words the integer k which labels the asymptotic data of $A_\mu(x)$ is the same integer which labels the bundle P to which $A_\mu(x)$ belongs.

10.4 INSTANTONS AND ABSOLUTE MINIMA

The next important fact that we shall prove is that the self-dual and anti-delf-dual connections are the most important extrema of the action S. We mean by this that rather than their being local maxima or minima or saddle points, they are always absolute minima for a given value of K. The absolute minima of the action are generally more important than the other extrema. To prove this we turn at once to the theory of characteristic classes applied to P. Since the group of the bundle is $SU(2)$ we are concerned with Chern classes rather than Pontrjagin classes. We have already calculated the two non-trivial Chern classes for P in Chapter 7 where we found that†

$$\begin{aligned} c_1(\mathbf{F}) &= 0 \\ c_2(\mathbf{F}) &= -\frac{1}{16\pi^2}\mathbf{F}^a \wedge \mathbf{F}^a = +\frac{1}{8\pi^2} tr(\mathbf{F} \wedge \mathbf{F}) \end{aligned} \tag{10.34}$$

If $c_2(\mathbf{F})$ is integrated over S^4 we obtain an integer since $c^2(\mathbf{F})$ as well as being a real cohomology class is actually an integral cohomology class, and we are using the fact that $H^4(S^4; \mathbf{Z}) = \mathbf{Z}$. In summary we have

$$C_2(P) = -k = +\frac{1}{8\pi^2}\int_{S^4} tr(\mathbf{F} \wedge \mathbf{F}) \tag{10.35}$$

a reappearance of the familiar integer k‡. This equation is what is needed to obtain the result about absolute minima. To derive this result we decompose our arbitrary $\mathbf{F}$ into its self-dual and anti-self-dual parts:

$$\mathbf{F} = \mathbf{F}^+ + \mathbf{F}^- \tag{10.36}$$

(It is easy to see that this is always possible since we can solve (10.36) for $\mathbf{F}^+$ and $\mathbf{F}^-$ obtaining $\mathbf{F}^+ = \frac{1}{2}(\mathbf{F} + {}^*\mathbf{F})$, and $\mathbf{F}^- = \frac{1}{2}(\mathbf{F} - {}^*\mathbf{F})$.) We can now write the action S as

$$S = -tr\int_{S^4} (\boldsymbol{F}^+ + \boldsymbol{F}^-) \wedge {}^*(\boldsymbol{F}^+ + \boldsymbol{F}^-)$$

$$= -\left[tr\int_{S^4} (\boldsymbol{F}^+ \wedge {}^*\boldsymbol{F}^+) + tr\int_{S^4} \boldsymbol{F}^+ \wedge {}^*\boldsymbol{F}^-\right] + tr\int_{S^4} \mathbf{F}^- \wedge {}^*\mathbf{F}^+ + tr\int_{S^4} \mathbf{F}^- \wedge {}^*\mathbf{F}^-$$

† In general one can regard $c_1(\mathbf{F}) = 0$ as a necessary condition that an arbitrary $Gl(2, \mathbf{C})$-bundle should reduce to an $SU(2)$-bundle.

‡ We shall prove below that the various definitions of k coincide.

$$= -tr\int F^+ \wedge F^+ - tr\int \mathbf{F}^- \wedge \mathbf{F}^- + 0 \tag{10.37}$$

where we have used $^*\mathbf{F}^\pm = \pm\mathbf{F}^\pm$ and the fact that $tr[A, B] = 0$ for any matrices A and B. Each of the two terms in (10.37) is separately positive, to emphasise this we write (10.37) as

$$S = \|\mathbf{F}^+\|^2 + \|\mathbf{F}^-\|^2 \geq 0 \tag{10.38}$$

where we have used (10.11) and the obvious notation

$$\|\boldsymbol{\omega}\|^2 = -tr\langle\boldsymbol{\omega}, \boldsymbol{\omega}\rangle = -tr\int \boldsymbol{\omega} \wedge {}^*\boldsymbol{\omega} \tag{10.39}$$

Applying the same decomposition (10.36) to (10.35), we obtain in exactly similar fashion

$$8\pi^2 k = \|\mathbf{F}^+\|^2 - \|\mathbf{F}^-\|^2 \tag{10.40}$$

Now since for any pair of real numbers a and b

$$a^2 + b^2 \geq |a^2 - b^2| \tag{10.41}$$

then (10.38, 40) imply that

$$S \geq 8\pi^2 |k| \tag{10.42}$$

so the absolute minima of S are given by $8\pi^2|k|$ for each k. When the absolute minimum of S is attained there are three cases to consider

$$\begin{aligned} &\text{i. } k = 0\\ &\text{ii. } k > 0\\ &\text{iii. } k < 0 \end{aligned} \tag{10.43}$$

If $k = 0$, then $S = 0$, and $\mathbf{F} = \mathbf{F}^+ = \mathbf{F}^- = \mathbf{0}$, so this is a trivial case corresponding to a flat connection $A_\mu(x)$. If $k > 0$, then

$$\begin{aligned} &S = 8\pi^2 k\\ &\Rightarrow \|\mathbf{F}^+\|^2 + \|\mathbf{F}^-\|^2 = \|\mathbf{F}^+\|^2 - \|\mathbf{F}^-\|^2\\ &\Rightarrow \|\mathbf{F}^-\|^2 = 0\\ &\Rightarrow \mathbf{F}^- = 0 \end{aligned}$$

i.e.

$$\mathbf{F} = \mathbf{F}^+ \quad \text{or} \quad \mathbf{F} = {}^*\mathbf{F} \tag{10.44}$$

if $k < 0$, then

$$S = 8\pi^2|k| = -8\pi^2 k \tag{10.45}$$

so

$$\|\mathbf{F}^{+}\|^{2}+\|\mathbf{F}^{-}\|^{2}=-\|\mathbf{F}^{+}\|^{2}+\|\mathbf{F}^{-}\|^{2}$$
$$\Rightarrow\|\mathbf{F}^{+}\|^{2}=0$$
$$\Rightarrow\mathbf{F}=-{}^{*}\mathbf{F} \tag{10.46}$$

Thus if we exclude the trivial case $k=0$, then the absolute minima of S are given by

$$\mathbf{F}=\varepsilon(k)^{*}\mathbf{F} \tag{10.47}$$

with $\varepsilon(k)=|k|/k$

10.5 THE INSTANTON SOLUTION

We are now ready to examine our first solution of (10.47). We shall look, to begin with, at the simplest case [4] where $k=\mp1$. What is done is to look for the simplest form of solution that is conceivable by requiring $\mathbf{A}$ to be spherically symmetric. This reduces the partial differential equation (10.47) to a simple soluble non-linear ordinary differential equation. For this simple case it is also more direct to work on $\mathbf{R}^4$ with the boundary condition (10.21) rather than on S^4. With these remarks in mind we make the ansatz

$$A_\mu(x)=f(x^2)g^{-1}(x)\partial_\mu g(x) \tag{10.48}$$

for $A_\mu(x)$ where f is a function of x^2 only and is to be determined subject to $f(x^2)\to1$ for $x^2\to\infty$. For $g(x)$, we make the further ansatz

$$g(x)=g_\alpha(x^2)\tau^\alpha \tag{10.49}$$

where τ^α are 2×2 matrices given by

$$\begin{aligned}\tau^0&=I\\ \tau_j&=i\sigma_j,\qquad j=1,2,3\end{aligned} \tag{10.50}$$

τ_j being the Pauli matrices, the τ^α are a convenient basis for the space of 2×2 matrices. The unknown function $g_\alpha(x^2)$ is easily determined by the requirement that $g(x)\in SU(2)$:

$$\begin{aligned}g^{+}(x)g(x)&=I\\ \operatorname{Det} g(x)&=1\end{aligned} \tag{10.51}$$

The first part of (10.51) gives

$$g_\alpha(x^2)g_\beta(x^2)(\tau^\alpha)(\tau^\beta)=1$$

($g_\alpha(x^2)$ is assumed to be real). Since

$$\begin{aligned}(\tau^0)^+ &= \tau^0 \\ (\tau_j)^+ &= -\tau_j, \qquad j = 1, 2, 3\end{aligned} \tag{10.52}$$

and we assume

$$\begin{aligned}g_\alpha(x^2) &= \partial_\alpha h(x^2) \\ &= 2x_\alpha h'(x^2)\end{aligned} \tag{10.53}$$

for some function $h(x^2)$, then we obtain

$$\begin{aligned}&4(h'(x^2))^2 x_\alpha x_\beta (\tau^\alpha)^+ \tau^\beta = 1 \\ &\Rightarrow 2(h'(x^2))^2 x_\alpha x_\beta \{(\tau^\alpha)^+, \tau^\beta\} = I \\ &\Rightarrow 4(h'(x^2))^2 x^2 = I \text{ using (10.52)} \\ &\Rightarrow h(x^2) = (x^2)^{1/2} + c\end{aligned} \tag{10.54}$$

Thus

$$g(x) = \frac{x_\alpha \tau^\alpha}{(x^2)^{1/2}} \tag{10.55}$$

and one can also verify that det $g = 1$, actually this was satisfied automatically by our assumption that $g_\alpha(x^2)$ was real. Our ansatz for $A_\mu(x^2)$ now takes the form

$$\begin{aligned}A_\mu(x^2) &= f(x^2)\frac{x_\alpha(\tau^\alpha)^+}{(x^2)^{1/2}}\left(\frac{\tau^\mu}{(x^2)^{1/2}} + \frac{x_\beta x_\mu}{(x^2)^{3/2}}\tau^\beta\right) \\ &= \frac{f(x^2)}{(x^2)^{1/2}}\left\{\frac{x_\alpha(\tau^\alpha)^+}{(x^2)^{1/2}}\tau^\mu + \frac{x_\mu}{(x^2)^{1/2}} I\right\}\end{aligned} \tag{10.56}$$

Note that the term proportional to the identity I in (10.56) contributes zero to $F_{\mu\nu}$. We shall therefore delete it when calculating $F_{\mu\nu} - \frac{1}{2}\varepsilon_{\mu\nu\alpha\beta}F^{\alpha\beta}$. We now calculate this quantity choosing a connection A_μ given by

$$\begin{aligned}A_\mu(x) &= \frac{f(x^2)}{x^2} x_\alpha(\tau^\alpha)^+ \tau^\mu \\ &= a(x^2) b^+ b_{,\mu}\end{aligned} \tag{10.57}$$

where we have defined for simplicity

$$\begin{aligned}a(x^2) &= f(x^2)/x^2 \\ b(x) &= (x_\alpha \tau^\alpha), \quad \text{note } b^+ b = x^2 I\end{aligned} \tag{10.58}$$

Using (10.57) we obtain

$$F_{\mu\nu}=a_{,\mu}b^{+}b_{,\nu}-a_{,\nu}b^{+}b_{,\mu}+a(b^{+}_{,\mu}b_{,\nu}-b^{+}_{,\nu}b_{,\mu})+a^{2}[b^{+}b_{,\mu},b^{+}b_{,\nu}] \quad (10.59)$$

If we differentiate $b^{+}b=x^{2}I$ and insert this information into (10.59), then we find that

$$\begin{aligned}F_{\mu\nu}&=(a_{,\mu}+2x_{\mu}a^{2})b^{+}b_{,\nu}-(a_{,\nu}+2x_{\nu}a^{2})b^{+}b_{,\mu}\\&\quad+(a-x^{2}a^{2})(b^{+}_{,\mu}b_{,\nu}-b^{+}_{,\nu}b_{,\mu})\end{aligned} \quad (10.60)$$

It is straightforward and quick to check that since $a_{,\mu}=\tau_{\mu}$, then the combination $b^{+}_{,\mu}b_{,\mu}-b^{+}_{,\nu}b_{,\mu}$ *is* self-dual but that the first two terms in (10.60) are *not*. This therefore requires $a(x^{2})$ to satisfy the simple non-linear equation:

$$a_{,\mu}+2x_{\mu}a^{2}=0 \quad (10.61)$$

This gives at once

$$a(x^{2})=\frac{1}{x^{2}+\lambda^{2}},\quad \lambda^{2}\text{ constant} \quad (10.62)$$

Thus our self-dual connection $A_{\mu}(x)$ is given by

$$\begin{aligned}A_{\mu}(x)&=f(x^{2})g^{+}g_{,\mu}\\&=a(x^{2})b^{+}b_{,\mu}\\&=\frac{1}{x^{2}+\lambda^{2}}(x_{\alpha}\tau^{\alpha})^{+}\tau_{\mu}\end{aligned} \quad (10.63)$$

and a, f, g and b are given by

$$\begin{aligned}f(x^{2})&=x^{2}a(x^{2})=\frac{x^{2}}{x^{2}+\lambda^{2}}\\g(x)&=\frac{b(x)}{\sqrt{x^{2}}}=\frac{(x_{\alpha}\tau^{\alpha})}{\sqrt{x^{2}}}\end{aligned} \quad (10.64)$$

Note that $f(x^{2})\to 1$ at infinity as desired. The origin of the term instanton comes from the fact that the connection $A_{\mu}(x)$ given in (10.63) attains its maximum value at $\mathbf{x}=0$, $x_{0}=0$, and so it dies quickly away before and after the instant in time given by $x_{0}=0$.

This instanton has $k=1$, instantons with $|k|>1$ are called multi-instantons. Let us verify that $k=1$ for this solution, i.e. let us show that

$$-\frac{1}{8\pi^{2}}tr\int \mathbf{F}\wedge\mathbf{F}=-\frac{1}{32\pi^{2}}\varepsilon_{\mu\nu\alpha\beta}tr\int_{R^{4}}F^{\mu\nu}F^{\alpha\beta}\,d^{4}x=1 \quad (10.65)$$

(recall that $\mathbf{F}=F_{\mu\nu}/2\,dx^{\mu}\wedge dx^{\nu}$). Since from (10.60, 62), we have that

$$F_{\mu\nu}=(a-x^{2}a^{2})(\tau^{+}_{\mu}\tau_{\nu}-\tau^{+}_{\nu}\tau_{\mu}) \quad (10.66)$$

then we find that

$$k=-\frac{1}{8\pi^2}\int \mathbf{F}\wedge\mathbf{F}=A\int_{R^4}(a-x^2a^2)^2\,\mathrm{d}^4x \tag{10.67}$$

with

$$\begin{aligned}A&=\frac{1}{32\pi^2}\varepsilon_{\mu\nu\alpha\beta}tr(\overset{+}{\tau}_\mu\tau_\nu-\overset{+}{\tau}_\nu\tau_\mu)(\overset{+}{\tau}_\alpha\overset{+}{\tau}_\beta-\overset{+}{\tau}_\beta\overset{+}{\tau}_\alpha)\\&=\frac{\varepsilon_{\mu\nu\alpha\beta}}{8\pi^2}tr(\overset{+}{\tau}_\mu\tau_\nu\overset{+}{\tau}_\beta\tau_\alpha)\end{aligned} \tag{10.68}$$

To evaluate A, we make use of the identity $\overset{+}{\tau}_\beta\tau_\alpha=\frac{1}{2}[\overset{+}{\tau}_\beta,\tau_\alpha]+\frac{1}{2}\{\overset{+}{\tau}_\beta,\tau^\alpha\}$ twice and hence find

$$A=\frac{6}{\pi^2} \tag{10.69}$$

where the commutator $[\overset{+}{\tau}_\beta,\tau_\alpha]$ and the anti-commutator $\{\overset{+}{\tau}_\beta,\tau_\alpha\}$ are easily calculated and are found to be

$$\begin{aligned}&[\overset{+}{\tau}_\beta\tau_\alpha]=2\varepsilon_{0\beta\alpha\gamma}\tau_\gamma\\&\{\overset{+}{\tau}_\beta,\tau_\alpha\}=2\delta_{\beta\alpha}I+2(\delta_{0\beta}\tau_\alpha-\delta_{0\alpha}\tau_\beta)\end{aligned} \tag{10.70}$$

The integral in (10.67) is standard, c.f. appendix reference [17], the result is

$$\begin{aligned}\int(a-x^2a^2)^2\,\mathrm{d}^4x&=\int\frac{\lambda^4}{(x^2+\lambda^2)^4}\,\mathrm{d}^4x\\&=\frac{\pi^2}{6}\end{aligned} \tag{10.71}$$

Thus, $k=1$ as claimed and this integer is the value of the second Chern class $C^2(P)$ when evaluated on S^4 or equivalently on $\mathbf{R}^4$. To obtain an instanton with $k=-1$, one need only interchange $g(x)$ and $g^{-1}(x)$ in (10.48). We have given two other ways of defining the integer k; one in terms of sphere S^3 at infinity, and the other in terms of an equatorial S^3 on S^4. We are now in a position to demonstrate the equivalence of these definitions. The key to this equivalence is the identity

$$Tr(\mathbf{F}\wedge\mathbf{F})=\mathrm{d}\{tr(\mathbf{F}\wedge\mathbf{A}-\tfrac{1}{3}\mathbf{A}\wedge\mathbf{A}\wedge\mathbf{A})\} \tag{10.72}$$

We leave the short proof of this identity to the reader and provide the following technical reminders as aid in the proof;

i. $d(\omega \wedge \eta) = d\omega \wedge \eta + (-1)^r \omega \wedge d\eta$

ii. Since we often deal with matrix-valued forms, then for such forms the identity

$$\omega \wedge \eta = (-1)^{rs} \eta \wedge \omega$$

is no longer true, its nearest analogue is the identity

$$tr(\omega \wedge \eta) = (-1)^{rs} tr(\eta \wedge \omega) \tag{10.73}$$

iii. Use the Bianchi identities $D\mathbf{F} = \mathbf{0}$ and remember that $tr(\mathbf{A} \wedge \mathbf{A} \wedge \mathbf{A} \wedge \mathbf{A}) = 0$ because of (ii) above.

10.6 THE INSTANTON NUMBER AND THE SECOND CHERN CLASS

To make use of identity (10.72) we have to think about how the non-triviality of the instanton bundle P over S^4 affects the form of the connection $\mathbf{A}$. We pointed out in Chapter 7 that a local change of a section of P was a gauge transformation. For our bundle P we know that P is trivial over the hemispheres A and B separately, hence there is a local section S_A of P over A and S_B of P over B. The two sections are connected by the gauge transformation g_{AB}; $S_A = g_{AB} S_B$. Hence, we can use S_A to give local coordinates for $\mathbf{A}$ above A and we can use S_B to give local coordinates for $\mathbf{A}$ above B. In the intersection region $A \cap B$, the two forms for $\mathbf{A}$ differ by a gauge transformation $g_{AB} \equiv g$, i.e.

$$\mathbf{A}^A = g^{-1}\mathbf{A}^B g + g^{-1}\,dg \tag{10.74}$$

in an obvious notation. Now we return to identity (10.72) and integrate it over S^4 having first split S^4 into A and B and contracted $A \cap B$ down to S^3. This means that we can write

$$\begin{aligned}
-8\pi^2 k = tr\int_{S^4} (\mathbf{F} \wedge \mathbf{F}) &= tr\int_A (\mathbf{F}^A \wedge \mathbf{F}^A) + tr\int_B (\mathbf{F}^B \wedge \mathbf{F}^B) \\
&= \int_A d\{tr(\mathbf{F}^A \wedge \mathbf{A}^A - \tfrac{1}{3}\mathbf{A}^A \wedge \mathbf{A}^A \wedge \mathbf{A}^A)\} \\
&\quad + \int_B d\{tr(\mathbf{F}^B \wedge \mathbf{A}^B - \tfrac{1}{3}\mathbf{A}^B \wedge \mathbf{A}^B \wedge \mathbf{A}^B)\} \\
&= \int_{S^3} tr(\mathbf{F}^A \wedge \mathbf{A}^A - \tfrac{1}{3}\mathbf{A}^A \wedge \mathbf{A}^A \wedge \mathbf{A}^A) \\
&\quad - \int_{S^3} tr(\mathbf{F}^B \wedge \mathbf{A}^B - \tfrac{1}{3}\mathbf{A}^B \wedge \mathbf{A}^B \wedge \mathbf{A}^B)
\end{aligned} \tag{10.75}$$

where we have used Stokes' theorem† and the fact that $\partial A = \partial B = S^3$, the minus sign in the second S^3 integral is due to the fact that the boundaries of A and B have opposite orientations. To further simplify (10.75) we note that (10.74) implies that

$$\mathbf{F}^A = g^{-1}\mathbf{F}^B g \tag{10.76}$$

If we insert (10.74, 76) in (10.75), and bear in mind (10.73), then we obtain the result that

$$\begin{aligned} -8\pi^2 k &= \frac{-1}{3}\int_{S^3} tr\{(g^{-1}\,dg)\wedge(g^{-1}\,dg)\wedge(g^{-1}\,dg)\} - \int_{S^3} tr\, d(g^{-1}A^A \wedge dg) \\ &= \frac{-1}{3}\int_{S^3} tr\{(g^{-1}\,dg)\wedge(g^{-1}\,dg)\wedge(g^{-1}\,dg)\} \end{aligned} \tag{10.77}$$

so

$$k = \frac{1}{24\pi^2}\int_{S^3} tr(g^{-1}\,dg)\wedge(g^{-1}\,dg)\wedge(g^{-1}\,dg) \tag{10.78}$$

Equation (10.78) is the result that we were after. The integral on the RHS can be seen to have the property that it counts the number of times the group G covers the sphere under the mapping $g: S^3 \to G$, i.e. it evaluates the integer k to which g corresponds in $\pi_3(G)$. A short argument is required to establish this intuitively reasonable claim. We show (i) that the RHS of (10.78) is constant for g of the same homotopy type, (ii) that (10.78) takes integer values which change precisely when the homotopy type of g changes, and that when this integer is k the homotopy type of g is also k.

i. Let

$$\omega = \frac{1}{24\pi^2} tr\{(g^{-1}\,dg)\wedge(g^{-1}\,dg)\wedge(g^{-1}\,dg)\},$$

now ω is a 3-form on S^3 and is really the pullback under the map $g: S^3 \to G$ of a 3-form η on G; (η is in fact the form ω regarded as a form on G so that $dg \equiv a_\alpha\, dg^\alpha$ where dg^α are differentials on G, the map g provides G with S^3 valued coordinates so that now $g \equiv g(x)$, $x \in S^3$ and $dg \equiv b_\beta\, dx^\beta$ where dx^β are differentials on S^3), in view of this we have

$$\omega = g^*\eta$$

† We could not use Stokes' theorem on the identity (10.72) for this would imply that $\int_{S^4} tr\, \mathbf{F}\wedge\mathbf{F} = 0$. This is not allowed because the quantity $tr\{\mathbf{F}\wedge\mathbf{A} - \frac{1}{3}\mathbf{A}\wedge\mathbf{A}\wedge\mathbf{A}\}$, unlike $tr(\mathbf{F}\wedge\mathbf{F})$ is not gauge invariant and hence is affected by a gauge transformation. This is indeed why we shall find (10.75) to be non-zero for $k \neq 0$.

and

$$\int_{S^3} \omega = \int_{S^3} g^* \eta \tag{10.79}$$

But because S^3 has a dimension 3 and ω is a closed, non-exact 3-form on S^3 then ω determines a cohomology class $[\omega]$ where $[\omega] \in H^3(S^3; \mathbf{Z})$ (note we use integer coefficients here). We saw in Chapter 6 that if g and g' were homotopic maps then $g^*\eta$ determine the same cohomology class, i.e. $[g^*\eta] = [(g')^*\eta]$. The integral

$$\int_{S^3} g^* \eta \tag{10.80}$$

is therefore constant on each homotopy class of g. For if $g \simeq h$, then

$$\begin{aligned} &[g^*\eta] = [h^*\eta] \\ &\Rightarrow g^*\eta = h^*\eta + d\phi \\ &\Rightarrow \int_{S^3} g^*\eta = \int_{S^3} h^*\eta \end{aligned} \tag{10.81}$$

since $\partial S^3 = \phi$. Thus we have established that $\int \omega$ is constant on each homotopy class of g.

ii. Now let us take two non-cohomologous ω, i.e. take

$$\omega = g^*\eta$$

$$\omega' = h^*\omega'$$

where

$$[\omega] \neq [\omega'] \tag{10.82}$$

This assumption implies of course that g and h are non-homotopic. Let us further assume that $[\omega]$ corresponds to the first cohomology class and that $[\omega']$ corresponds to the n-th cohomology class, $n > 1$. Then, since $H^3(S^3; \mathbf{Z}) = \mathbf{Z}$, we know that, although $\omega' - \omega$ is not exact, the combination $\omega' - n\omega$ is exact or

$$\omega' - n\omega = \mathrm{d}\phi \tag{10.83}$$

If we now integrate we find that

$$\int_{S^3} \omega' = n \int_{S^3} \omega \tag{10.84}$$

Thus, we have established so far that

$$\int_{S^3} \omega = \int_{S^3} g^* \eta$$

is constant on both the homotopy class of g and the cohomology class of $g^*\eta$. It also is true [28] that there is a one to one correspondence between these classes, we do not prove this here. This means that the integer n labelling the homotopy class of g and the cohomology class of $g^*\eta$ can be taken to be the same. Finally, if one chooses $G = SU(2)$, then a $g(x)$ which covers $SU(2)$ just once, is $g(x) = x_\alpha \tau^\alpha/\sqrt{x^2}$. For this case the integral (10.78) was already evaluated in (10.67, 71) and found to be unity. This completes the last step in the argument.

For the case of $\mathbf{R}^4$ and the sphere S^3 at infinity, the definition of k can be shown to coincide with the other two by using a similar argument, which is a simple exercise for the reader, and by using an $\mathbf{R}^4$ version of the identity (10.72) namely:

$$\tfrac{1}{2}tr(F_{\mu\nu}{}^*F_{\mu\nu}) = \varepsilon_{\alpha\beta\gamma\delta} tr\, \partial_\alpha (A_\beta \partial_\gamma A_\delta + \tfrac{2}{3} A_\beta A_\gamma A_\delta) \qquad (10.85)$$

10.7 MULTI-INSTANTONS

Next our interest turns to multi-instantons. These are characterized by $\mathbf{F} = \mp{}^*\mathbf{F}$ with $|k|>1$. The multi-instanton connection $\mathbf{A}$, therefore, sits on a bundle characterized by $|k|>1$. So far we have constructed bundles with $k=\mp 1$ using $g(x) = x_\alpha \tau^\alpha/\sqrt{x^2}$. If we replace $g(x)$ by $g^k(x)$, it turns out [29] that $g^k(x)$ gives rise to a bundle P with $C_2(P) = -k$. (This is because [29] multiplication in the group g induces addition in the homotopy group $\pi_3(G)$.) Accepting this result one sees that the construction of examples of bundles P with arbitrary values of k is no trouble. The difficult part is to find self-dual or anti-self-dual connections $\mathbf{A}$ on P, and more, to find all such connections for each value of k. Happily all this work has been done, but the proofs take us into the subject of algebraic geometry, and so we shall not prove all the results that we give on instantons. We shall however make plausible the general nature of the results, and shall be able nevertheless to give quite a satisfactory account of instantons within these limitations. The references to the papers which make the proofs complete will also be given.

10.8 QUATERNIONS AND SU(2) CONNECTIONS

It turns out that there are two apparently quite different methods of constructing $SU(2)$ multi-instantons. The first uses linear algebra of matrices with quaternionic entries, and in doing so makes use of the fact that $SU(2)$ may be regarded as the group of unit quaternions. The second uses techniques of twistor theory to translate the problem into one where there is a one to one correspondence between self-dual connections on S^4 and certain bundles over $\mathbf{CP}^3$. These bundles have then to be constructed. The connection between the two approaches is more conveniently brought to the fore by showing that the quaternionic approach gives all self-dual connections, this being far from self-evident in the quaternionic approach. We deal first with the quaternionic approach and then with the twistor approach.

To establish our notation, we need some preliminaries about quaternions. The space of quaternions is denoted, as is customary, by $\mathbf{H}$, (H standing for Hamilton). A quaternion x is specified by four real numbers according to:

$$x = x_1 + x_2\mathbf{i} + x_3\mathbf{j} + x_4\mathbf{k} \tag{10.86}$$

In (10.86), $x_1, \ldots, x_4$ are real and $\mathbf{i}$, $\mathbf{j}$, and $\mathbf{k}$ obey the equations

$$\begin{gathered} \mathbf{i}^2 = \mathbf{j}^2 = \mathbf{k}^2 = -1 \\ \mathbf{ij} = -\mathbf{ji} = \mathbf{k}; \qquad \mathbf{jk} = -\mathbf{kj} = \mathbf{i}; \qquad \mathbf{ki} = -\mathbf{ik} = \mathbf{j} \end{gathered} \tag{10.87}$$

The space $\mathbf{H}$ of quaternions is therefore isomorphic to $\mathbf{R}^4$. It is also possible to express a quaternion x as an ordered pair of complex numbers thus dwelling on the isomorphism $\mathbf{H} \simeq \mathbf{C}^2$. To do this we write

$$x = z_1 + z_2\mathbf{j}$$

with

$$z_1 = x_1 + x_2\mathbf{i}, \qquad z_2 = x_3 + x_4\mathbf{i} \tag{10.88}$$

Equation (10.88) is equivalent to (10.86) if we use (10.87). We define the norm $|x|$ of a quaternion by

$$\begin{aligned} |x| &= \sqrt{(x_1^2 + x_2^2 + x_3^2 + x_4^2)} \\ &= \sqrt{(x\bar{x})} \end{aligned}$$

where

$$\bar{x} = x_1 - x_2\mathbf{i} - x_3\mathbf{j} - x_4\mathbf{k} \tag{10.89}$$

The quaternion $\bar{x}$ is called the conjugate quaternion of x, and in further analogy with the complex numbers x_1 is called the real part of x and $x_2\mathbf{i} + x_3\mathbf{j} + x_4\mathbf{k}$ is called the imaginary part. If we insist that the norm of x be unity then the components of x satisfy

$$x_1^2 + x_2^2 + x_3^2 + x_4^2 = 1$$

i.e. they lie on an S^3. The unit quaternions, since they are always non-zero, also form a group—and this group is isomorphic to $SU(2)$. To see this explicitly one exploits the isomorphism of $\mathbf{H} \simeq \mathbf{C}^2$ and represents quaternionic multiplication by the action of 2×2 complex matrices. Let $x, y \in \mathbf{H}$

$$\begin{aligned} x &= z_1 + z_2\mathbf{j} \\ y &= w_1 + w_2\mathbf{j} \end{aligned} \tag{10.90}$$

Then the product xy is given by

$$xy = (z_1 w_1 - z_2 \bar{w}_2) + (z_1 w_2 + z_2 \bar{w}_1)\mathbf{j} \tag{10.91}$$

This means that the vector $(z_1, z_2) \in \mathbf{C}^2$ is multiplied on the *right* by the matrix

$$\begin{pmatrix} w_1 & w_2 \\ -\bar{w}_2 & \bar{w}_1 \end{pmatrix} \tag{10.92}$$

and this is a representation of y. If we restrict to quaternions of unit norm, then

$$|y| = \mathrm{Det}\begin{pmatrix} w_1 & w_2 \\ -\bar{w}_2 & \bar{w}_1 \end{pmatrix} = 1 \tag{10.93}$$

so that (10.92) is indeed an $SU(2)$ matrix. The Lie algebra of the group of unit quaternions is represented by the 2×2 matrices which correspond to the units $\mathbf{i}$, $\mathbf{j}$ and $\mathbf{k}$; i.e. it is given by

$$\mathbf{i} = \begin{pmatrix} i & 0 \\ 0 & -i \end{pmatrix}, \quad \mathbf{j} = \begin{pmatrix} 0 & 1 \\ -1 & 0 \end{pmatrix}, \quad \mathbf{k} = \begin{pmatrix} 0 & i \\ i & 0 \end{pmatrix} \tag{10.94}$$

These three matrices being just $-i\sigma$ where σ is one of the three Pauli matrices. Thus the Lie algebra of the group of unit quaternions is given simply by the pure imaginary quaternions.

Now we turn to an $SU(2)$-connection $\mathbf{A}$. $\mathbf{A}$ takes values in the Lie algebra of $SU(2)$ and can therefore be expressed also as taking values in the space of pure imaginary quaternions. Let us do this. Let

$$\mathbf{A} = A_\mu(x)\, \mathrm{d}x^\mu \tag{10.95}$$

in the usual notation. To express $\mathbf{A}$ in quaternionic form we use quaternionic differentials which we write as $\mathrm{d}x$ where

$$\mathrm{d}x = \mathrm{d}x^1 + \mathrm{d}x^2\,\mathbf{i} + \mathrm{d}x^3\,\mathbf{j} + \mathrm{d}x^4\,\mathbf{k} \tag{10.96}$$

and we write (10.95) as

$$\mathbf{A} = \mathrm{Im}\, A(x)\, \mathrm{d}x = A_\mu\, \mathrm{d}x^\mu$$

with

$$A(x) = \tilde{A}_1(x) + \tilde{A}_2(x)\mathbf{i} + \tilde{A}_3(x)\mathbf{j} + \tilde{A}_4(x)\mathbf{k} \tag{10.97}$$

Let us evaluate (10.97) so as to understand clearly the notation. Using the rules of multiplication given in (10.87) we find that

$$\begin{aligned} A(x)\,\mathrm{d}x &= (\tilde{A}_1+\tilde{A}_2\mathbf{i}+\tilde{A}_3\mathbf{j}+\tilde{A}_4\mathbf{k})(\mathrm{d}x^1+\mathrm{d}x^2\,\mathbf{i}+\mathrm{d}x^3\,\mathbf{j}+\mathrm{d}x^4\,\mathbf{k}) \\ &= (\tilde{A}_1+\tilde{A}_2\mathbf{i}+\tilde{A}_3\mathbf{j}+\tilde{A}_4\mathbf{k})\,\mathrm{d}x^1 \\ &\quad +(-\tilde{A}_2+\tilde{A}_1\mathbf{i}+\tilde{A}_4\mathbf{j}-\tilde{A}_3\mathbf{k})\,\mathrm{d}x^2 \\ &\quad +(-\tilde{A}_3-\tilde{A}_4\mathbf{i}+\tilde{A}_1\mathbf{j}+\tilde{A}_2\mathbf{k})\,\mathrm{d}x^3 \\ &\quad +(-\tilde{A}_4+\tilde{A}_3\mathbf{i}-\tilde{A}_2\mathbf{j}+\tilde{A}_1\mathbf{k})\,\mathrm{d}x^4 \end{aligned} \tag{10.98}$$

Thus

$$\begin{aligned} \mathbf{A} &= A_\mu\,\mathrm{d}x^\mu = \operatorname{Im} A(x)\,\mathrm{d}x \\ &= (\tilde{A}_2\mathbf{i}+\tilde{A}_3\mathbf{j}+\tilde{A}_4\mathbf{k})\,\mathrm{d}x^1+(\tilde{A}_1\mathbf{i}+\tilde{A}_4\mathbf{j}-\tilde{A}_3\mathbf{k})\,\mathrm{d}x^2 \\ &\quad +(-\tilde{A}_4\mathbf{i}+\tilde{A}_1\mathbf{j}+\tilde{A}_2\mathbf{k})\,\mathrm{d}x^3+(\tilde{A}_3\mathbf{i}-\tilde{A}_2\mathbf{j}+\tilde{A}_1\mathbf{k})\,\mathrm{d}x^4 \end{aligned} \tag{10.99}$$

Thus the $A_\mu(x)$ are given in terms of the quaternionic coefficients $\tilde{A}_\mu$ by:

$$\begin{aligned} A_1(x) &= \tilde{A}_2\mathbf{i}+\tilde{A}_3\mathbf{j}+\tilde{A}_4\mathbf{k} \\ A_2(x) &= \tilde{A}_1\mathbf{i}+\tilde{A}_4\mathbf{j}-\tilde{A}_3\mathbf{k} \\ A_3(x) &= -\tilde{A}_4\mathbf{i}+\tilde{A}_1\mathbf{j}+\tilde{A}_2\mathbf{k} \\ A_4(x) &= \tilde{A}_3\mathbf{i}-\tilde{A}_2\mathbf{j}+\tilde{A}_1\mathbf{k} \end{aligned} \tag{10.100}$$

Next, we need an example of a calculation of the curvature form **F** in this method. Let

$$\mathbf{A} = \operatorname{Im} A(x)\,\mathrm{d}x \tag{10.101}$$

then

$$\begin{aligned} \mathbf{F} &= D\mathbf{A} = \mathrm{d}\mathbf{A}+\mathbf{A}\wedge\mathbf{A} \\ &= d(\operatorname{Im} A(x)\,\mathrm{d}x)+(\operatorname{Im} A(x)\,\mathrm{d}x)\wedge(\operatorname{Im} A(x)\,\mathrm{d}x) \end{aligned} \tag{10.102}$$

But, rather than working with equation (10.102) where one first computes $\operatorname{Im} A(x)\,\mathrm{d}x$ and then computes $D\mathbf{A}$, one can instead use the fact that

$$\mathbf{F} = \operatorname{Im} D(A(x)\,\mathrm{d}x) = \operatorname{Im}\{d(A(x)\,\mathrm{d}x)+(A(x)\,\mathrm{d}x)\wedge(A(x)\,\mathrm{d}x)\} \tag{10.103}$$

In other words the process of evaluating the imaginary part of $A(x)\,\mathrm{d}x$ and computing its curvature commute. This can be seen on general grounds by noting that the algebra formed by the quaternions before taking the imaginary part is just the Lie algebra of $SU(2)$ plus a one dimensional space spanned by x_1; and this one dimensional space may be projected out either before or after formation of the curvature. In view of these remarks we have the simpler formula for the curvature

$$\mathbf{F}=\mathrm{Im}\,\{\mathrm{d}A\wedge \mathrm{d}x+(A(x)\,\mathrm{d}x)\wedge(A(x)\,\mathrm{d}x\} \tag{10.104}$$

10.9 THE $k=1$ INSTANTON IN TERMS OF QUATERNIONS

As an instructive exercise we shall now write the $k=1$ instanton down using quaternions. Equation (10.100) gives the relation between the $A_\mu(x)$ and the quaternionic $\tilde{A}_\mu(x)$, and this linear relation can be solved to give the $A_\mu(x)$ in terms of the $\tilde{A}_\mu(x)$. Instead of doing this it is actually quite easy to use the representation of **i**, **j** and **k** given in (10.94) and compare this with the $k=1$ expression $\mathbf{A}(x)=f(x^2)g^{-1}(x)\,\mathrm{d}g(x)$, simple inspection then gives the components $\tilde{A}_\mu(x)$ as:

$$\begin{aligned}
\tilde{A}_1(x)&=\frac{(x_2\mathbf{i}+x_3\mathbf{j}+x_4\mathbf{k})}{(|x|^2+\lambda^2)}\\
\tilde{A}_2(x)&=\frac{(-x_1\mathbf{i}-x_4\mathbf{j}+x_3\mathbf{k})}{(|x|^2+\lambda^2)}\\
\tilde{A}_3(x)&=\frac{(x_4\mathbf{i}-x_1\mathbf{j}-x_2\mathbf{k})}{(|x|^2+\lambda^2)}\\
\tilde{A}_4(x)&=\frac{(-x_3\mathbf{i}+x_2\mathbf{j}-x_1\mathbf{k})}{(|x|^2+\lambda^2)}
\end{aligned} \tag{10.105}$$

which we obtained from the elegant formula

$$\mathbf{A}(x)=\mathrm{Im}\left\{\frac{x\,\mathrm{d}\bar{x}}{(|x|^2+\lambda^2)}\right\} \tag{10.106}$$

The curvature **F** of **A** is also expressible using quaternions

$$\mathbf{F}=\mathrm{Im}\left\{\mathrm{d}\left(\frac{x}{|x|^2+\lambda^2}\right)\wedge \mathrm{d}\bar{x}+\frac{(x\,\mathrm{d}\bar{x})\wedge(x\,\mathrm{d}\bar{x})}{(|x|^2+\lambda^2)^2}\right\}. \tag{10.107}$$

To handle the expression on the RHS one only needs to set $|x|^2=\bar{x}x$ and one then finds that

$$\begin{aligned}
\mathrm{d}\left(\frac{x}{|x|^2+\lambda^2}\right)\wedge \mathrm{d}\bar{x}&=\frac{\mathrm{d}x\wedge \mathrm{d}\bar{x}}{(|x|^2+\lambda^2)}-\frac{x}{(|x|^2+\lambda^2)^2}\,\mathrm{d}(\bar{x}x)\wedge \mathrm{d}\bar{x}\\
&=\frac{\mathrm{d}x\wedge \mathrm{d}\bar{x}}{(|x|^2+\lambda^2)}-\frac{(x\,\mathrm{d}\bar{x})\wedge(x\,\mathrm{d}\bar{x})}{(|x|^2+\lambda^2)^2}-\frac{|x|^2\,\mathrm{d}x\wedge \mathrm{d}\bar{x}}{(|x|^2+\lambda^2)^2}
\end{aligned} \tag{10.108}$$

One discovers that **F** is already pure imaginary and is given by the elegant expression

$$\mathbf{F} = \frac{\lambda^2}{(|x|^2 + \lambda^2)} \, \mathrm{d}x \wedge \mathrm{d}\bar{x} \tag{10.109}$$

From this we make the observation that the quaternionic 2-form $\mathrm{d}x \wedge \mathrm{d}\bar{x}$ is self-dual and pure imaginary and its presence guarantees at once the required self-duality of **F**. To change from instanton to anti-instanton is also easy in this method one simply exchanges x and $\bar{x}$ in the formula for the potential **A** and so therefore in **F**. This equivalent to the interchange of $g(x)$ and $g^{-1}(x)$ mentioned already since on S^4 $g(x)$ is given by

$$\begin{aligned} g &= x, \qquad |x|^2 = 1 \\ g^{-1} &= \bar{x} \end{aligned} \tag{10.110}$$

Since the three matrices representing **i**, **j** and **k** are traceless, then it is easy to see that the formula for the integer k becomes, in the quaternionic notation

$$k = -\frac{Re}{8\pi^2} \int (\mathbf{F} \wedge \mathbf{F}) \tag{10.111}$$

When one works on S^4 one needs, for the general case when $k \neq 0$, more than one coordinate patch on S^4 to express the potential **A**. Let us examine this explicitly for the $k = 1$ instanton. We saw in Chapter 2 that 2 charts or patches are sufficient for coordinates on a sphere; one patch can be taken to be a neighbourhood of the south pole, and the other (overlapping) patch can be a neighbourhood of the north pole. These would correspond in $\mathbf{R}^4$ to the open sets $0 \leq |x| < +\varepsilon$, and $R \leq |x| < \infty$. Our expression for **A** will have to undergo a gauge change in the overlap region. To see this let us take the expression for **A**:

$$\mathbf{A} = \mathrm{Im} \frac{x \, \mathrm{d}\bar{x}}{(|x|^2 + \lambda^2)} \tag{10.112}$$

This is non-singular in the vicinity of the North Pole, i.e. $|x| \to \infty$, for there **A** tends to

$$\mathrm{Im} \frac{x \, \mathrm{d}\bar{x}}{\bar{x}x} = \bar{x}^{-1} \, \mathrm{d}\bar{x} \tag{10.113}$$

which is pure gauge. However, we cannot also use the form (10.112) near the South Pole, i.e. $|x| \to 0$. For suppose we can, then, if $y = x^{-1}$, **A** becomes

$$\mathrm{Im} \frac{y^{-1} \, \mathrm{d}\bar{y}^{-1}}{(|x|^{-2} + \lambda^2)} = -\mathrm{Im} \frac{\mathrm{d}\bar{y} y^{-1}}{(1 + \lambda^2 |y|^2)} \tag{10.114}$$

Now if we try to move from the South Pole $y=\infty$ back to the North Pole $y=0$, then near $y=0$, $\mathbf{A}$ tends to the singular expression

$$-\frac{1}{y^2}\,\mathrm{Im}\,\frac{\mathrm{d}\bar{y}}{|y|^2 y} \tag{10.115}$$

which appears to contradict (10.113). So before getting to the North Pole one must perform a gauge transformation on $\mathbf{A}$ and then one will be free of contradiction. Indeed, if in the overlap region $R \leq |x| \leq R+\varepsilon$, one gauge transforms $\mathbf{A}$ by the gauge transformation $g=\bar{x}^{-1}$. Then $\mathbf{A}$ becomes the imaginary part of

$$\begin{aligned} g^{-1}\mathbf{A}g+g^{-1}\,\mathrm{d}g &= \frac{\bar{x}(x\,\mathrm{d}\bar{x})}{(|x|^2+\lambda^2)}\bar{x}^{-1}+\bar{x}\,\mathrm{d}\bar{x}^{-1} \\ &= \frac{|x|^2\,\mathrm{d}\bar{x}\bar{x}^{-1}}{(|x|^2+\lambda^2)}+\bar{x}\,\mathrm{d}\bar{x}^{-1} \\ &= \frac{(|x|^2+\lambda^2-\lambda^2)\,\mathrm{d}\bar{x}\bar{x}^{-1}}{(|x|^2+\lambda^2)}+\bar{x}\,\mathrm{d}\bar{x}^{-1} \\ &= \frac{-\lambda^2\,\mathrm{d}\bar{x}\bar{x}^{-1}}{(|x|^2+\lambda^2)} \quad \text{since } \mathrm{d}\bar{x}\bar{x}^{-1}+\bar{x}\,\mathrm{d}\bar{x}^{-1}=0 \\ &= \frac{\lambda^2 y\,\mathrm{d}\bar{y}}{(1+\lambda^2|y|^2)}, \qquad y=x^{-1} \end{aligned} \tag{10.116}$$

This gauge transformed $\mathbf{A}$ is now non-singular at $y=0$ as it should be. Of course, this necessity to change gauge reflects the non-triviality of the bundle P, which in turn comes from the fact that $k=1$. If $k=0$, then one would be able to choose $\mathbf{A}=0$, or equivalently, choose $\mathbf{A}$ as a pure gauge globally on S^4.

10.10 INSTANTONS WITH |k|>1 AND QUATERNIONS

Turning now to instantons with $|k|>1$, we proceed by making an ansatz for $\mathbf{A}$ and discuss how it gives rise to a multi-instanton. For this purpose we introduce the space $\mathbf{H}^k$ of k-dimensional quaternionic vectors. If $\mathbf{v}\in\mathbf{H}^k$, then

$$\mathbf{v}=\begin{pmatrix} v_1 \\ v_2 \\ \vdots \\ v_k \end{pmatrix}$$

where $v_1, \ldots, v_k$ are quaternions. Since $\mathbf{H} \simeq \mathbf{R}^4$, then $\mathbf{H}^k = \mathbf{R}^{4k}$ so $\mathbf{H}^k$ is a $4k$-dimensional real vector space or a k-dimensional quaternionic vector space. In this space we make the simplest ansatz which is analagous to the $k = 1$ case. This is

$$\mathbf{A} = \mathrm{Im}\left\{\frac{\bar{\mathbf{v}}^T\, d\mathbf{v}}{1+|\mathbf{v}|^2}\right\} \tag{10.117}$$

where $\bar{\mathbf{v}}^T$ is the transpose conjugate of $\mathbf{v}$, i.e.

$$\bar{\mathbf{v}}^T = [\bar{v}_1 \bar{v}_2, \ldots, \bar{v}_k]$$

$\bar{v}_i$ is the quaternion conjugate to v_i. Also

$$|\mathbf{v}| = \sum_{i=1}^{k} \bar{v}_i v_i = \bar{\mathbf{v}}^T \cdot \mathbf{v} \tag{10.118}$$

The quaternionic components $v_1, \ldots, v_k$ of the vector $\mathbf{v}$ are chosen to be functions of $x \in \mathbf{H} \simeq \mathbf{R}^4$. Our ansatz will only give self-dual connections if $\mathbf{v}(x)$ is suitably chosen. To motivate our choice for $\mathbf{v}(x)$, we return once again to the $k = 1$ instanton, and by considering how to generate further $k = 1$ instantons one is led to a choice for $\mathbf{v}(\mathbf{x})$.

So far, we only have one instanton, however, it is easy to see that if one replaces the quaternion x by the linear combination $ax + b$ where $a \in \mathbf{R}$, $b \in H$, then the connection $\mathbf{A}$ of equation (10.112) remains self-dual. The curvature $\mathbf{F}$ of this connection is easily calculated and is given by (we have set $\lambda = 1$)

$$\mathbf{F} = \frac{a^2\, dx \wedge d\bar{x}}{|(ax+b)|^2 + 1} \tag{10.119}$$

Since this curvature is proportional to the self-dual 2-form $dx \wedge d\bar{x}$, it is automatically self-dual. Further, for each choice of a, b one obtains a new $k = 1$ self-dual connection $\mathbf{A}$, i.e. no two such $\mathbf{A}$ are gauge transforms of each other. This can be seen immediately by examining the curvature $\mathbf{F}$ in (10.119). The curvature transforms tensorially under gauge transformations ($\mathbf{F} \to g^{-1}\mathbf{F}g$), thus the maximum of $\mathbf{F}$ is a gauge invariant object. From (10.119) we see that the maximum of $\mathbf{F}$ occurs at $x = -b/a$ and has a size proportional to a. A gauge transformation cannot change these geometrically based quantities and cannot therefore connect two $\mathbf{A}$s with differing values of a and b. We can tidy this up a little by redefining a and b so as to replace $ax + b$ by the combination $a(x + b)$, this change in notation anticipates some notation that will be used shortly. It is clear that, physically speaking a and b correspond to a change in size and a translation respectively. The quantities a and b represent (since $a \in \mathbf{R}, b \in \mathbf{H}$), five real parameters. Hence we have not just one self-dual connection $\mathbf{A}$ but a 5-parameter family of self-dual connections with $k=1$. It is natural to ask if there are any more $k=1$ self-dual connections? The answer is no. This is one of the statements that will not be completely proved. However, it will be made more plausible when we describe the twistor theory approach to instantons. A further piece of evidence supporting this answer is that, for arbitrary k, a theorem known as the Atiyah-Singer index theorem can be used to show that the space of multi-instantons which are gauge inequivalent has dimension $8k-3$. So, for $k=1$, this space has dimension 5. This is at least consistent with our 5 parameter family. Of course it does not rule out the possibility that our 5 parameter family is only a subspace of the space of instantons with $k=1$. For a proof of this assertion c.f. reference [2] Chapter 2, and references therein.

Returning to multi-instantons, we make, for $\mathbf{v}(x)$, the ansatz

$$\mathbf{v}(x)=\bar{\mathbf{w}}^T \tag{10.120}$$

with $\mathbf{w}(x)=\boldsymbol{\lambda}(B-x)$. This $\mathbf{v}(x)$ is a matrix analogue of $ax+b$, in (10.120), B is a $k\times k$ symmetric matrix with quaternionic entries, x is the matrix xI, I being the $k\times k$ identity matrix, and $\boldsymbol{\lambda}$ is a k-dimensional vector with quaternion components. Thus when $k=1$, $\boldsymbol{\lambda}$ up to a sign corresponds to a, and B to b. The transpose T in (10.120) ensures that this is an ansatz for a multi-instanton rather than a multi-anti-instanton. The matrix B, and the vector $\boldsymbol{\lambda}$ are not arbitrary, and must be constrained if $\mathbf{v}(x)$ is to give rise to a multi-instanton. There are two constraints on $\mathbf{v}(x)$ and both of them are algebraic. They are

i. $B^{\dagger}B+\bar{\boldsymbol{\lambda}}^T\boldsymbol{\lambda}$ is a real matrix

ii. $$\left.\begin{aligned}(B-x)\mathbf{z}&=\mathbf{0}\\ \boldsymbol{\lambda}\cdot\mathbf{z}&=0\end{aligned}\right\}\Rightarrow\mathbf{z}=0 \tag{10.121}$$

The second condition in (10.121) can be restated as

$$A\mathbf{z}=0\Rightarrow\mathbf{z}=\mathbf{0}$$

where A is the non-square $(k+1)\times k$ matrix

$$A=\begin{pmatrix}b-x\\ \boldsymbol{\lambda}\end{pmatrix}_{(k+1)\times k} \tag{10.122}$$

This condition is therefore the statement that the rank of A has the value k, its maximal value. We cannot, at the moment, further clarify the meaning of this condition. We pass on to the first condition in (10.121). The statement that a $k\times k$ quaternionic matrix is real is just the statement that its entries are real. In any case this reality condition is the one which ensures that the connection A given by (10.117) is self-dual.

10.11 EXAMPLE OF INSTANTONS WITH $|\kappa|>1$

An example of a multi instanton can now be given. The matrix B is chosen to be diagonal.

$$B=\begin{bmatrix} b_1 & & & \\ & b_2 & & 0 \\ & & \ddots & \\ 0 & & & b_k \end{bmatrix}_{k\times k} \quad ; \qquad b_i \in \mathbf{H} \tag{10.123}$$

and the vector $\boldsymbol{\lambda}$ is chosen to have positive, real components $\lambda_1, \ldots, \lambda_k$, i.e. its quaternionic part is zero. Condition (i) of (10.121) becomes the condition that the diagonal matrix C should have real entries where

$$C=\begin{bmatrix} \bar{b}_1 b_1+\lambda_1^2 & & & \\ & \bar{b}_2 b_2+\lambda_2^2 & & 0 \\ & & \ddots & \\ 0 & & & \bar{b}_k b_k+\lambda_k^2 \end{bmatrix}_{k\times k} \tag{10.124}$$

This condition is evidently satisfied. Condition (ii) requires the matrix A to have maximal rank where now

$$A=\begin{pmatrix} b_1-x & & & \\ & b_2-x & & 0 \\ & & \ddots & \\ 0 & & & b_k-x \\ \lambda_1 & \lambda_2 & \cdots & \lambda_k \end{pmatrix}_{(k+1)\times k} \tag{10.125}$$

Requiring $A\mathbf{z}=\mathbf{0}$ gives $(k+1)$ quaternionic simultaneous equations for $\mathbf{z}$ which have as solution $\mathbf{z}=\mathbf{0}$ provided no two of $b_1, \ldots, b_k$ are equal. Thus our only restriction on the $b_1, \ldots, b_k$ is that they should be distinct. We can now compute the connection $\mathbf{A}$ using equation (10.117), we obtain

$$\mathbf{A}=\operatorname{Im}\left[\frac{\{-\lambda_1^2(b_1-x)-\lambda_2^2(b_2-x), \ldots, -\lambda_k^2(b_k-x)\}\,d\bar{x}}{(1+\lambda_1^2|b_1-x|^2+\ldots\lambda_k^2|b_k-x|^2)}\right] \tag{10.126}$$

This multi-instanton looks like k single instantons with scales and centres determined by $\lambda_1, \ldots, \lambda_k$ and $b_1, \ldots, b_k$ respectively. It is straightforward to compute $\mathbf{F}$ and thus verify that $\mathbf{F}={}^*\mathbf{F}$. This solution was found first without quaternionic or twistor methods. Unfortunately though not all

k-instantons are of the form given in (10.126). Direct calculation shows readily enough that (10.126) has instanton number k. After reading the rest of this chapter the reader may also deduce this from a more general viewpoint, see reference [2]. Now, however, we wish to count the parameters involved in our ansatz for **A**. We shall be able to show that when this count is done correctly, the number of parameters is the required number $8k-3$. This is encouraging and shows at once that we have a large family of multi-instantons. The description of the proof that all multi-instantons are given by (10.117) subject to (10.120, 121) is not given in this book, but cf. reference [2]. Returning to our parameter count, if we make the following transformations on $\boldsymbol{\lambda}$ and B, then the connection **A** is unchanged.

There are two of these transformations,

i.
$$\boldsymbol{\lambda} \mapsto \boldsymbol{\lambda} T$$
$$B \mapsto T^{-1} B T; \quad T \in O(k) \tag{10.127}$$

(T is a real orthogonal matrix as the notation indicates).

ii.
$$\boldsymbol{\lambda} \mapsto u\boldsymbol{\lambda}, \quad u \in \mathbf{H}, \quad |u|^2 = 1 \tag{10.128}$$

(u is just a unit quaternion).

Actually, what happens more precisely is that under the transformations given in (i) **A** does change but it only undergoes a gauge transformation as can be immediately checked; and under (ii) **A** is unchanged. There are no more transformations which either leave **A** unchanged or gauge transform it. Thus if we count the number of parameters in our original ansatz, and subtract from this total the number of parameters in (10.127, 128), then we shall have the dimension of gauge inequivalent **A**s provided by our ansatz. Let us begin. The k-dimensional quaternionic vector $\boldsymbol{\lambda}$ has $4k$ parameters, the symmetric quaternionic matrix B has $4k(k+1)/2$ parameters, the matrix T is specified by $k(k-1)/2$ parameters, and the unit quaternion u is specified by 3 parameters. However, the ansatz is also required to satisfy the constraints (10.121) and this further reduces the number of parameters. This reduction is simply by the number of (real) equations in (10.121) which is just the number of such equations in (10.121.i). This number is easily enough seen to be the number of pure imaginary components in a $k \times k$ self-adjoint quaternionic matrix (though this sounds at first far from easy to see), in any case this number is $3 \cdot k(k-1)/2$. Thus the grand total is

$$4k + \frac{4k(k+1)}{2} - \frac{k(k-1)}{2} - 3 - \frac{3k(k-1)}{2} = 8k - 3 \tag{10.129}$$

as promised.

10.12 TWISTOR METHODS AND INSTANTONS

The starting point for this section is the observation that if **F** is self-dual, then the components of **F** vanish when contracted with those of an anti-self dual tensor $\boldsymbol{\omega}$ rank 2. More explicitly let $F_{\mu\nu}$ and $\omega_{\mu\nu}$ be the components of **F** and $\boldsymbol{\omega}$ referred to local coordinates. Then by assumption

$$\mathbf{F} = {}^*\mathbf{F}, \quad \boldsymbol{\omega} = -{}^*\boldsymbol{\omega}$$

or

$$\tfrac{1}{2}\varepsilon_{\mu\nu\lambda\sigma}F_{\lambda\sigma} = F_{\mu\nu}, \; -\tfrac{1}{2}\varepsilon_{\mu\nu\rho\sigma}\omega_{\rho\sigma} = \omega_{\mu\nu} \tag{10.129}$$

Thus we have immediately that

$$F_{\mu\nu}\omega_{\mu\nu} = 0 \tag{10.130}$$

Similarly if **F** is anti-self-dual and $\boldsymbol{\omega}$ is self-dual then equation (10.130) holds in this case also. The reason that this observation is important in this context, is that in twistor theory, such $\boldsymbol{\omega}$ arise naturally and define certain families of planes which we shall describe in due course. Before pursuing this matter we wish to introduce some of the elements of twistor theory.

10.13 THE PROJECTIVE TWISTOR SPACE

The general idea in twistor theory is to replace Minkowski space-time *Mk* by a complex manifold of 3-complex dimensions denoted by *PT*—the projective twistor space. One can think of this as a transform: a problem in Minkowski space is transformed into an equivalent one on *PT* where the powerful methods of complex analysis are available; after the investigations are completed the results can be transformed back to Minkowski space. The space *PT* referred to above is just $\mathbf{C}P^3$—complex projective 3-space or the space of all lines through the origin in $\mathbf{C}^4$. The first step in obtaining *PT* is to complexify Minkowski space *Mk*. This complexification is just $\mathbf{C}^4$. $\mathbf{C}^4$ is also, of course, the complexification of $\mathbf{R}^4$ the space of interest for the instanton problem. In fact Euclidean space $\mathbf{R}^4$, and Minkowski space, *Mk*, are both real slices of $\mathbf{C}^4$.

The usual intention in twistor theory is pass back and forth between *Mk* and *PT*, we intend to pass back and forth between $\mathbf{R}^4$ and *PT*. The details of the geometry in the two above cases differ, but if one case is understood the other may be readily constructed. Having obtained $\mathbf{C}^4$ by the complexification just described consider the space of all complex lines through the origin in $\mathbf{C}^4$, i.e. consider the 3-dimensional complex projective space $\mathbf{C}P^3$. Now since the quaternions **H** are isomorphic to $\mathbf{C}^2$, then $\mathbf{C}^4 \simeq \mathbf{H}^2$

(cf. 10.87). Now each element of $\mathbf{C}P^3$, that is to say each complex line through the origin in $\mathbf{C}^4$, gives rise to a quaternionic line through the origin in $\mathbf{C}^4 \simeq \mathbf{H}^2$. For example if we use local coordinates then a line in $\mathbf{C}P^3$ is given by the four homogeneous coordinates (of which of course only 3 are independent)†:

$$(z_1, z_2, z_3, z_4) \tag{10.131}$$

To this line in $\mathbf{C}^4$ we associate the quaternionic line:

$$(z_1 + z_2 j, z_3 + z_4 j) \tag{10.132}$$

which are recognised as homogeneous coordinates for a line in $\mathbf{H}^2$, i.e. an element of 1-*dimensional quaternionic projective space* which we denote by $\mathbf{H}P^1$. The space $\mathbf{H}P^1$ is actually isomorphic to $\mathbf{S}^4$. To see this consider the unit quaternionic sphere $S^1_{\mathbf{H}}$ in $\mathbf{H}^2$. It is given by all pairs of quaternions (q_1, q_2) subject to $|q_1|^2 + |q_2|^2 = 1$ and is therefore isomorphic to a sphere in $\mathbf{R}^8$, i.e. to S^7. Now let a quaternionic line l intersect $S^1_{\mathbf{H}} \simeq S^7$ at the point $(\overset{0}{q}_1, \overset{0}{q}_2)$. Since l is a line, and $S^1_{\mathbf{H}}$ a unit sphere, then l will also intersect $S^1_{\mathbf{H}}$ at the points $(a\overset{0}{q}_1, a\overset{0}{q}_2)$ where a is a unit quaternion. The unit quaternions a form an $S^3 \simeq SU(2)$. This makes $\mathbf{H}P^1$, the space of such lines l, isomorphic to the quotient $S^7/SU(2)$ which is in turn isomorphic to S^4. (Exactly similar reasoning can be used to show that $\mathbf{R}P^1 \simeq S^1$ and $\mathbf{C}P^1 \simeq S^2$.) This means that the association (10.131, 132) of elements of $\mathbf{C}P^3$ to elements of $\mathbf{H}P^1$ is a map π from $\mathbf{C}P^3$ to S^4, in fact $\mathbf{C}P^3$ is a bundle over $S^4 \simeq \mathbf{H}P^1$. This being the case we are interested in the fibres of this bundle, i.e. we wish to know how many lines in $\mathbf{C}^4$ are mapped by π onto the same line $l \in \mathbf{H}P^1$. In fact each line l is a copy of $\mathbf{H} \simeq \mathbf{C}^2$, and $\mathbf{C}^2$ can contain a whole space of complex lines namely the space $\mathbf{C}P^1 \simeq S^2$. In other words it is these complex lines that get mapped onto l and the fibres of this bundle are therefore copies of $\mathbf{C}P^1 \simeq S^2$. We can now describe the connection between twistor space PT and $\mathbf{R}^4$: first we add the point at infinity to $\mathbf{R}^4$ so as to obtain S^4. Then we make use of the bundle that we have just described

$$\begin{array}{ccc} \mathbf{C}P^3 & = & PT \\ \pi\big\downarrow & & \big\downarrow\pi \\ \mathbf{H}P^1 & = & S^4 \end{array} \qquad \text{fibre } \mathbf{C}P^1 = S^2 \tag{10.133}$$

† In homogeneous coordinates the points (z_1, z_2, z_3, z_4) and $(\lambda z_1, \lambda z_2, \lambda z_3, \lambda z_4)$, $\lambda \in \mathbf{C}$ represent the same line hence only three of the z_i are independent.

Above each point $x \in S^4$ lies its fibre which is a 2-sphere S_x^2. This means that there is the correspondence

$$\text{(points } x \text{ in } S^4) \textit{ correspond to } \text{(the spheres } S_x^2 \text{ in } PT) \quad (10.134)$$

10.14 TWISTOR SPACE AND PLANES IN $\mathbf{C}^4$

At this point we can return to the curvature **F** and the vanishing of $F_{\mu\nu}\omega_{\mu\nu}$ obtained in (10.130) for self-dual **F** and anti-self-dual $\boldsymbol{\omega}$. Examples of such $\boldsymbol{\omega}$ are provided by considering certain families of planes in $\mathbf{C}^4$. To construct such a plane first consider any plane P in $\mathbf{C}^4$ together with any two orthogonal tangent vectors **U** and **V** to P. With **U** and **V** form the tensor or bivector

$$\boldsymbol{\omega} = \mathbf{V} \otimes \mathbf{U} - \mathbf{U} \otimes \mathbf{V} \quad (10.135)$$

The tensor $\boldsymbol{\omega}$ is, by construction, antisymmetric, if it is also anti-self-dual, $^*\boldsymbol{\omega} = -\boldsymbol{\omega}$, then P is a plane of the required type. Anti-self-dual complex 2-planes such as P are usually called β-planes in twistor theory. If the tensor $\boldsymbol{\omega}$ is self-dual rather than anti-self-dual then P is called an α-plane, clearly α-planes have the property that the components of an anti-self-dual **F** vanish when contracted with $\boldsymbol{\omega}$. In general P will be neither an α- nor a β-plane. There are however families of each type of plane in $\mathbf{C}^4$. To see this let $\mathbf{e}_1, \ldots, \mathbf{e}_4$ be a (complex) orthonormal basis for $\mathbf{C}^4$. Then the plane P_0 defined by the points (z_1, z_2) is a β-plane where

$$\begin{aligned} z_1 &= \lambda_1(\mathbf{e}_1 + i\mathbf{e}_3), \qquad \lambda_1 \in \mathbf{C} \\ z_2 &= \lambda_2(\mathbf{e}_2 + i\mathbf{e}_4), \qquad \lambda_2 \in \mathbf{C} \end{aligned} \quad (10.136)$$

The plane P_0 passes through the origin in $\mathbf{C}^4$ and it is a straightforward and a recommended exercise to construct a general $\boldsymbol{\omega}$ according to (10.135) and verify that $^*\boldsymbol{\omega} = -\omega$. For example if one takes **U** in the z_1 direction and V in the z_2 direction then one calculates that:

$$\begin{aligned} \omega_{12} &= -\omega_{34} = \lambda_1\lambda_2 \\ \omega_{13} &= \omega_{24} = 0 \\ \omega_{14} &= -\omega_{23} = i\lambda_1\lambda_2 \end{aligned} \quad (10.137)$$

which together with the fact that $\boldsymbol{\omega}$ is antisymmetric by construction verifies that $\boldsymbol{\omega}$ is anti-self-dual. Using P_0 we can construct all α-planes and all β-planes. Consider an element E of the complex orthogonal group $O(4, \mathbf{C})$; E, being orthogonal, will transform P_0 to another plane P_0^E, furthermore, if $\operatorname{Det} E = 1$ P_0^E will remain a β-plane, on the other hand, if

$\operatorname{Det} E = -1$ P_0^E will become an α-plane since such transformations change orientation and interchange self-dual and anti-self-dual objects. Thus P_0^E for all $E \in O(4, \mathbf{C})$ with $\operatorname{Det} E = 1$ corresponds to all β-planes passing through the origin in $\mathbf{C}^4$, and P_0^E for all $E \in O(4, \mathbf{C})$ with $\operatorname{Det} E = -1$ corresponds to all α-planes passing through the origin in $\mathbf{C}^4$. To obtain all α-planes and all β-planes in $\mathbf{C}^4$ it is just necessary to consider displacements of the origin. This leads to the following characterisation of self-dual and anti-self-dual $\mathbf{F}$'s:

i. If $\mathbf{F}$ vanishes when restricted to all α-planes and all β-planes, then $\mathbf{F} = \mathbf{0}$
ii. $\mathbf{F} = {}^*\mathbf{F} \Leftrightarrow \mathbf{F}$ vanishes when restricted to all β-planes.
iii. $\mathbf{F} = -{}^*\mathbf{F} \Leftrightarrow \mathbf{F}$ vanishes when restricted to all α-planes.

Statements (ii) and (iii) are obtained as an immediate consequence of (i) if one decomposes $\mathbf{F}$ as $\mathbf{F} = \mathbf{F}^+ + \mathbf{F}^-$ as was done in (10.36). The procedure for proving (10.138.i) is as follows: both α-planes and β-planes have the property that their tangent vectors $\mathbf{V}$ and $\mathbf{W}$ are always null vectors, we shall prove this below. Now take a point $p \in \mathbf{C}^4$ and consider all the α-planes and β-planes through p. Each pair of α- and β-planes intersect in a necessarily null line through p. The collection of these null lines form a null cone at p. In any case $\mathbf{F}(x)$, $(x = p)$, vanishes along any null line and since one can always take as a basis for $\mathbf{C}^4$ four linearly independent null vectors, then $\mathbf{F}$ vanishes at p, and so $\mathbf{F}$ vanishes everywhere in $\mathbf{C}^4$ since p is a general point. To see that displacements in α-planes and β-planes are always null we go to local coordinates. Then we have for displacements $\mathbf{V}$ and $\mathbf{W}$

$$\tfrac{1}{2}\varepsilon_{\alpha\beta\gamma\delta}(V^\gamma W^\delta - W^\gamma V^\delta) = \pm(V_\alpha W_\beta - W_\alpha V_\beta) \tag{10.139}$$

(the choice of sign in (10.139) corresponds to the choice of an α-plane or a β-plane). If we contract (10.139) successively with $\mathbf{V}$ and $\mathbf{W}$ then we find that

$$\begin{aligned} \mathbf{V}^2 W_\beta - \mathbf{V}\cdot\mathbf{W}\, V_\beta &= 0 \\ (\mathbf{W}\cdot\mathbf{V})\, W_\beta - \mathbf{W}^2 V_\beta &= 0 \end{aligned} \tag{10.140}$$

and so, we have $\mathbf{V}^2 = \mathbf{W}^2 = 0$.

Because of the importance of α- and β-planes in characterizing the duality properties of $\mathbf{F}$ it is essential to know some more about them. A single fact underlies the further properties of α- and β-planes. This is that the collection of all α-planes is isomorphic to $\mathbf{C}P^3$, and that the collection of all β-planes is isomorphic to another $\mathbf{C}P^3$ which we denote by $\mathbf{C}P^3_*$. The connection between the first $\mathbf{C}P^3$ and $\mathbf{C}P^3_*$ is a duality relationship where we are now using duality in the sense of projective geometry. We show first how $\mathbf{C}P^3$ arises and then explain the meaning of $\mathbf{C}P^3_*$.

Consider $\mathbf{C}P^3$ and endow it with homogeneous coordinates (z_1, z_2, z_3, z_4), we rewrite these coordinates as $(\omega^0, \omega^1, \pi_{0'}, \pi_{1'})$ to conform with the notation used in twistor theory. Then the equation of a complex 2-plane in $\mathbf{C}^4$ is given below

$$\omega^A = X^{AA'}\pi_{A'}; \qquad A, A' = 0, 1 \tag{10.141}$$

where $X^{AA'}$ is a 2×2 matrix representing the point $x \in \mathbf{C}^4$ according to the definition†

$$X^{AA'} = \begin{bmatrix} x^0 + ix^1 & ix^2 + x^3 \\ ix^2 - x^3 & x^0 - ix^1 \end{bmatrix} \tag{10.142}$$

In equation (10.141) it is easy to check that the points $X^{AA'}$ are constrained to lie on a plane. For example if we set $\omega^A = 0$ so that the plane passes through the origin, and choose $\pi_{0'} = \pi_{1'} = 1$. Then equation (10.141) becomes

$$\begin{aligned} x^0 + ix^1 + ix^2 + x^3 &= 0 \\ x^0 - ix^1 + ix^2 - x^3 &= 0 \end{aligned} \tag{10.143}$$

so that $x^2 = ix^0$ and $x^3 = -ix^1$. In other words we have a plane P defined by (z_1, z_2) where

$$\begin{aligned} z_1 &= \lambda_1(\mathbf{e}_0 + i\mathbf{e}_2) \\ z_2 &= \lambda_2(\mathbf{e}_1 - i\mathbf{e}_3) \end{aligned} \tag{10.144}$$

If we compare P with the plane P_0 defined in (10.136) then we see that P is obtained from P_0 by application of an element E of $O(4, \mathbf{C})$ with $\det E = -1$. Thus since P_0 is a β-plane then P is an α-plane. More precisely

$$P = P_0^E, \qquad E = \begin{bmatrix} 1 & 0 & 0 & 0 \\ 0 & 1 & 0 & 0 \\ 0 & 0 & 1 & 0 \\ 0 & 0 & 0 & -1 \end{bmatrix} \tag{10.145}$$

It should now be clear that all α-planes arise as solutions of (10.141) and also that non-zero ω^A correspond to α-planes which do not pass through the origin. In summary the homogeneous coordinates $(\omega^A, \pi_{A'})$ of $\mathbf{C}P^3$ parametrize all α-planes. It is also clear that all β-planes are parametrized by a $\mathbf{C}P^3$ for example the equation

$$\omega_*^A = X_*^{AA'}\pi_{*A'} \tag{10.146}$$

† A useful property of (10.142) is the relation $\det(X^{AA'}) = x^2$, note then that null vectors are characterized by $\det(X^{AA'}) = 0$.

has as its solutions all β-planes, where $(\omega_*^A, \pi_{*A'})$ are the homogeneous coordinates of another $\mathbf{C}P^3$, and $X_*^{AA'}$ is defined by

$$X_*^{AA'} = \begin{bmatrix} x^0 + ix^1 & ix^2 - x^3 \\ ix^2 + x^3 & x^0 - ix^1 \end{bmatrix} \tag{10.147}$$

The only difference between $X_*^{AA'}$ and $X^{AA'}$ is that x^3 has changed sign, this corresponds to the action of the $E \in O(4, \mathbf{C})$ defined in (10.145) and is the reason for the changeover from α-planes to β-planes. As we have mentioned above there is a dual connection between $\mathbf{C}P^3$ and $\mathbf{C}P_*^3$. This connection comes from projective geometry and we shall briefly explain its origin. Consider a line l in the $x-y$ plane it has an equation of the form

$$Ax + By = C \tag{10.148}$$

To each such line l we may associate the pair of independent numbers (A, B). Thus the lines l can be thought of as points in another $\mathbf{R}^2$ which we can denote by $\mathbf{R}_*^2$. $\mathbf{R}_*^2$ is then said to be dual to $\mathbf{R}^2$ because it is clearly also true that the lines l of $\mathbf{R}_*^2$ correspond to points of $\mathbf{R}^2$. In general it is easy to check that in $\mathbf{R}^n$ we have a similar correspondence between points of $\mathbf{R}^n$ and hyperplanes or $(n-1)$-planes in $\mathbf{R}^n$. Thus $\mathbf{R}_*^n$ is the space of hyperplanes in $\mathbf{R}^n$ and vice-versa†. A further consequence of this duality is that any r-dimensional geometrical object O^r in $\mathbf{R}^n$ is dual to an $(n-4-r)$ dimensional object $O_*^{(n-r-1)}$ in $\mathbf{R}_*^n$. If we apply this to α-planes and β-planes in $\mathbf{C}^4$ then we can say that points in $\mathbf{C}P^3$ are dual to 2-planes in $\mathbf{C}P^3$. Thus if $\mathbf{C}P^3$ is the space of α-planes then β-planes are either given by points of the dual $\mathbf{C}P_*^3$ are by planes in $\mathbf{C}P^3$. $\mathbf{C}P_*^3$ is then called the dual projective twistor space. For more details on this point cf. reference [22].

10.15 α-PLANES AND ANTI-SELF-DUAL CONNECTIONS

We shall now, for the sake of being specific, only deal with $PT = \mathbf{C}P^3 =$ the space of α-planes. It remains for us to describe how PT leads to a solution of the instanton problem. In fact we shall show how twistor methods lead to the construction of anti-instantons rather than instantons, this compliments our quaternionic discussion which described instantons. As we have

† It is also usual to compactify $\mathbf{R}^n$ by adding the point at infinity.

observed before the difference between instantons and anti-instantons is merely that of a difference in orientation. If, in what follows, we were to substitute β-planes for α-planes or, PT_* for PT, then we would obtain instantons rather than anti-instantons. Consider then $\mathbf{C}^4$ with a curvature $\mathbf{F}(x)$ such that the components of $\mathbf{F}$ vanish in all α-planes so that $\mathbf{F} = {}^*\mathbf{F}$. Let $\mathbf{A}(x)$ be the anti-self-dual connection for $\mathbf{F}$. $\mathbf{A}(x)$, $x \in \mathbf{C}^4$, is constructed as follows: $\mathbf{F}(x)$ has vanishing components in any α-plane, hence if we restrict x to lie in an α-plane P, $\mathbf{A}(x)$ is a flat connection for $x \in P$, therefore $\mathbf{A}(x) = g^{-1}(x)\, dg(x)$, $x \in P$. The collection of all α-planes P through x form a null cone, and the restriction of $\mathbf{F}$ to this cone is flat. Thus we know $\mathbf{A}(x)$ for the null directions along this cone. That is to say we know

$$A_\mu(x) n^\mu, \; n^2 = 0 \tag{10.149}$$

Further, if one wishes one can take a null basis for $\mathbf{C}^4$ thus we know $A_\mu(x)$. On the other hand if we do not restrict x to lie on α-planes then $\mathbf{F}$ is no longer zero and $A(x)$ is no longer flat, more specifically, $\mathbf{F}$ is not also flat on β-planes, (unless it is identically zero which we obviously exclude). Next we must find $\mathbf{A}(x)$ on β-planes. This is by using the fact that $A_\mu(x)$ is holomorphic and homogeneous of degree one along the null cone, and such a function always extends in a unique way to β-planes. This will be clarified in the course of what follows. The description given above of how to obtain anti-self-dual $\mathbf{A}(x)$ is the basis for the breakthrough made by Ward in his seminal paper of reference [30].

10.16 THE EQUIVALENCE BETWEEN INSTANTONS AND HOLOMORPHIC VECTOR BUNDLES

We now wish to give further details of the instanton problem in the following way: the idea is to describe how an anti-self-dual $\mathbf{F}$ on S^4 corresponds in a one to one fashion with a certain kind of vector bundle $\tilde{E}$ on $\mathbf{C}P^3$. These bundles $\tilde{E}$ can all be calculated and their construction amounts to the solution of a problem in algebraic geometry. The ability to perform this construction and the one to one nature of the correspondence means that all anti-self-dual connections on S^4 are found.

First we translate anti-self-duality on S^4 into a simple holomorphic property for bundles on $\mathbf{C}P^3$: consider $\mathbf{F}$ defined, not on S^4, but on $\mathbf{R}^4$. Identify $\mathbf{R}^4$ with $\mathbf{C}^2$ and replace the four real variables x_1, x_2, x_3, x_4 by two complex variables z_1, z_2, e.g. we can choose $z_1 = x_1 + ix_2$, $z_2 = x_3 + ix_4$. Evidently the dx^μ can be expressed in a formal way in terms of dz^1 and

dz^2 by use of complex conjugation:

$$\begin{aligned} dx^1 &= \tfrac{1}{2}(dz^1 + d\bar{z}^1) \\ dx^2 &= \frac{-i}{2}(dz^1 - d\bar{z}^1) \\ dx^3 &= \tfrac{1}{2}(dz^2 + d\bar{z}^2) \\ dx^4 &= \frac{-i}{2}(dz^2 - d\bar{z}^2) \end{aligned} \tag{10.150}$$

This means that any form, ω say, may be expressed in terms of dzs and $d\bar{z}$s for

$$\begin{aligned} \omega = \omega_{\mu\nu}\, dx^\mu \wedge dx^\nu = \tilde{\omega}^{(2,0)}_{ab}\, dz^a \wedge dz^b \\ +\tilde{\omega}^{(1,1)}_{ab}\, dz^a \wedge d\bar{z}^b \\ +\tilde{\omega}^{(0,2)}_{ab}\, d\bar{z}^a \wedge d\bar{z}^b \end{aligned} \tag{10.151}$$

In (10.151) the Latin letters $a, b = 1, 2$ and the superscripts on the components $\tilde{\omega}^{(i,j)}_{ab}$ of the form indicate that the form contains i factors of dz and j factors of $d\bar{z}$. Using an obvious notation we can write ω as

$$\omega = \omega^{(2,0)} + \omega^{(1,1)} + \omega^{(0,2)} \tag{10.152}$$

Suppose ω is anti-self-dual then it is a routine calculation to decompose ω according to (10.152), one then discovers that $\omega^{(2,0)} = \omega^{(0,2)} = 0$ and we say that ω is of type (1, 1). On the other hand if ω is of type (1, 1) then ω is not always anti-self-dual, but ω does decompose into an anti-self-dual piece and a self-dual piece:

$$\omega = \omega^{(1,1)} = \omega^+ + \omega^- \tag{10.153}$$

To verify this last statement take as an example any particular anti-self-dual ω^- which will then be of type (1, 1) and add to it the self-dual form ω^+, where $\omega^+ = dz_1 \wedge d\bar{z}_1 + dz_2 \wedge d\bar{z}_2$. The sum $\omega^+ + \omega^-$ is then of type (1, 1).

However there is an arbitrariness in the definition of z_1 and z_2 in terms of $x_1, \ldots, x_4$; a permutation of $x_1, \ldots, x_4$ would produce a different pair of complex variables. If, then, a 2-form ω is of type (1, 1) for all† possible such complex structures in $\mathbf{R}^4$ then ω is exactly anti-self-dual. This is because the space of such ω is invariant under $SO(4)$ as is the space of anti-self-dual ω. In summary then we can characterize anti-self-dual ω by

† A technical point here is that a complex structure endows $\mathbf{R}^4$ with an orientation, and that the definition of the * operation endows it with a metric; and we leave these two things fixed. Thus we consider all complex structures and all anti-self-dual ω which are compatible with the metric and the orientation.

the statement:

$$\boldsymbol{\omega} \text{ anti-self-dual on } \mathbf{R}^4 \Leftrightarrow \boldsymbol{\omega} \text{ is always of type } (1, 1) \tag{10.154}$$

(in (10.154) the word always stands for, all complex structures). Now we move from $\mathbf{R}^4$ to S^4. As we saw in Chapter 7, S^n has no complex structure for $n = 4$. However we can lift from S^4 to $\mathbf{C}P^3$ since S^4 is the base space of the twistor fibration given in (10.133). More specifically, if π is the projection from $\mathbf{C}P^3$ to S^4, and $\boldsymbol{\omega}$ is a 2-form on S^4, then if U is an open set of S^4 the pull back $\pi^*\boldsymbol{\omega}$ is a 2-form on an open set $\tilde{U}$ say of $\mathbf{C}P^3$. We call this form the lift of $\boldsymbol{\omega}$ to $\mathbf{C}P^3$ and denote it by $\tilde{\boldsymbol{\omega}}$. The space $\mathbf{C}P^3$ has quite an intimate connection with the possible complex structures on $\mathbf{R}^4$. The possible complex structures on $\mathbf{R}^4$ are parametrized by the fibre coordinates in the fibration of $\mathbf{C}P^3$ over S^4. This statement comes from the following reasoning: take a point $x \in S^4$. Above x is a fibre F_x, let $f \in F_x$ be a point in the fibre F_x. Then since f also belongs to $\mathbf{C}P^3$ (recall that a fibre bundle is the union of all its fibres), consider the tangent space at f, $T_f(\mathbf{C}P^3)$. A subspace of $T_f(\mathbf{C}P^3)$ is the vertical subspace $V_f(\mathbf{C}P^3)$ of tangents to the fibre F_x (cf. Chapter 7). These are complex vector spaces and their quotient $T_f(\mathbf{C}P^3)/V_f(\mathbf{C}P^3$ is a complex vector space of complex dimension $\dim T_f - \dim V_f = 3 - 1 = 2$, i.e. of real dimension 4. So for $x \in S^4$, and for each f belonging to the fibre F_x above x, one can identify the quotient T_f/V_f, which is a complex vector space, with the ordinary real tangent space $T(S^4)$ to S^4 at X. These quotient spaces correspond therefore, for each $f \in F_x$ to possible complex structures on $\mathbf{R}^4$. The reason that one writes $\mathbf{R}^4$ above instead of S^4 is that one must exclude from the above discussion the point at infinity in $\mathbf{R}^4$. In any case the fibre coordinate† f parametrizes the possible complex structures on $\mathbf{R}^4$, incidentally these are already naturally endowed with an orientation induced from $\mathbf{C}P^3$ and projected onto $\mathbf{R}^4$ by π.

The statement (10.154) translates to a version on CP^3:

$$\begin{matrix} \boldsymbol{\omega} \text{ anti-self-dual} \\ \text{on } U \subset S^4 \end{matrix} \quad \Leftrightarrow \quad \begin{matrix} \tilde{\boldsymbol{\omega}} \text{ is of type } (1, 1) \\ \text{on } \tilde{U} \subset \mathbf{C}P^3 \end{matrix} \tag{10.155}$$

where the word always is no longer needed in view of our preceding discussion.

Now let F be the curvature of an anti-self-dual connection on S^4. We wish to bring together the requirements for $\tilde{\mathbf{F}}$ to be of type (1, 1) on $\mathbf{C}P^3$ and the requirement for $\mathbf{F}$ to be $su(2)$-valued. To bring these requirements together let $\tilde{E}$ be a holomorphic vector bundle of rank 2 over $\mathbf{C}P^3$

† The fibre $F_x \simeq S^2$, so the possible complex structures on $\mathbf{R}^4$ are parametrized by points on a 2-sphere.

(by holomorphic we mean that the fibre F_z above the point z varies holomorphically with z) and let $\mathbf{A}$ be a connection on $\tilde{E}$. Since $\mathbf{A}$ is a 1-form then, in local coordinates, it decomposes into the sum of two pieces: one of type $(1, 0)$, and one of type $(0, 1)$. (Now $a = 1, 2, 3$)

$$\mathbf{A} = A'_a \, dz^a + A''_a \, d\bar{z}^a \tag{10.156}$$

The covariant derivative D_a is then written as

$$\left(\partial'_a \equiv \frac{\partial}{\partial z^a}, \ \partial''_a \equiv \frac{\partial}{\partial \bar{z}^a}\right) \qquad D_a = \partial'_a + \partial''_a + A'_a + A''_a \tag{10.157}$$

If it happens that $A''_a = 0$, then $\mathbf{A}$ is said to be *compatible* with the holomorphic structure on $\tilde{E}$. This can be more easily comprehended by realizing that if $S(z)$ is a section of $\tilde{E}$ which is also holomorphic (i.e. $S(z)$ depends* on z but not on $\bar{z}$. or, the 1-form dS is of type $(1, 0)$), then $A''_a = 0$ is equivalent to the statement

$$D''_a S(z) = 0 \tag{10.158}$$

where $D''_a = \partial_a - A''_a$. In other words this connection allows one to parallel transport in a holomorphic manner. We further require $\mathbf{A} \in su(2)$ the Lie algebra of $SU(2)$, this is the requirement that $\mathbf{A}^{\dagger} = -\mathbf{A}$ and that $tr\mathbf{A} = 0$. In terms of sections this is the requirement that there exists a gauge transformation $g(z)$ from the section $S(z)$ to another section $S'(z)$, and that the gauge transformed connection, $\mathbf{A}_g$ say, belongs to $su(2)$. (From now on we replace $SU(2)$ by $U(2)$, we rectify this later.)

Now we can rephrase the foregoing discussion by saying that we require‡ a connection $\mathbf{A}$ on $\tilde{E}$ that is

a. Compatible with the holomorphic structure on $\tilde{E}$.
b. Compatible with the unitary structure§ on $\tilde{E}$. (10.159)
c. Endowed with a curvature $\mathbf{F}$ of type $(1, 1)$

Let $\mathbf{A}$ satisfy (a) i.e. $A''_a = 0$. Then let g be a gauge transformation such that $\mathbf{A}^{\dagger}_g = -\mathbf{A}_g$. In other words we have

$$\mathbf{A}_g = \mathbf{A}'_g + \mathbf{A}''_g,$$

---

* Recall that for holomorphic functions $f(z)$ of complex variable the holomorphicity of $f(z)$ expressed by the Cauchy–Riemann equations is equivalent to $df/d\bar{z} = 0$.

‡ It turns out, and we shall now indicate the reasons, that given (a) then there is a connection satisfying (b) and (c). Further there is an important converse (which we do not prove but cf. refs. [3, 18]) that given (b) and (c) there is a unique holomorphic structure satisfying (a).

§ The use of the word unitary structure is the following: unitary matrices are those which preserve a positive definite Hermitian inner product. The bundle $\tilde{E}$ is said to have a unitary structure if there exists a positive definite inner product on the fibres.

with

$$\begin{aligned}\mathbf{A}''_g &= g^{-1}(d''+\mathbf{A}'')g \\ &= g^{-1}d''g, \text{ since } \mathbf{A}''=0\end{aligned} \tag{10.160}$$

and we want $\mathbf{A}_g$ to satisfy

$$(\mathbf{A}'_g+\mathbf{A}''_g)^{\dagger} = -\mathbf{A}'_g - \mathbf{A}''_g \tag{10.161}$$

To satisfy (10.161) let us simply choose $\mathbf{A}'_g = -(\mathbf{A}''_g)^{\dagger}$ so that

$$\mathbf{A}'_g = -(g^{-1}d''g)^{\dagger} + g^{-1}d''g \tag{10.162}$$

where g is our (unitary) gauge transformation. The connection $\mathbf{A}_g$ just defined has already a curvature $\mathbf{F}$ of type $(1, 1)$. For, if we use the holomorphic section $S(z)$ then, since $\mathbf{A}''=0$,

$$\mathbf{F} = d'\mathbf{A}' + d''\mathbf{A}' + \mathbf{A}' \wedge \mathbf{A}' \tag{10.163}$$

Now in (10.163) the term $d''\mathbf{A}'$ is of type $(1, 1)$ but the combination $d'\mathbf{A}' + \mathbf{A}' \wedge \mathbf{A}'$ is of type $(2, 0)$. However this latter term is actually zero because in the unitary gauge we have $\mathbf{F}_g^{\dagger} = -\mathbf{F}_g$ with $\mathbf{F}_g = g^{-1}\mathbf{F}g$, thus we also have $\mathbf{F}^{\dagger} = -\mathbf{F}$. This means that $\mathbf{F}$ has no term of type $(2, 0)$ for under Hermitian conjugation this would become a term of type $(0, 2)$. We have described how given (10.159.a) we can satisfy (10.159.b, c). We remind the reader of the converse given above. Given (10.159.a, b, c) and its discussion, and given also (10.155), then we can amalgamate (10.155, 159) to provide the statement that

E a bundle over S^4 with an anti-self-dual curvature $\mathbf{F}$	$\Leftrightarrow$	$\tilde{E}$ over $\mathbf{C}P^3$ has the unique holomorphic structure determined by (10.159.a, b, c)

(10.164)

To characterize all anti-self-dual connections on S^4 we simply have to produce a converse of (10.164). This we can do and it corresponds to considering all holomorphic bundles over $\mathbf{C}P^3$ and restricting them so that they project down to an appropriate anti-self-dual bundle E on S^4. To see how this works consider the lifted bundle $\tilde{E}$ of (10.164). The curvature $\tilde{\mathbf{F}}$ on $\tilde{E}$ is the lift of $\mathbf{F}$ under the projection $\pi: \mathbf{C}P^3 \to S^4$. It is easy to check that this means that $\tilde{F}_{\mu\nu}(z)$, $z \in \mathbf{C}P^3$, is zero if (i) one or both of the components μ and ν correspond to a fibre direction, and (ii) z is restricted to a fibre F_x of $\mathbf{C}P^3$. In terms simply of bundles, the lifted bundle $\tilde{E}$ over $\mathbf{C}P^3$ is trivial when restricted to a fibre F_x of the fibration $\pi: \mathbf{C}P^3 \to S^4$.

Conversely, given a holomorphic bundle B over $\mathbf{C}P^3$ with a connection $\mathbf{A}$ on B, the condition that the curvature $\mathbf{F}$ of $\mathbf{A}$ projects onto the curvature of a connection on S^4 is also that $F_{\mu\nu}(z)$ be zero when one or both of μ and ν correspond to fibre directions and $z \in F_x$ a fibre of $\mathbf{C}P^3$. This property of $F_{\mu\nu}(z)$ is not necessarily guaranteed just by requiring that B restricted to fibres F_x be trivial, the fact that it is true is connected with the fact that B is a complex vector bundle over $\mathbf{C}P^3$ rather than just over an arbitrary complex manifold. We do not go in to this here since a treatment of this point requires the use of sheaf cohomology—a topic that we have unfortunately not included, mainly for considerations of space, in this book, cf. however reference [2]. In any case this point corresponds, in the discussion following (10.149), to the extension of $\mathbf{A}(x)$ from α-planes to β-planes. A piece of terminology that we shall use is that if B is a complex vector bundle over $\mathbf{C}P^3$ which is trivial when restricted to the fibres F_x of the twistor fibration $\pi : \mathbf{C}P^3 \to S^4$; and if there is a positive Hermitian form on $B|_{F_x}$, the restriction of B to F_x, then B is said to have a positive real form (cf. also reference [2]).

The characterization of all anti-self-dual $U(2)$ connections on S^4 can now be given. It says that: there is *one to one correspondence* between

a. holomorphic vector bundles $\tilde{E}$ of rank 2 over $\mathbf{C}P^3$ having a positive real form. (10.165)
b. anti-self-dual $U(2)$ connections $\mathbf{A}$ on S^4

Note that once the holomorphic structure in (a) is specified that there is no need to discuss connections in (a)—one simply produces holomorphic bundles $\tilde{E}$ satisfying the required conditions. To convert $U(2)$ connections into $SU(2)$ connections one needs to restrict to those bundles which have first Chern class zero, i.e. $C_1(\tilde{E}) = 0$. To see this note that in (7.252) we found that $C_1(\mathbf{F}) = i/2\pi\, tr(\mathbf{F})$. The second Chern class $C_2(\tilde{E}) = k$, the instanton number. Thus all one needs to do is to construct all these holomorphic bundles $\tilde{E}$ over $\mathbf{C}P^3$ satisfying the condition (10.164.a) and the condition $C_1(\tilde{E}) = 0$. Fortunately, at about the same time as physicists were considering the instanton problem, various mathematicians completed work in which all these bundles $\tilde{E}$ were constructed. We are unable to describe this work here. One point is worth mentioning though. It is this: every holomorphic bundle $\tilde{E}$ over $\mathbf{C}P^3$ is actually algebraic, not just holomorphic, this follows from basic work of Serre [26] on analytic and algebraic geometry. The point that we wish to mention is that a consequence of this algebraic nature of the $\tilde{E}$s is that there always exists a gauge in which the connection $\mathbf{A}(x)$ is a rational function of the x_μ. Another detail of interest here is that if one wishes to construct $SU(n)$ instantons with $n > 2$ then this is done by simply demanding that $\tilde{E}$ have rank n instead of rank 2.

10.17 CONSTRUCTION OF AN INSTANTON GIVEN A HOLOMORPHIC VECTOR BUNDLE

To conclude our discussion of instantons we wish to explain how given $\tilde{E}$ of rank 2 one may extract from it an anti-self-dual connection $A_\mu(x)$ on S^4. Let us specify a bundle $\tilde{E}$ over $\mathbf{C}P^3$ by giving its transition functions $g_{\alpha\beta}(z)$, and from $g_{\alpha\beta}(z)$ we shall extract $A_\mu(x)$. First of all how many coordinate patches are needed to cover $\mathbf{C}P^3$? Since $\mathbf{C}P^3$ has homogeneous coordinates (z_1, z_2, z_3, z_4) then the four patches $z_i \neq 0$, $i = 1, \ldots, 4$, will cover $\mathbf{C}P^3$. For example in the patch $z_1 \neq 0$ the three complex coordinates on $\mathbf{C}P^3$ are given by $(z_2/z_1, z_3/z_1, z_4/z_1)$. The point in $\mathbf{C}P^3$ corresponding to $z_i = 1$, $i = 1, \ldots, 4$ lies in all four patches hence their four-fold intersection is non-empty. This means that six transition functions are needed to specify $\tilde{E}$. Let us use the twistor notation $(\omega^0, \omega^1, \pi_{0'}, \pi_{1'})$ of (10.141) instead of (z_1, z_2, z_3, z_4). Denote the transition function between two patches by $g(\omega^A, \pi_{A'})$; because g is defined on $\mathbf{C}P^3$ it must be homogeneous of degree zero in ω^A, $\pi_{A'}$.

Now $\tilde{E}$ is trivial when restricted to fibres $F_x \in \mathbf{C}P^3$ of the twistor fibration $\pi : \mathbf{C}P^3 \to S^4$. $F_x \simeq \mathbf{C}P^1$ and $\mathbf{C}P^1$ needs only two patches. Let us fix an x and denote these two patches by $\pi_{0'} \neq 0$, $\pi_{1'} \neq 0$. From our discussion of the triviality of transition functions in Chapter 7, it follows that the transition function $g(X^{AB'}\pi_{B'}, \pi_{A'})$ for these patches factorizes into a product

$$g(X^{AB'}\pi_{B'}, \pi_{A'}) = \lambda(X^{AB'}\pi_{B'}, \pi_{A'})\tilde{\lambda}^{-1}(X^{AB'}\pi_{B'}, \pi_{A'}) \qquad (10.166)$$

where λ and $\tilde{\lambda}$ are 2×2 matrices of homogeneous functions homogeneous of degree zero. λ is holomorphic for $\pi_{0'} \neq 0$ and $\tilde{\lambda}$ is holomorphic for $\pi_{1'} \neq 0$. A more compact notation is

$$a(x, \zeta) = \lambda(X^{AB'}\pi_{B'}, \pi_{A'})$$

$$\tilde{a}(x, \zeta) = \tilde{\lambda}(X^{AB'}\pi_{B'}, \pi_{A'})$$

where ζ stands for $\pi_{0'}/\pi_{1'}$, thus a is holomorphic for $\zeta \neq 0$ and $\tilde{a}$ is holomorphic for $\zeta \neq \infty$. The anti-self-dual connection $\mathbf{A}(x)$ is given by:

$$\mathbf{A}(x) = g^{-1}\,\mathrm{d}g = A_{PQ'}\,\mathrm{d}X^{PQ'} = A_\mu\,\mathrm{d}x^u, \qquad dX^{PQ'}\pi_{Q'} = 0 \qquad (10.167)$$

the latter equation following from $\omega^P = X^{PQ'}\pi_{Q'}$. Note that $dX^{PQ'}\pi_{PQ'} = 0$ implies that $\mathrm{d}X^{P1'} = -\zeta\,\mathrm{d}X^{P0'}$, this in turn implies that

$$\left(\frac{\partial}{\partial X^{P0'}} - \frac{\zeta\partial}{\partial X^{P1'}}\right)g = 0 \qquad (10.168)$$

If we now substitute $g = a\tilde{a}^{-1}$ into (10.168) we obtain the result

$$a^{-1}\left(\frac{\partial}{\partial X^{P0'}} - \frac{\zeta\partial}{\partial X^{P1'}}\right)a = \tilde{a}^{-1}\left(\frac{\partial}{\partial X^{P0'}} - \frac{\zeta\partial}{\partial X^{P1'}}\right)\tilde{a}$$
$$= f_p(x, \zeta), P = 0, 1. \qquad (10.169)$$

Because a and $\tilde{a}$ are homogenous of degree zero in ζ it follows that $f_p(x, \zeta)$ is homogeneous of degree one in ζ, i.e. $f_p = A\zeta + B$. The quantities A and B are actually components of the connection as may be seen from the following argument: since a is holomorphic and invertible, so that det $a \neq 0$, then so is a^{-1}, a similar remark applies to $\tilde{a}$. Thus, because a is holomorphic for $\zeta \neq 0$ and $\tilde{a}$ is holomorphic for $\zeta \neq \infty$, then if we bear in mind the explicit linear factor of ζ in (10.169) we can reason as follows: The equality of $f_p(x, \zeta)$ to the first term in (10.169) means that $f_p(x, \zeta) - A\zeta$ is holomorphic except for $\zeta = 0$ where A is some constant. On the other hand the equality of $f_p(x, \zeta)$ to the second term in (10.169) means that $f_p - A\zeta$ is holomorphic except for $\zeta = \infty$. Hence Liouvilles theorem implies immediately that $f_p - A\zeta$, being holomorphic for all ζ, is a constant: $f_p - A\zeta = B$ with B independent of ζ but dependent on x. In fact if we define $A_{PQ'}$ by

$$f_p(x, \zeta) = A_{P0'} - \zeta A_{P1'}, \qquad p = 0, 1 \qquad (10.170)$$

then $A_{PQ'}$ are functions of x only and are the components of the anti-self-dual connection given by (10.168). The anti-self-duality of $A_{PQ'}$ is of course guaranteed but can be verified by direct calculation. Thus we see that if we can find easily all transition functions g one can find all anti-self-dual connections. Atiyah and Ward [3] and also Corrigan *et al.* [8] do essentially this. In reference [3] Atiyah and Ward give a sequence of ansatze for instantons which they denote by A_l, $l = 1, 2, \ldots$, all instantons can be constructed by using larger and larger values of l, further it is sufficient to consider only transition functions which are of a certain triangular form:

$$g = \begin{bmatrix} 1 & a \\ 0 & b \end{bmatrix} \qquad (10.171)$$

In reference [3] the ansatze for $l > 3$ are not explicitly given. Corrigan *et al.* [8] in an important sequel to reference 12 give explicit formula for A_l for $l > 3$. In reference [8] g is written as:

$$g = \begin{bmatrix} \zeta^l & \rho(\omega, \pi) \\ 0 & \zeta^{-l} \end{bmatrix} \qquad (10.172)$$

where ρ is a function satisfying certain properties. There is then [8] a formula for $A_\mu(x)$ in terms of an integral over ρ. Thus the problem is narrowed down to a search for all such ρ. In conclusion we remark that the factorization $g = a\tilde{a}^{-1}$ is by no means unique, different factorizations produce connections which are gauge transforms of one another. That is to say a choice of factors a, $\tilde{a}$ is a choice of gauge. Also with regard to explicit formulae for instanton connections $A_\mu(x)$ we remind the reader that there are also the quaternionic formulae of (10.117, 120, 121). Unfortunately to check in specific cases the constraints of (10.121) leads to non-trivial problems in quaternionic linear algebra. For further work on instantons including the quaternionic approach cf. reference [2].

10.18 THE MINKOWSKI CASE

Recall from (10.19) that in Minkowski space Mk the instanton problems corresponds to solving

$$^*\mathbf{F} = \mp i\mathbf{F} \tag{10.173}$$

Let the gauge group be $G = SL(2, \mathbf{C})$ so that $\mathbf{F}$ is g-valued. We must make a few remarks about this case. In the Euclidean case we compactify $\mathbf{R}^4$ to S^4, thus we would also like to compactify Mk. This can be done [21] the resulting compactification is homeomorphic to $S^1 \times S^3$ and is endowed [21] with the flat Lorentz metric $dt^2 - d\mathbf{x}^2$. For an account of the geometry in the Minkowski case cf. reference [10], and for information about specific solutions of (10.173) cf. reference [19].

10.19 MONOPOLES

Monopoles may also be constructed using similar geometrical methods to the instanton case. We describe first the Abelian monopoles of Maxwell theory which were first considered by Dirac [9], for a more modern treatment cf. also Wu *et al.* [33]. The problem, mathematically speaking, arises from a static magnetic monopole situated at the origin in $\mathbf{R}^3$. Since the gauge group of Maxwell theory is $U(1)$ this corresponds to a principal-$U(1)$-bundle P over $\mathbf{R}^3 - \{0\}$. But $\mathbf{R}^3 - \{0\}$ can be contracted or retracted to S^2 without changing the topology of the bundle P. Thus P is equivalent to a $U(1)$-bundle over S^2. P is therefore classified by elements of $\pi_1(U(1)) = \pi_1(S^1) = \mathbf{Z}$. Also the integer n corresponding to the element of $\pi_1(S^1)$ is given by evaluating the first Chern class of P, $C_1(P)$. Using the formulae of Chapter 7 we have $C_1(\mathbf{F}) = -tr(\mathbf{F}/2\pi)$ so that

$$C_1 = -tr\int_{S^2}\frac{\mathbf{F}}{2\pi} = n \qquad (10.174)$$

where **F** is the curvature of the Maxwell field. Recall that if Maxwell's equations are

$$\text{div}\,\mathbf{B} = 0 \qquad (10.175)$$

to be solved on $\mathbf{R}^3 - \{0\}$ then since $\mathbf{R}^3 - \{0\}$ is not contractible there are solutions which are not of the form $\mathbf{B} = \nabla \times \mathbf{A}$. In fact these other solutions are parametrized by $H^2(\mathbf{R}^3 - \{0\}; \mathbf{R})$ since **B** is a 2-form. But, as we showed in Chapter 6, $H^2(\mathbf{R}^3 - \{0\}; \mathbf{R}) = H^2(S^2; \mathbf{R})$, and $H^2(S^2; \mathbf{R}) = \mathbf{R}$ which is non-zero. The integer n which classifies solutions to Maxwell's equations in $\mathbf{R}^3 - \{0\}$ is called the magnetic charge. An example of a monopole with charge n is given below [11]

$$\mathbf{A}_{\mp} = \frac{n}{2r}(z \mp r)(x\,dy - y\,dx)$$

with

$$\mathbf{F} = \frac{n}{2r^3}(x\,dy \wedge dz + y\,dz \wedge dx + z\,dx \wedge dy)$$

$$\mathbf{B} = \frac{n\hat{\mathbf{r}}}{2r^2}, \qquad r = (x^2 + y^2 + z^2)^{1/2} \qquad (10.176)$$

Having described some monopoles in Maxwell theory we would like to change the guage group from $U(1)$ to $SU(2)$ and consider Yang–Mills monopoles. We shall only discuss what are called Yang–Mills–Higgs monopoles. The monopoles are static objects and are purely magnetic and are solutions of an equation known as the Bogomolny equation [5, 6]. A static monopole of the above type corresponds to a $su(2)$-valued pair $(\mathbf{A}, \phi)$ where **A** is a connection and ϕ is a scalar field. The energy of the monopole system is E where

$$E = \tfrac{1}{2}\int d^3x[\mathbf{B}^2 + (D\phi)^2 + \lambda(\phi^2 - C^2)^2] \qquad (10.177)$$

The pair $(\mathbf{A}, \phi)$ is defined on $\mathbf{R}^3$ corresponding to the property that the system is static. Also **B** and $D\phi$ are defined by

$$\mathbf{B} = \nabla \times \mathbf{A} + [\mathbf{A}, \mathbf{A}], \qquad D\phi = \nabla\phi + [\mathbf{A}, \phi] \qquad (10.178)$$

The field equations for a static system with energy E are

$$\mathbf{D}\cdot\mathbf{B}=0, \qquad \mathbf{D}\times\mathbf{B}=[\phi, \mathbf{D}\phi]$$
$$D^2\phi=4\lambda\phi^3-2\lambda C^2\phi^2 \tag{10.179}$$

These equations are second order and difficult to solve. If they are considered in the limit where the scalar potential term is zero i.e. $\phi^2=C^2$, then a first order equation results and minimizes the energy E. This equation is

$$\mathbf{B}=\mathbf{D}\phi \tag{10.180}$$

and is called the Bogomolny equation. So far there has been a minimum of geometry. Geometry begins to enter in an essential way if we define ϕ to be a 4th component A_4 of the connection $\mathbf{A}$. Thus the pair $(\mathbf{A}, \phi)$ can be identified with a static four-dimensional connection or gauge potential $\mathbf{A}$. It is easy to check that having done this the Bogomolny equation becomes the self-dual equation

$$F_{\mu\nu}=\tfrac{1}{2}\varepsilon_{\mu\nu\rho\sigma}F_{\rho\sigma}$$

where

$$F_{ab}=\tfrac{1}{2}\varepsilon_{abc}B_c, \qquad F_{a4}=D_a\phi; \qquad a, b, c=1, 2, 3 \tag{10.181}$$

Thus a monopole corresponds to a *static* self-dual gauge potential. The Bogomolny equations are to be solved subject to the conditions

i. $E<\infty$

ii. $\phi^2\to 1-\dfrac{\mathrm{m}}{(x^2)^{1/2}}+O(\mathbf{x}^{-2});\qquad m\in R$

$$\mathbf{x}^2=x_1^2+x_2^2+x_3^2$$

$$\mathbf{x}^2\to\infty \tag{10.182}$$

The number m appearing in (ii) above is actually an integer and is called the magnetic charge. A topological argument establishes this: the condition (ii) above provides a map from a sphere S^2 at infinity to those ϕ in $SU(2)$ satisfying $\phi^2=1$. Since $SU(2)\simeq S^3$ the condition $\phi^2=1$ reduces S^3 to S^2 and thus we have a map from S^2 to S^2, i.e. an element of $\pi_2(S^2)=\mathbf{Z}$. If S^2_∞ denotes the sphere at infinity then the integer n corresponding to the homotopy class of ϕ is given by [1] the formula

$$n=\frac{1}{8\pi}\int_{S^2_\infty}\xi_i d^2s^i$$

where

$$\xi_j = \varepsilon_{jkl}\varepsilon_{abc}\hat{\phi}^a \partial_k \hat{\phi}^b \partial_l \hat{\phi}^c, \ \hat{\phi}^a = \frac{\phi^a}{(\phi^2)^{1/2}} \tag{10.183}$$

and ϕ^a are the $su(2)$ components of ϕ. If we now use the form of ϕ^2 for large x^2 we find that $n = m$.

For the case $n = 1$ there is a well understood solution [25] to the Bogomolny equation. As in the $k = 1$ instanton case the solution is spherically symmetric. For $n > 1$ solutions are harder to find and it is here that the geometrical methods come into their own. These are two main ways in which one may apply these methods. The first way is to take the Atiyah–Ward ansatze $A_1, A_2, \ldots$ for the instanton problem, and to specialize these quite general ansatze for self-dual connections to the monopole problem. That is to say to impose the conditions (10.182), and to impose the topological constraint (10.183).

The second way in which one may apply geometrical methods to the Bogomolny equation is to try and set up a one-to-one correspondence between solutions of the Bogomolny equations and certain holomorphic bundles over some appropriate complex manifold. We shall say a little about both these uses of geometrical methods.

To return to the first method we consider the spherically symmetric monopole with $n = 1$, then take the ansatz A_1 so that

$$g = \begin{bmatrix} \zeta & \rho(\omega, \pi) \\ 0 & \zeta^{-1} \end{bmatrix} \tag{10.184}$$

and let ρ be given by

$$\rho = \frac{\exp[2\mu - 2\nu]}{\gamma} \tag{10.185}$$

with μ, ν, and γ defined by

$$\mu = \frac{i\omega_2}{\pi_2}, \qquad \nu = i\omega_1/\pi_1, \qquad \gamma = \mu - \nu \tag{10.186}$$

Then this choice of g gives [25] the spherically symmetric $n = 1$ monopole, to obtain a monopole with $n = 2$ Ward showed [31] that ansatz A_2 is sufficient where g is given by

$$g = \begin{bmatrix} \zeta^2 & \dfrac{e^{2\mu} + e^{2\nu}}{\gamma^2 + 4c^2} \\ 0 & \zeta^{-2} \end{bmatrix} \tag{10.187}$$

where c is a real constant. Since then there has been considerable further work cf. for example references [7, 12, 16, 20, 24, 31] and works cited therein.

As regards the second method Hitchin [14] has provided the one-to-one correspondence alluded to above. Hitchin describes a complex manifold T and constructs, for the $su(2)$ case, holomorphic bundles $\tilde{E}$ of rank 2 over T. If these bundles $\tilde{E}$ satisfy certain conditions reminiscent of those in the instanton case, then there is a one-to-one correspondence between such $\tilde{E}$ and solutions of the Bogomolny equation. Further, the Atiyah–Ward ansatze are sufficient to generate all the monopole solutions.

The manifold T is the space of all *oriented* straight lines or geodesics in $\mathbf{R}^3$. This space actually can carry a complex structure. In fact T has real dimension 4 and hence complex dimension 2. T is not however compact and so the bundles $\tilde{E}$ can not be expected to be algebraic as they are for instantons. There is a link between the twistor space $PT = CP^3$ and T: so if one takes S^4 and removes the point at infinity as to return to $\mathbf{R}^4$ then in $\mathbf{C}P^3$ this corresponds to the removal of the projective line at infinity; in other words the twistor fibration $\pi : \mathbf{C}P^3 \to S^4$ becomes a fibration $\pi' : (\mathbf{C}P^3 - \mathbf{C}P^1) \to \mathbf{R}^4$. Now since monopoles are static, Hitchin [14] takes the quotient of $(\mathbf{C}P^3 - \mathbf{C}P^1)$ by the action of time translation this we write as $(\mathbf{C}P^3 - \mathbf{C}P^1)/\mathbf{X}_0$, where $\mathbf{X}_0$ is the (holomorphic) vector field $\partial/\partial x_0$. The space T and this quotient space are the same:

$$T = (\mathbf{C}P^3 - \mathbf{C}P^1)/\mathbf{X}_0 \tag{10.188}$$

The construction resembles in spirit the twistor construction for instantons for there is a correspondence between null planes in $\mathbf{C}^3$, the complexification of $\mathbf{R}^3$, and points of T. For further details on $\tilde{E}$ and the conditions imposed on it cf. reference [14].

10.20 THE BOHM–AHARANOV EFFECT

We wish to give here with some brief remarks on the physical significance of the connection as compared with the curvature. We deal with Maxwell theory and consider the consequence of assuming the field $\mathbf{F}$ to be identically zero in some region Ω. At first one may think that there will be no physically measurable electromagnetic effects in such a region Ω. This is not so, effects may arise if the topology of Ω is non-trivial, e.g. if Ω is not simply connected. Take then $\mathbf{F} = 0$ in a region Ω

$$\mathbf{F} = \mathrm{d}\mathbf{A} = 0 \tag{10.189}$$

then if $H^1(\Omega; \mathbf{R}) \neq 0$ one cannot conclude that $\mathbf{A} = \text{const.}$ and hence physically trivial. As an example let Ω be $\mathbf{R}^3 - L$ where L is an infinite straight line through the origin. $H^1(\mathbf{R}^3 - L; \mathbf{R}) = H^1(S^1; \mathbf{R}) = \mathbf{R}$ so that Ω is not simply connected as indeed one can see directly. In the Bohm–Aharonov effect the region Ω is created by choosing L to be an infinite zero width solenoid. Since the field $\mathbf{B}$ is zero outside a solenoid we have

$$\mathbf{F} = 0 \tag{10.190}$$

in Ω. An experiment to confirm the presence of a physical effect can be done [32].

Figure 10.2 depicts electrons being diffracted by Young's slits and the consequent diffraction pattern being detected on a screen. A black dot representing a solenoid is also shown, the diffraction pattern is found to *change* when the field is switched on even though the electrons are passing through a region of zero field. This is the Bohm–Aharonov effect. In terms of parallel transport one says that zero curvature does not imply trivial parallel transport if the region in which the curvature is zero is not simply connected. This underlines the fact that there is a sense in which the connection is a more fundamental object than the curvature, even though a connection is gauge dependent and not directly measurable.

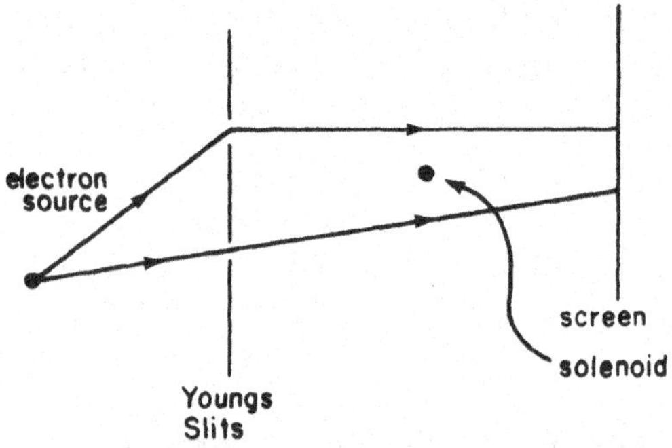

Figure 10.2

We close this chapter with a remark about the quantization of Yang–Mills theories. The quantization of Yang–Mills theories is most conveniently done using the Feynman path-integral [17]. The integration space is the functional space of connections and this space is gauge dependent and infinite dimensional. Physical scattering amplitudes are gauge independent, and thus the integration must be modified to integrate over equivalence classes of gauge equivalent connections. This amounts to a kind of choice of gauge in the functional space of connections. Gribov [13] and also Singer [27] showed that a global choice of gauge condition is not always possible.

This is because the appropriate space to integrate over is a non-trivial infinite dimensional principal bundle which does not therefore admit a global gauge. This is an interesting insight into the Yang–Mills functional integral and raises interesting technical questions.

REFERENCES

1. ARAFUNE, J., FREUND, P. G. O. and GOEBEL, C. J., *J. Math. Phys.* **16**, 433 (1975).
2. ATIYAH, M. F., "The Geometry of Yang–Mills Fields". Lezione Fermiane, Scuola Normale Superiore Pisa, 1979.
3. ATIYAH, M. F. and WARD, R. S., *Commun. Math. Phys.* **55**, 117 (1977).
4. BELAVIN, A. A., POLYAKOV, A. M., SCHWARTZ, A. S. and TYUPKIN, Yu. S., *Phys. Lett.* **59B**, 85 (1975).
5. BOGOMOLNY, E., *Sov. J. Nucl. Phys.* **24**, 449 (1976).
6. COLEMAN, S., PARKE, S., NEVEU, A. and SOMMERFIELD, C., *Phys. Rev.* **D15**, 544 (1976).
7. CORRIGAN, E. and GODDARD, P., *Commun. Math. Phys.* **80**, 575 (1981).
8. CORRIGAN, E.F., FAIRLIE, D.B., YATES, R.G. and GODDARD, P., *Commun. Math. Phys.* **58**, 223 (1978).
9. DIRAC, P. A. M., *Proc. Roy. Soc. London* **A133**, 60 (1931).
10. DOUADY, A., *In* "Les Equations de Yang–Mills". Seminaire E.N.S. 1977–78 Asterique 71-2, Société Mathématique de France, 1980.
11. EGUCHI, T., GILKEY, P. B. and HANSON, A. J., *Phy. Rep.* **66**, 213 (1980).
12. FORGÁCS, P., HORVÁTH, Z. and PALLA, L., *Nucl. Phys.* **B192**, 2182 (1981).
13. GRIBOV, V. N., *Nucl. Phys.* **B139**, 1 (1978).
14. HITCHIN, N. J., *Commun. Math. Phys.* **83**, 579 (1982).
15. IYANAGA, S. and KAWADA, Y., For a comprehensive list of the homotopy groups of Lie groups see Appendix, *In* "Encyclopedic Dictionary of Mathematics". MIT Press, 1980.
16. JAFFE, A. and TAUBES, C., "Vortices and Monopoles". Birkauser Boston, 1980.
17. NASH, C., "Relativistic Quantum Fields". Academic Press, 1978 (the *i* in the relevant formulae should be deleted for Euclidean space).
18. NEWLANDER, A. and NIRENBERG, L., *Ann. Math.* **65**, 391 (1957).
19. NEWMAN, E. T., *Phys. Rev.* **D22**, 3023 (1980).
20. O'RAIFEARTAIGH, L. and ROUHANI, S., *Acta Physika Austriaca* (Suppl. **23**) 525 (1981).
21. PENROSE, R., *In* "Group Theory in Non-Linear Problems" (Barut, A. O., Ed). Reidel Publishers Co., 1974.
22. PENROSE, R., *In* "Quantum Gravity" (Isham, C. J., Penrose, R. and Sciama, D. W., Eds). Clarendon Press, 1975.
23. PENROSE, R., The twistor programme. *Rep. Math. Phys.* **12**, 65 (1977).
24. PRASAD, M. K. and ROSSI, P., *Phys. Rev.* **D24**, 2182 (1981).
25. PRASAD, M. K. and SOMMERFIELD, C. M., *Phys. Rev. Lett.* **35**, 760 (1975).
26. SERRE, J. P., *Ann. Inst. Fourier* **VI**, 1 (1956).
27. SINGER, I. M., *Commun. Math. Phys.* **60**, 7 (1978).
28. SPANIER, E. H., "Algebraic Topology". p. 398. McGraw-Hill, 1966.

29. STEENROD, N., "The Topology of Fibre Bundles". p. 88. Princeton University Press, 1970.
30. WARD, R. S., *Phys. Lett.* **61A**, 81 (1977).
31. WARD, R. S., *Commun. Math. Phys.* **79**, 317 (1981).
32. WERNER, F. G. and BRILL, D. R., *Phys. Rev. Lett.* **4**, 344 (1960).
33. WU, T. T. and YANG, C. N., *Phys. Rev.* **D12**, 3845 (1975).

BIBLIOGRAPHY

For further reading we give here an alphabetic list of books and papers, some of which are referenced to in the text.

CHERN, S., "Complex Manifolds without Potential Theory". Van Nostrand, 1967.

DOLD, A., "Lectures on Algebraic Topology". Springer–Verlag, 1972.

EGUCHI, T., GILKEY, P. B. and HANSON, A. J., *Phys. Rep.* **66**, 213 (1980).

EILENBERG, S. and STEENROD, N., "Foundations of Algebraic Topology". Princeton University Press, 1952.

GREUB, W., HALPERIN, S. and VANSTONE, R., "Connections, Curvature and Cohomology". Vols 1–3. Academic Press, 1972, 1973, 1978.

HILTON, P. J. "An Introduction to Homotopy Theory". Cambridge University Press, 1953.

HILTON, P. J. and WYLIE, S., "Homology Theory". Cambridge University Press, 1962.

HOCKING, J. G. and YOUNG, G. J., "Topology". Addison Wesley, 1961.

HU, S. T., "Homotopy Theory". Academic Press, 1959.

HUSEMOLLER, D., "Fibre Bundles". Springer-Verlag, 1966.

IYANAGA, S. and KAWADA, Y. (Eds), "Encyclopaedic Dictionary of Mathematics". MIT Press, 1980.

KABAYASHI, S. and NOMIZU, K., "Foundations of Differential Geometry". Vols 1–2. Interscience, 1963, 1969.

MASSEY, W. S., "Algebraic Topology: An Introduction". Springer-Verlag, 1977.

MASSEY, W. S., "Singular Homology". Springer-Verlag, 1980.

MAUNDER, C. R. F., "Algebraic Topology". Van Nostrand Rheinhold Co., 1972.

MILNOR, J. W., "Morse Theory". Princeton University Press, 1963.

MILNOR, J. W. and STASHEFF, J. D., "Characteristic Classes". Princeton University Press, 1974.

MORSE, M., "The Calculus of Variations in the Large". American Mathematical Society, 1964.

SPANIER, E. H., "Algebraic Topology". McGraw-Hill, 1966.

SPIVAK, M., "Differential Geometry". Vols 1–5. Publish or Perish Inc., 1979 (in Volume 5 of this series there is an excellent critical bibliography).

STEENROD, N., "The Topology of Fibre Bundles". Princeton University Press, 1970.

WELLS, R. O., "Differential Analysis on Complex Manifolds". Springer-Verlag, 1979.

Subject Index